The Myth of the Machine:

TECHNICS AND HUMAN DEVELOPMENT

BOOKS BY LEWIS MUMFORD

THE STORY OF UTOPIAS   1922

STICKS AND STONES   1924

THE GOLDEN DAY   1926

HERMAN MELVILLE   1929

THE BROWN DECADES   1931

TECHNICS AND CIVILIZATION   1934

THE CULTURE OF CITIES   1938

MEN MUST ACT   1939

FAITH FOR LIVING   1940

THE SOUTH IN ARCHITECTURE   1941

THE CONDITION OF MAN   1944

CITY DEVELOPMENT   1945

VALUES FOR SURVIVAL   1946

GREEN MEMORIES   1947

THE CONDUCT OF LIFE   1951

ART AND TECHNICS   1952

IN THE NAME OF SANITY   1954

FROM THE GROUND UP   1956

THE TRANSFORMATIONS OF MAN   1956

THE CITY IN HISTORY   1961

THE HIGHWAY AND THE CITY   1963

THE URBAN PROSPECT   1968

THE MYTH OF THE MACHINE

  Vol. I: TECHNICS AND HUMAN
    DEVELOPMENT   1967

  Vol. II: THE PENTAGON OF POWER   1970

INTERPRETATIONS AND FORECASTS:
  1922—1972   1973

The Myth of the Machine

TECHNICS AND HUMAN DEVELOPMENT

by LEWIS MUMFORD

HARCOURT BRACE JOVANOVICH, INC., NEW YORK
A HARVEST BOOK

Pictures 1a, 9a, and 9c are from *The Eternal Present* .... .ollingen
XXXV.6.1, distributed by Pantheon Books, copyright 1962 by the Trustees of the
National Gallery of Art, Washington, D.C. By permission of Achille Weider.

CONTENTS

ILLUSTRATIONS
Between pages 166 and 167

In April 1962 I had the honor of inaugurating at my Alma Mater, the City College of New York,

THE JACOB C. SAPOSNEKOW LECTURES

These lectures were established by his sisters in memory of a devoted scholar, an alert citizen, and a loyal alumnus of the College. Some of the main themes of this book were first lightly sketched out in those three lectures and I am grateful to both the donors, the Misses Sadie and Rebecca Saposnekow, and to the College, for their prior consent to my incorporating that material in this larger work, on which I was then already engaged.

The Myth of the Machine:

TECHNICS AND HUMAN DEVELOPMENT

CHAPTER ONE

Prologue

Ritual, art, poesy, drama, music, dance, philosophy, science, myth, religion are all as essential to man as his daily bread: man's true life consists not alone in the work activities that directly sustain him, but in the symbolic activities which give significance both to the processes of work and their ultimate products and consummations.

THE CONDITION OF MAN (1944)

The last century, we all realize, has witnessed a radical transformation in the entire human environment, largely as a result of the impact of the mathematical and physical sciences upon technology. This shift from an empirical, tradition-bound technics to an experimental mode has opened up such new realms as those of nuclear energy, supersonic transportation, cybernetic intelligence and instantaneous distant communication. Never since the Pyramid Age have such vast physical changes been consummated in so short a time. All these changes have, in turn, produced alterations in the human personality, while still more radical transformations, if this process continue unabated and uncorrected, loom ahead.

In terms of the currently accepted picture of the relation of man to technics, our age is passing from the primeval state of man, marked by his invention of tools and weapons for the purpose of achieving mastery over the forces of nature, to a radically different condition, in which he will have not only conquered nature, but detached himself as far as possible from the organic habitat.

With this new 'megatechnics' the dominant minority will create a uniform, all-enveloping, super-planetary structure, designed for automatic operation. Instead of functioning actively as an autonomous personality, man will become a passive, purposeless, machine-conditioned animal whose proper functions, as technicians now interpret man's role, will either be fed into the machine or strictly limited and controlled for the benefit of de-personalized, collective organizations.

My purpose in this book is to question both the assumptions and the predictions upon which our commitment to the present forms of technical and scientific progress, treated as if ends in themselves, has been based. I shall bring forward evidence that casts doubts upon the current theories of man's basic nature which over-rate the part that tools once played—and machines now play—in human development. I shall suggest that not only was Karl Marx in error in giving the material instruments of production the central place and directive function in human development, but that even the seemingly benign interpretation of Teilhard de Chardin reads back into the whole story of man the narrow technological rationalism of our own age, and projects into the future a final state in which all the possibilities of human development would come to an end. At that 'omega-point' nothing would be left of man's autonomous original nature, except organized intelligence: a universal and omnipotent layer of abstract mind, loveless and lifeless.

Now, we cannot understand the role that technics has played in human development without a deeper insight into the historic nature of man. Yet that insight has been blurred during the last century because it has been conditioned by a social environment in which a mass of new mechanical inventions had suddenly proliferated, sweeping away ancient processes and institutions, and altering the traditional conception of both human limitations and technical possibilities.

Our predecessors mistakenly coupled their particular mode of mechanical progress with an unjustifiable sense of increasing moral superiority. But our own contemporaries, who have reason to reject this smug Victorian belief in the inevitable improvement of all other human institutions through command of the machine, nevertheless concentrate, with manic fervor, upon the continued expansion of science and technology, as if they alone magically would provide the only means of human salvation. Since our present over-commitment to technics is in part due to a radical misinterpretation of the whole course of human development, the first step toward recovering our balance is to bring under review the main stages of man's emergence from its primal beginnings onward.

Just because man's need for tools is so obvious, we must guard ourselves against over-stressing the role of stone tools hundreds of thousands of years before they became functionally differentiated and efficient. In treating tool-making as central to early man's survival, biologists and anthropologists for long underplayed, or neglected, a mass of activities in which many other species were for long more knowledgeable than man. Despite the contrary evidence put forward by R. U. Sayce, Daryll Forde, and André Leroi-Gourhan, there is still a tendency to identify tools and machines with technology: to substitute the part for the whole.

Even in describing only the material components of technics, this practice overlooks the equally vital role of containers: first hearths, pits, traps, cordage; later, baskets, bins, byres, houses, to say nothing of still later collective containers like reservoirs, canals, cities. These static components play an important part in every technology, not least in our own day, with its high-tension transformers, its giant chemical retorts, its atomic reactors.

In any adequate definition of technics, it should be plain that many insects, birds, and mammals had made far more radical innovations in the fabrication of containers, with their intricate nests and bowers, their geometric beehives, their urbanoid anthills and termitaries, their beaver lodges, than man's ancestors had achieved in the making of tools until the emergence of *Homo sapiens*. In short, if technical proficiency alone were sufficient to identify and foster intelligence, man was for long a laggard, compared with many other species. The consequences of this perception should be plain: namely, that there was nothing uniquely human in tool-making until it was modified by linguistic symbols, esthetic designs, and socially transmitted knowledge. At that point, the human brain, not just the hand, was what made a profound difference; and that brain could not possibly have been just a hand-made product, since it was already well developed in four-footed creatures like rats, which have no free-fingered hands.

More than a century ago Thomas Carlyle described man as a "tool-using animal," as if this were the one trait that elevated him above the rest of brute creation. This overweighting of tools, weapons, physical apparatus, and machines has obscured the actual path of human development. The definition of man as a tool-using animal, even when corrected to read 'tool-making,' would have seemed strange to Plato, who attributed man's emergence from a primitive state as much to Marsyas and Orpheus, the makers of music, as to fire-stealing Prometheus, or to Hephaestus, the blacksmith-god, the sole manual worker in the Olympic pantheon.

Yet the description of man as essentially a tool-making animal has become so firmly embedded that the mere finding of the fragments of little primate skulls in the neighborhood of chipped pebbles, as with the Australopithecines of Africa, was deemed sufficient by their finder, Dr. L. S. B. Leakey, to identify the creature as in the direct line of human ascent, despite marked physical divergences from both apes and later men. Since Leakey's sub-hominids had a brain capacity about a third of *Homo sapiens* —less indeed than some apes—the ability to chip and use crude stone tools plainly neither called for nor by itself generated man's rich cerebral equipment.

If the Australopithecines lacked the beginning of other human characteristics, their possession of tools would only prove that at least one other species outside the true genus *Homo* boasted this trait, just as parrots and magpies share the distinctly human achievement of speech, and the bower bird that for colorful decorative embellishment. No single trait, not even tool-making, is sufficient to identify man. What is specially and uniquely human is man's capacity to combine a wide variety of animal propensities into an emergent cultural entity: a human personality.

If the exact functional equivalence of tool-making with utensil-making had been appreciated by earlier investigators, it would have been plain that there was nothing notable about man's hand-made stone artifacts until far along in his development. Even a distant relative of man, the gorilla, puts together a nest of leaves for comfort in sleeping, and will throw a bridge of great fern stalks across a shallow stream, presumably to keep from wetting or scraping his feet. Five-year-old children, who can talk and read and reason, show little aptitude in using tools and still less in making them: so if tool-making were what counted, they could not yet be identified as human.

In early man we have reason to suspect the same kind of facility and the same ineptitude. When we seek for proof of man's genuine superiority to his fellow creatures, we should do well to look for a different kind of evidence than his poor stone tools alone; or rather, we should ask ourselves what activities preoccupied him during those countless years when with the same materials and the same muscular movements he later used so skillfully he might have fashioned better tools.

The answer to this question I shall spell out in detail in the first few chapters; but I shall briefly anticipate the conclusion by saying that there was nothing specifically human in primitive technics, apart from the use and preservation of fire, until man had reconstituted his own physical organs by employing them for functions and purposes quite different from those they had originally served. Probably the first major displacement was the transformation of the quadruped's fore-limbs from specialized organs of locomotion to all-purpose tools for climbing, grasping, striking, tearing, pounding, digging, holding. Early man's hands and pebble tools played a significant part in his development, mainly because, as Du Brul has pointed out, they facilitated the preparatory functions of picking, carrying, and macerating food, and *thus liberated the mouth for speech.*

If man was indeed a tool-maker, he possessed at the beginning one primary, all-purpose tool, more important than any later assemblage: his own mind-activated body, every part of it, including those members that made clubs, hand-axes or wooden spears. To compensate for his extremely primitive working gear, early man had a much more important asset that extended his whole technical horizon: he had a far richer biological equipment than any other animal, a body not specialized for any single activity, and a brain capable of scanning a wider environment and holding all the different parts of his experience together. Precisely because of his extraordinary plasticity and sensitivity, he was able to use a larger portion of both his external environment and his internal, psychosomatic resources.

Through man's overdeveloped and incessantly active brain, he had more mental energy to tap than he needed for survival at a purely animal level; and he was accordingly under the necessity of canalizing that energy, not just into food-getting and sexual reproduction, but into modes of living that would convert this energy more directly and constructively into appropriate cultural—that is, symbolic—forms. Only by creating cultural outlets could he tap and control and fully utilize his own nature.

Cultural 'work' by necessity took precedence over manual work. These new activities involved far more than the discipline of hand, muscle, and eye in making and using tools, greatly though they aided man: they likewise demanded a control over all man's natural functions, including his organs of excretion, his upsurging emotions, his promiscuous sexual activities, his tormenting and tempting dreams.

With man's persistent exploration of his own organic capabilities, nose, eyes, ears, tongue, lips, and sexual organs were given new roles to play. Even the hand was no mere horny specialized work-tool: it stroked a lover's body, held a baby close to the breast, made significant gestures, or expressed in shared ritual and ordered dance some otherwise inexpressible sentiment about life or death, a remembered past, or an anxious future. Tool-technics, in fact, is but a fragment of biotechnics: man's total equipment for life.

This gift of free neural energy already showed itself in man's primate ancestors. Dr. Alison Jolly has recently shown that brain growth in lemurs derived from their athletic playfulness, their mutual grooming, and their enhanced sociability, rather than from tool-using or food-getting habits; while man's exploratory curiosity, his imitativeness, and his idle manipulativeness, with no thought of ulterior reward, were already visible in his simian relatives. In American usage, 'monkey-shines' and 'monkeying' are popular identifications of that playfulness and non-utilitarian handling of objects. I shall show that there is even reason to ask whether the standardized patterns observable in early tool-making are not in part derivable from the strictly repetitive motions of ritual, song, and dance, forms that have long existed in a state of perfection among primitive peoples, usually in far more finished style than their tools.

Only a little while ago the Dutch historian, J. Huizinga, in 'Homo Ludens' brought forth a mass of evidence to suggest that play, rather than work, was the formative element in human culture: that man's most serious activity belonged to the realm of make-believe. On this showing, ritual and mimesis, sports and games and dramas, released man from his insistent animal attachments; and nothing could demonstrate this better, I would add, than those primitive ceremonies in which he played at being another kind of animal. Long before he had achieved the power to transform the natural environment, man had created a miniature environment, the symbolic field of play, in which every function of life might be re-fashioned in a strictly human style, as in a game.

So startling was the thesis of 'Homo Ludens' that his shocked translator deliberately altered Huizinga's express statement, that all culture was a form of play, into the more obvious conventional notion that play is an element in culture. But the notion that man is neither *Homo sapiens* nor *Homo ludens,* but above all *Homo faber,* man the maker, had taken such firm possession of present-day Western thinkers that even Henri Bergson held it. So certain were nineteenth-century archeologists about the primacy of stone tools and weapons in the 'struggle for existence' that when the first paleolithic cave paintings were discovered in Spain in 1879, they were denounced, out of hand, as an outrageous hoax, by 'competent authorities' on the ground that Ice Age hunters could not have had the leisure or the mind to produce the elegant art of Altamira.

But mind was exactly what *Homo sapiens* possessed in a singular degree: mind based on the fullest use of all his bodily organs, not just his hands. In this revision of obsolete technological stereotypes, I would go even further: for I submit that at every stage man's inventions and transformations were less for the purpose of increasing the food supply or controlling nature than for utilizing his own immense organic resources and expressing his latent potentialities, in order to fulfill more adequately his superorganic demands and aspirations.

When not curbed by hostile environmental pressures, man's elaboration of symbolic culture answered a more imperative need than that for control over the environment—and, one must infer, largely predated it and for long outpaced it. Among sociologists, Leslie White deserves credit for giving due weight to this fact by his emphasis on 'minding' and 'symboling,' though he has but recovered for the present generation the original insights of the father of anthropology, Edward Tylor.

On this reading, the evolution of language—a culmination of man's more elementary forms of expressing and transmitting meaning—was incomparably more important to further human development than the chipping of a mountain of hand-axes. Besides the relatively simple coordinations required for tool-using, the delicate interplay of the many organs needed for the creation of articulate speech was a far more striking advance. This effort must have occupied a greater part of early man's time, energy, and mental activity, since the ultimate collective product, spoken language, was infinitely more complex and sophisticated at the dawn of civilization than the Egyptian or Mesopotamian kit of tools.

To consider man, then, as primarily a tool-using animal, is to overlook the main chapters of human history. Opposed to this petrified notion, I shall develop the view that man is pre-eminently a mind-making, self-mastering, and self-designing animal; and the primary locus of all his activities lies first in his own organism, and in the social organization through which it finds fuller expression. Until man had made something of himself he could make little of the world around him.

In this process of self-discovery and self-transformation, tools, in the narrow sense, served well as subsidiary instruments, but not as the main operative agent in man's development; for technics has never till our own age dissociated itself from the larger cultural whole in which man, as man, has always functioned. The classic Greek term 'tekhne' characteristically makes no distinction between industrial production and 'fine' or symbolic art; and for the greater part of human history these aspects were inseparable, one side respecting the objective conditions and functions, the other responding to subjective needs.

At its point of origin, technics was related to the whole nature of man, and that nature played a part in every aspect of industry: thus technics, at the beginning, was broadly life-centered, not work-centered or power-centered. As in any other ecological complex, varied human interests and purposes, different organic needs, restrained the overgrowth of any single component. Though language was man's most potent symbolic expression, it flowed, I shall attempt to show, from the same common source that finally produced the machine: the primeval repetitive order of ritual, a mode of order man was forced to develop, in self-protection, so as to control the tremendous overcharge of psychal energy that his large brain placed at his disposal.

So far from disparaging the role of technics, however, I shall rather demonstrate that once this basic internal organization was established, technics supported and enlarged the capacities for human expression. The discipline of tool-making and tool-using served as a timely correction, on this hypothesis, to the inordinate powers of invention that spoken language gave to man—powers that otherwise unduly inflated the ego and tempted man to substitute magical verbal formulae for efficacious work.

On this interpretation, the specific human achievement, which set man apart from even his nearest anthropoid relatives, was the shaping of a new self, visibly different in appearance, in behavior, and in plan of life from his primitive animal forebears. As this differentiation widened and the number of definitely human 'identification marks' increased, man speeded the process of his own evolution, achieving through culture in a relatively short span of years changes that other species accomplished laboriously through organic processes, whose results, in contrast to man's cultural modes, could not be easily corrected, improved, or effaced.

Henceforth the main business of man was his own self-transformation, group by group, region by region, culture by culture. This self-transformation not merely rescued man from permanent fixation in his original animal condition, but freed his best-developed organ, his brain, for other tasks than those of ensuring physical survival. The dominant human trait, central to all other traits, is this capacity for conscious, purposeful self-identification, self-transformation, and ultimately for self-understanding.

Every manifestation of human culture, from ritual and speech to costume and social organization, is directed ultimately to the remodelling of the human organism and the expression of the human personality. If it is only now that we belatedly recognize this distinctive feature, it is perhaps because there are widespread indications in contemporary art and politics and technics that man may be on the point of losing it—becoming not a lower animal, but a shapeless, amoeboid nonentity.

In recasting the stereotyped representations of human development, I have fortunately been able to draw upon a growing body of biological and anthropological evidence, which has not until now been correlated or fully interpreted. Yet I am aware, of course, that despite this substantial support the large themes I am about to develop, and even more their speculative subsidiary hypotheses, may well meet with justifiable skepticism; for they have still to undergo competent critical scrutiny. Need I say that so far from starting with a desire to dispute the prevailing orthodox views, I at first respectfully accepted them, since I knew no others? It was only because I could find no clue to modern man's overwhelming commitment to his technology, even at the expense of his health, his physical safety, his mental balance, and his possible future development, that I was driven to re-examine the nature of man and the whole course of technological change.

In addition to discovering the aboriginal field of man's inventiveness, not in his making of external tools, but primarily in the re-fashioning of his own bodily organs, I have undertaken to follow another freshly blazed trail: to examine the broad streak of irrationality that runs all through human history, counter to man's sensible, functionally rational animal inheritance. As compared even with other anthropoids, one might refer without irony to man's superior irrationality. Certainly human development exhibits a chronic disposition to error, mischief, disordered fantasy, hallucination, 'original sin,' and even socially organized and sanctified misbehavior, such as the practice of human sacrifice and legalized torture. In escaping organic fixations, man forfeited the innate humility and mental stability of less adventurous species. Yet some of his most erratic departures have opened up valuable areas that purely organic evolution, over billions of years, had never explored.

The mischances that followed man's quitting mere animalhood were many, but the rewards were great. Man's proneness to mix his fantasies and projections, his desires and designs, his abstractions and his ideologies, with the commonplaces of daily experience were, we can now see, an important source of his immense creativity. There is no clean dividing line between the irrational and the super-rational; and the handling of these ambivalent gifts has always been a major human problem. One of the reasons that the current utilitarian interpretations of technics and science have been so shallow is that they ignore the fact that this aspect of human culture has been as open to both transcendental aspirations and demonic compulsions as any other part of man's existence—and has never been so open and so vulnerable as today.

The irrational factors that have sometimes constructively prompted, yet too often distorted, man's further development became plain at the moment when the formative elements in paleolithic and neolithic cultures united in the great cultural implosion that took place around the Fourth Millennium B.C.: what is usually called 'the rise of civilization.' The remarkable fact about this transformation technically is that it was the result, not of mechanical inventions, but of a radically new type of social organization: a product of myth, magic, religion, and the nascent science of astronomy. This implosion of sacred political powers and technological facilities cannot be accounted for by any inventory of the tools, the simple machines, and the technical processes then available. Neither the wheeled wagon, the plow, the potter's wheel, nor the military chariot could of themselves have accomplished the mighty transformations that took place in the great valleys of Egypt, Mesopotamia, and India, and eventually passed, in ripples and waves, to other parts of the planet.

The study of the Pyramid Age I made in preparation for writing 'The City in History' unexpectedly revealed that a close parallel existed between the first authoritarian civilizations in the Near East and our own, though most of our contemporaries still regard modern technics, not only as the highest point in man's intellectual development, but as an entirely new phenomenon. On the contrary, I found that what economists lately termed the Machine Age or the Power Age, had its origin, not in the so-called Industrial Revolution of the eighteenth century, but at the very outset in the organization of an archetypal machine composed of human parts.

Two things must be noted about this new mechanism, because they identify it throughout its historic course down to the present. The first is that the organizers of the machine derived their power and authority from a heavenly source. Cosmic order was the basis of this new human order. The exactitude in measurement, the abstract mechanical system, the compulsive regularity of this 'megamachine,' as I shall call it, sprang directly from astronomical observations and scientific calculations. This inflexible, predictable order, incorporated later in the calendar, was transferred to the regimentation of the human components. As against earlier forms of ritualized order, this mechanized order was external to man. By a combination of divine command and ruthless military coercion, a large population was made to endure grinding poverty and forced labor at mind-dulling repetitive tasks in order to insure "Life, Prosperity, and Health" for the divine or semidivine ruler and his entourage.

The second point is that the grave social defects of the human machine were partly offset by its superb achievements in flood control and grain production, which laid the ground for an enlarged achievement in every area of human culture: in monumental art, in codified law, in systematically pursued and permanently recorded thought, in the augmentation of all the potentialities of the mind by the assemblage of a varied population, with diverse regional and vocational backgrounds in urban ceremonial centers. Such order, such collective security and abundance, such stimulating cultural mixtures were first achieved in Mesopotamia and Egypt, and later in India, China, Persia, and in the Andean and Mayan cultures: and they were never surpassed until the megamachine was reconstituted in a new form in our own time. Unfortunately these cultural advances were largely offset by equally great social regressions.

Conceptually the instruments of mechanization five thousand years ago were already detached from other human functions and purposes than the constant increase of order, power, predictability, and, above all, control. With this proto-scientific ideology went a corresponding regimentation and degradation of once-autonomous human activities: 'mass culture' and 'mass control' made their first appearance. With mordant symbolism, the ultimate products of the megamachine in Egypt were colossal tombs, inhabited by mummified corpses; while later in Assyria, as repeatedly in every other expanding empire, the chief testimony to its technical efficiency was a waste of destroyed villages and cities, and poisoned soils: the prototype of similar 'civilized' atrocities today. As for the great Egyptian pyramids, what are they but the precise static equivalents of our own space rockets? Both devices for securing, at an extravagant cost, a passage to Heaven for the favored few.

These colossal miscarriages of a dehumanized power-centered culture monotonously soil the pages of history from the rape of Sumer to the blasting of Warsaw and Rotterdam, Tokyo and Hiroshima. Sooner or later, this analysis suggests, we must have the courage to ask ourselves: Is this association of inordinate power and productivity with equally inordinate violence and destruction a purely accidental one?

In the working out of this parallel and in the tracing of the archetypal machine through later Western history, I found that many obscure irrational manifestations in our own highly mechanized and supposedly rational culture became strangely clarified. For in both cases, immense gains in valuable knowledge and usable productivity were cancelled out by equally great increases in ostentatious waste, paranoid hostility, insensate destructiveness, hideous random extermination.

This survey will bring the reader to the threshold of the modern world: the sixteenth century in Western Europe. Though some of the implications of such a study cannot be fully worked out until the events of the last four centuries are re-examined and re-appraised, much that is necessary for understanding—and eventually redirecting—the course of contemporary technics will be already apparent, to a sufficiently perceptive mind, from the earliest chapters on. This widened interpretation of the past is a necessary move toward escaping the dire insufficiencies of current one-generation knowledge. If we do not take the time to review the past we shall not have sufficient insight to understand the present or command the future: for the past never leaves us, and the future is already here.

CHAPTER TWO

The Mindfulness of Man

1: THE NEED FOR DISCIPLINED SPECULATION

Modern man has formed a curiously distorted picture of himself, by interpreting his early history in terms of his present interests in making machines and conquering nature. And then in turn he has justified his present concerns by calling his prehistoric self a tool-making animal, and assuming that the material instruments of production dominated all his other activities. As long as the paleoanthropologist regarded material objects—mainly bones and stones—as the only scientifically admissible evidence of early man's activities, nothing could be done to alter this stereotype.

I shall find it necessary as a generalist to challenge this narrow view. There is sound reason to believe that man's brain was from the beginning far more important than his hands, and its size could not be derived solely from his shaping or using of tools; that ritual and language and social organization which left no material traces whatever, although constantly present in every culture, were probably man's most important artifacts from the earliest stages on; and that so far from conquering nature or reshaping his environment primitive man's first concern was to utilize his overdeveloped, intensely active nervous system, and to give form to a human self, set apart from his original animal self by the fabrication of symbols—the only tools that could be constructed out of the resources provided by his own body: dreams, images and sounds.

The overemphasis on tool-using was the result of an unwillingness to consider any evidence other than that based on material finds, along with a decision to exclude much more important activities that have characterized all human groups, in every part of the world, at every known period. Although no one part of our present culture may, without risk of serious error, be taken as a clue to the past, our culture as a whole remains the living witness of all that man has undergone, whether recorded or unrecorded; and the very existence of grammatically complex and highly articulated languages at the onset of civilization five thousand years ago, when tools were still extremely primitive, suggests that the human race may have had even more fundamental needs than getting a living, since it might have continued to do the latter on the same terms as its hominid ancestors.

If this is so, what were these needs? These questions are waiting to be answered, or rather, they must first be asked; and they cannot be asked without a willingness to look freshly at the evidence, and to apply rational speculation, fortified with careful analogies, to the large blank spaces of prehistoric existence, when the character of man, as other than a mere animal, was first formed. So far both anthropologists and historians of technics have guarded themselves against speculative error by taking too much for granted, including their own premises; and this has led to greater errors of interpretation than those they have avoided.

The result has been a single-factor explanation of man's original development centered around the stone tool: an oversimplification in method that has now been abandoned elsewhere as inadequate for the general theory of evolution, and for the interpretation of better documented areas of human history.

What has limited scientific investigation, of course, is the fact that as concerns the unrecorded beginnings of man's life—all but the last one or two per cent of his whole existence—one can for the greater part only speculate. This is a hazardous business, whose difficulties are not lessened by scattered finds of fragmentary bones and artifacts, since without some imaginative insight and analogical interpretation these solid objects tell all too little. Yet to refrain from speculation may be even more stultifying, for it gives to man's later recorded history an appearance of singularity and suddenness, as if a different species had come into existence. In talking about the 'agricultural revolution' or the 'urban revolution' we forget how many foothills the race had climbed before it reached those peaks. Let me then present the case for speculation as a necessary instrument for arriving at adequate knowledge.

2 : DEDUCTIONS AND ANALOGIES

Now there are two ways in which the obscurity of man's early development may be partly overcome. The first is commonly used in all the sciences: deducing from the observed facts the unseen or unrecorded context. Thus if one finds a shell fishhook embedded in a datable site, one may infer, from that bit of evidence alone, not only the existence of water, even if the stream bed or lake has dried up, but likewise the presence of human beings who included fish in their diet, who selected the shell and shaped the hook after a model that could exist only in their own mind, who were ingenious enough to use guts or plant fiber for line, and who were sufficiently patient and skillful to catch fish by this method. Although various other animals, and birds, eat fish, no other species than man uses a hook.

These conclusions would be sound, although every trace of positive evidence, other than the hook, had vanished, along with the fisherman's own bones. If one remembered to allow for the possibility that the fishhook might have been transported from a distance, all these deductions would be firm and unshakable. Under similar limitations, with a similar liability to error, anatomists will derive the character of a whole human body from the size and shape of a broken skull and a few teeth—although the ghost of Piltdown Man may rise up to slay them, if they overestimate their powers.

Samuel Butler in his 'Notebooks' once speculated on "the finding of a lot of old photographs at Herculaneum; and they should turn out to be of no interest." But he forgot that such a singular find would in itself reveal many matters of extraordinary interest, which would cause a revolutionary rewriting of history: they would disclose that the Romans had invented photography; and this in turn would show that they had advanced beyond the Greeks in both chemistry and physics, knew the special chemical properties of the halogen group, probably had lenses and had done optical experiments, and had at their disposal metal, glass, or plastics, with smooth surfaces to use as backing for the precipitated image. Such firm knowledge as we have of prehistory rests precisely upon this kind of identification and inference, usually of commonplace, 'uninteresting' objects, like potsherds, animal bones, or pollen grains.

In the realm of prehistory the generalist has a special office, that of bringing together widely separated fields, prudently fenced in by specialists, into a larger common area, visible only from the air. Only by forfeiting the detail can the over-all pattern be seen, though once that pattern is visible new details, unseen even by the most thorough and competent field workers digging through the buried strata, may become visible. The generalist's competence lies not in unearthing new evidence but in putting together authentic fragments that are accidentally, or sometimes arbitrarily, separated, because specialists tend to abide too rigorously by a gentlemen's agreement not to invade each other's territory. Although this makes for safety and social harmony, it ignores the fact that the phenomena studied do not hold to the same principles. Such "No Trespassing" laws, if observed by the generalist, would halt his cross-country excursions, and prevent him from performing his own special function—one oddly similar to that of those Polynesian traders and interpreters who have a license to escape tribal taboos and wander freely over a wide area.

Nevertheless there are certain rules of the game that a generalist must keep, when he tries to fit the scattered pieces of evidence together in a more meaningful mosaic. Even when he seems on the verge of completing an emerging pattern, he must not surreptitiously chip a piece to make it fit, as in a jigsaw puzzle, nor yet must he manufacture any of the pieces in order to fill out the design—although he of course may look in unlikely places for them. He must likewise be ready to scrap any piece of evidence, however he may cherish it, as soon as one of his specialist colleagues discovers that it is suspect, or that it does not fit into the particular environment or the particular time sequence under discussion. When not enough parts exist, the generalist must wait until competent authorities find or fabricate them. But if, on the other hand, his design will not hold all the pieces the specialists present to him, then the pattern itself must be abandoned as faulty; and the generalist must begin all over again with a more adequate frame.

Yet even the specialized scholars who are most ready to decry speculation often succumb to it, chiefly by presenting purely speculative conclusions as if they were well-established facts, without allowing alternative hypotheses. Let me take a case sufficiently remote, I trust, to hurt no one's feelings. From the fact that the thigh bones of Peking Man in the Choukoutien caves were found cracked open, various anthropologists jumped to the conclusion that this creature was a cannibal. *Possibly he was.* But all we actually know is that the bones of unidentifiable humanoid creatures were cracked open, under special conditions that caused them to be preserved.

Apart from marks left by blows on the skull, which might have been made in a futile attempt after death to split it open, or even have happened earlier without causing death, we have no evidence to indicate whether these creatures were killed or died a natural death. If we suppose they were killed, we do not know if homicide was the custom of the country, or whether this was a particular case: certainly no statistically valid conclusions can be drawn on the basis of the few specimens discovered on a single site.

Nor yet do we know if they were killed by their own kind, by another group, or by some more gigantic predatory hominid of a vanished race whose huge teeth were also found in China.

Further, although the skulls indicate that the brains had been extracted through the base, we do not know if the rest of the flesh and marrow was eaten; and finally, even if cannibalism were firmly established, we still do not know if such victims were habitually slain for food, or whether this was done under pressure of starvation—something that has happened occasionally, as with the American pioneers at the Donner Pass, among people to whom cannibalism was abhorrent. Or again, was this extraction of marrow and brain like that of some later peoples, part of a sacrificial, magico-religious ceremony? And finally, was the marrow used as infant's food, or to help start a fire—both attested uses for marrow under primitive conditions?

Coldly appraised, the probability against cannibalism seems about as great as that for it. Few mammals kill their own kind for *food* under any conditions, and the likelihood is that if this perversion had been as common among early men as it was among many later savages, it would have worked against the survival of the groups practicing it, since the human population was extremely sparse and no one would have been safe against his neighbor's hunger. We know, from later evidence, that primitive hunting peoples feel guilty about taking the life of animals they need for food, and even pray the animal for forgiveness or rationalize its death as due to the animal's own wish. Is it so sure then that early man felt less sympathetic toward other human beings—except in surging moments of anger or fear?

Even plentiful examples of cannibalism among 'contemporary' savages —for long it was rife in Africa and New Guinea—do not establish it as common early practice. Just as primitive man was incapable of our own massive exhibitions of cruelty, torture, and extermination, so he may have been quite innocent of manslaughter for food. The assertion that man was always a killer, and a cannibal at that, once he had acquired a taste for flesh, must reckon with these many alternative possibilities. Any flatfooted assumption of man's aboriginal cannibalism rests on no sounder evidence than the contrary hypothesis, and should never have been presented as if it were unquestionable.

Such pitfalls do not rob deduction, scrupulously applied, of its value. All that this argument suggests is that when alternative explanations are equally plausible and may be equally valid, one must leave the question open, and hope some day to find a bit of positive evidence to clinch this or that hypothesis. But if the deduced traits exist in a kindred primate species, as cannibalism does not, and if it also emerges in later human groups, as with close and relatively durable marital attachments, one may with fair safety attribute it also to early man. I propose to adhere to this rule. But the fact that a question worth opening speculatively may have to remain open for an indefinite time is not a sufficient reason for not posing it at all. This holds for practically the entire sphere of human origins.

In short, Leslie White's point is well taken: "Scientists unhesitatingly tackle such problems as the origin of galaxies, stars, planetary systems, and life in general and in its many orders. . . . If the origin of the earth some two billion years ago, or the origin of life untold millions of years ago, can be and is a proper problem for science, why not the origin of culture a mere million years ago?"

The second method available for reading into the aboriginal nature of early man has equally serious drawbacks, so much so that many ethnologists during the last generation often dismissed it as unworthy of scientific consideration. This is the method of analogy: finding parallels between known practices and those that seem indicated by ancient artifacts. During the nineteenth century many primitive tribes, which had for long escaped direct encounters with civilized men, still made their living by grubbing and hunting alone, using stone tools and weapons similar to those that Boucher de Perthes had first uncovered in paleolithic remains in 1832. This led many observers to suppose that the traditions of these contemporary primitives could be directly traced back to ancestral stocks, and that differences in cultural development between groups even corresponded to differences in time.

That was a tempting error. The fallacy lay in forgetting that the surviving 'primitives,' even if they had long ago retreated into a safe niche, had nevertheless continued the process of cultural accumulation, modification, and elaboration: they had long ceased to be culturally naked, and had possibly, as Father Wilhelm Schmidt held in the case of religion, sometimes fallen from a higher early cultural level, through giving later fantasies or inventions free play. Between the language and ceremonial of the Australian aborigines and that of Mousterian culture was a spread of perhaps fifty thousand years: long enough to produce many salient differences, even if certain specific traits might nevertheless have persisted.

Yet once the processes of diversification and degeneration are allowed for, the parallels become suggestive and are sometimes highly illuminating. In fact, one cannot make any valid observation about otherwise unidentified stone tools without reference to similar later tools whose use is known. The Pygmies or the Bushmen of Africa, as they were 'discovered' by Europeans more than a century ago, hunted much the same kinds of animals, with much the same weapons as paleolithic man had used in other parts of the world, too, more than fifteen thousand years before; and the Bushmen had even earlier practiced the Magdalenian art of cave painting. Apart from the differences in climatic conditions and human stature, these people were far closer to their remote ancestral cultures than they were to contemporary European man. Though W. J. Sollas went too far in looking upon the Tasmanian, the Bushman, and the Eskimo as the lineal survivors of their respective paleolithic ancestors, early, middle, and late, their analogous activities give vital clues to earlier cultures.

Using the Eskimos' stone oil lamp, a paleolithic artifact, one can estimate the amount of light available to painters in caves where similar paleolithic lamps were found. From the Eskimos' thorough utilization of their meager natural resources, under climatic conditions similar to those of the Ice Ages, we can gather much information about the kind of economy that made survival possible and even provided a margin for positive cultural development. So, too, the weapons, the masks and costumes and ornaments, the rituals and ceremonies, give hints that illuminate comparable images found in the caves of Spain, France, and North Africa. Yet, as André Leroi-Gourhan has insisted in his recent monumental work, 'Préhistoire de l'Art Occidental,' these hints must not be taken as conclusive demonstrations: the fact that the footprints of boys and youths are found in certain paleolithic caves proves only that the young were permitted or encouraged to enter, not that they underwent an initiation rite. Even the arrows and marks of wounds on some ten per cent of the cave paintings are not free from ambiguity: if they reveal a magic hunting ceremony, they may also, he points out, symbolize the masculine and feminine principle: the penis spear, thrust into the vulva-wound.

One of the reasons that important clues to man's early development may have been missed is that the scientific tradition in the nineteenth century was—whatever the individual practices of some scientists—rationalist, utilitarian, and definitely skeptical about the value of any set of beliefs that tacitly denied science's own uncriticized assumptions. While magic was admitted as an early practice, perhaps interpretable, in James Frazer's terms, as an attempt to control natural forces that would, in the end, succumb to the scientific method, anything like the larger consciousness of cosmic forces that is associated with religion, was treated as negligible. That early man may have scanned the sky, and have reacted to the presence of the sun and moon, may even have identified the seemingly fixed pole star, as Zelia Nuttall suggested more than half a century ago, seemed as removed from possibility as the fact that he had produced works of art.

Yet from the moment *Homo sapiens,* at least, makes his appearance, we find evidences in his attitude toward death, toward ancestral spirits, toward future existence, toward sun and sky, that betray a consciousness that forces and beings, distant in space and time, unapproachable if not invisible, may nevertheless play a controlling part in man's life. This was a true intuition, although it may have taken hundreds of thousands of years before its full import and rational proof could be grasped by the human mind, which now ranges between invisible particles and equally mysterious retreating galaxies.

There seems a likelihood that the earliest peoples, perhaps even before language was available, had a dim consciousness of the mystery of their own being: a greater incentive to reflection and self-development than any pragmatic effort to adjust to a narrower environment. Some of this grave religious response is still present in the legends of creation among many surviving tribal cultures, and notably among the American Indians.

Here again we may judiciously make use of our knowledge of contemporary primitives to cast a fresh light on the beliefs and acts of early man. Take the mysterious imprints of human hands made upon the walls of caves as far apart as Africa and Australia. These imprints are all the more puzzling because so many of these hands show one or more finger joints missing. One would have no clue to this symbol were it not for the fact that there are still tribes equally widely separated where the sacrifice of a finger joint is a rite of mourning: a personal loss to emphasize a greater loss.

Is one not justified in concluding that the mutilated hand on the cave wall is probably a secondary symbol of grief, transferred for perpetuation from the short-lived primary symbol of flesh and bone to a stone surface? Such a symbolic hand may, even more sharply than a cairn of stones, count as the earliest public memorial to the dead. But it is also possible that this rite had an even deeper religious significance; for Robert Lowie describes the same mode of sacrifice among the Crow Indians as part of a truly religious retreat undergone in order to achieve communion with Deity.

In all these cases, the rite itself reveals an eminent human susceptibility to strong feeling about matters of ultimate concern, along with a desire to retain and transmit that feeling. This must have cemented family life and group loyalty, and thus have contributed quite as effectively to survival as any improvement in flaking flint tools. Although in many other species the parent will on occasion sacrifice its life to protect its mate or its young, this voluntary *symbolic* sacrifice of a finger joint is a distinctly human trait. Where such feeling is lacking, as so often in the whole routine of our mechanized, impersonal megalopolitan culture, the human ties become so weak that only stringent external regimentation will hold the group together. Witness the classic case of emotional frigidity and moral depravity in those

New York householders who heard a woman's cries for help in the night, and who watched her being murdered without even phoning for the police —as if they were watching a television program.

In short, to overlook these analogies would be as foolish as to be overconfident in our use of them. At a later stage, as Grahame Clark pointed out, it was contemporary mud-and-reed architecture in Mesopotamia that helped Leonard Woolley to interpret the traces of prehistoric architecture in Sumer; while the circular clay disks found on Minoan sites remained misidentified until Stephanos Xanthodides recognized them as the upper disks of potters' wheels, still in use on Crete. The fact that people in Mesopotamia were still, in the present century, using primitive boats made of bundles of reeds, like those of their ancestors five thousand years ago, as J. H. Breasted delightedly pointed out, gives support to the belief that other artifacts and even customs may have remained stationary over periods our own changeful age finds incredible.

Watchfully and delicately employed, then, analogy is indispensable for the interpretation of the behavior of other human beings, in other ages and cultures: it is wiser to assume, in any doubtful situation, that *Homo sapiens* fifty thousand years ago more closely resembled ourselves than any remoter animal ancestor.

3: STONES, BONES, AND BRAINS

The misleading notion that man is primarily a tool-making animal, who owes his inordinate mental development largely to his long apprenticeship in making tools, will not be easy to displace. Like other plausible conceits, it evades rational criticism, especially since it flatters the vanity of modern 'Technological Man,' that ghost clad in iron.

During the last half century, this short period has been described as the Machine Age, the Power Age, the Steel Age, the Concrete Age, the Air Age, the Electronic Age, the Nuclear Age, the Rocket Age, the Computer Age, the Space Age, and the Age of Automation. One would hardly guess from such characterizations that these recent technological triumphs constitute but a fraction of the immense number of highly diversified components that enter into present-day technology, and make up but an infinitesimal part of the entire heritage of human culture. If only one phase of the remote human past was blotted out—the cumulative inventions of paleolithic man, beginning with language—all these new achievements would be worthless. So much for our boasted one-generation culture.

The enlarged command of extra-human energy, which marks the recent period, along with the wholesale reconstruction of the human habitat, which began five thousand years ago, are both relatively minor events in the age-old transformation of man. Our chief reason for over-rating the importance of tools and machines is that man's most significant early inventions, in ritual, social organization, morals, and language, left no material remains, while stone tools can be associated with recognizable hominid bones for at least half a million years.

But if tools were actually central to mental growth beyond purely animal needs, how is it that those primitive peoples, like the Australian Bushmen, who have the most rudimentary technology, nevertheless exhibit elaborate religious ceremonials, an extremely complicated kinship organization, and a complex and differentiated language? Why, further, were highly developed cultures, like those of the Maya, the Aztecs, the Peruvians, still using only the simplest handicraft equipment, though they were capable of constructing superbly planned works of engineering and architecture, like the road to Machu Picchu and Machu Picchu itself? And how is it that the Maya, who had neither machines nor draught animals, were not only great artists but masters of abstruse mathematical calculations?

There is good reason to believe that man's technical progress was delayed until, with the advent of *Homo sapiens,* he developed a more elaborate system of expression and communication, and therewith a still more cooperative group life, embracing a larger number of members, than his primitive ancestors. But apart from the charcoal of ancient campfires, the only sure evidences of man's presence are the least animated parts of his existence, his bones and his stones—scattered, few in number and difficult to date, even in those later ages when urn-burial, mummification, or monumental inscription were practiced.

Material artifacts may stubbornly defy time, but what they tell about man's history is a good deal less than the truth, the whole truth, and nothing but the truth. If the only clue to Shakespeare's achievement as a dramatist were his cradle, an Elizabethan mug, his lower jaw, and a few rotted planks from the Globe Theatre, one could not even dimly imagine the subject matter of his plays, still less guess in one's wildest moments what a poet he was. Though we would still be far from justly appreciating Shakespeare, we should nevertheless have a better notion of his work through examining the known plays of Shaw and Yeats and reading backward.

So with early man. When we come to the dawn of history we find evidence that makes the pat identification of man with tools highly questionable, for by then many other parts of human culture were extremely well-developed while his tools were still crude. At the time the Egyptians and the Mesopotamians had invented the symbolic art of writing, they were still using digging-sticks and stone axes. But before this their languages had become complex, grammatically organized, delicate instruments, capable of articulating and transcribing a constantly enlarging area of human experience. This early superiority of language indicates, as I shall show later, if not a far longer history, then a more persistent and rewarding development.

Though it was by his symbols, not by his tools, that man's departure from a purely animal state was assured, his most potent form of symbolism, language, left no visible remains until it was fully developed. But when one finds red ochre on the bones of a buried skeleton in a Mousterian cave, both the color and the burial indicate a mind liberated from brute necessity, already advancing toward symbolic representation, conscious of life and death, able to recall the past and address the future, even to conceptualize the redness of blood as a symbol of life: in short, capable of tears and hopes. The burial of the body tells us more about man's nature than would the tool that dug the grave.

Yet because stone tools are such durable artifacts, past interpreters of early culture, with the significant exception of Edward Tylor, tended to attribute to them an importance out of all proportion to the rest of the culture that accompanied them, all the more because that culture remains so largely inaccessible. The mere survival of stone artifacts was enough to establish their pre-eminence. But as a matter of fact, this apparently solid evidence is full of holes; and its inadequacy has been covered over by speculations far more airy than any I shall dare to bring forward.

There is still a doubt, in some cases impossible to resolve, whether heaps of almost shapeless stones, once called eoliths, are the work of nature or man; and there is no tangible evidence to indicate what the so-called hand-axe, the chief tool of early paleolithic peoples for hundreds of thousands of years, was actually used for. Certainly it was not an axe in the modern sense: a specialized tool for felling trees. Even with a more shapely tool or weapon, like the mysterious instrument once called a 'baton de commande-ment,' the original function is still in doubt, though in later times the hole in this short staff was used as an arrow-straightener.

As against such material but tricky evidence, we have, on behalf of our mind-making thesis, an equally solid but also equally uncertain bit of testimony: the human skeleton, all too rarely available in complete form: in particular, the brain case. There is evidence, from other animals than man, cited by Bernhard Rensch, that the frontal lobe, the seat of the more discriminating, selective, and intelligent responses, grows faster than the rest of the brain; and that in man this part of the brain was always more developed than that of the nearest primates.

That development continued among intermediate human types until *Homo sapiens* emerged some fifty or a hundred thousand years ago, when the brain as a whole reached something like its present size and conformation. Unfortunately, the size and weight of the brain is only a rough indication of mental capacity, significant chiefly in comparing related species: the number of active layers, the complexity of neuron connections, the specialization and localization of functions, are even more important, since by bulk or weight a great scientist may have a smaller brain than a prize fighter. Here again what seems solid evidence gives a false sense of certainty.

Still, whatever else man may be, he was from the beginning pre-eminently a brainy animal. What is more, he stands indisputably at the climax of the vertebrate line, with its increasing specialization of the nervous system, which began with the development of the olfactory bulb and the brain stem and added progressively to the quantity and complexity of nervous tissue in the thalamus or 'old brain,' the ancestral seat of the emotions. With the massive growth of the frontal lobe a complete system was organized, capable of handling a larger portion of the environment than any other animal, enregistering sensitive impressions, inhibiting inappropriate responses, correcting unsuccessful reactions, making swift judgements and coherent responses, and not least storing the results in a capacious file of memories.

Given this original organic equipment, man 'minded' more of his environment than any other animal, and so has become the dominant species on the planet. What is even more important, perhaps, is that man began minding himself. The omnivorousness which gave him an advantage over more specialized feeders through many fluctuations in climate and food supply, likewise had its counterpart in his mental life, in his ceaseless searching, his indefatigable curiosity, his venturesome experimentalism. This was restricted at first no doubt to foods, but soon it turned to other resources, since the flint and obsidian that proved the best materials for tools and weapons are not to be found everywhere, and their finding and testing took time. Even primitive man often transported them from considerable distances. With his highly organized nervous equipment, this brainy creature could take more risks than any other animal could afford, because he eventually had something more than the dumb animal insight necessary for correcting his inevitable mistakes and aberrations. And he had, as no other animal shows any sign of, a potential capacity to relate the parts of his experience into organized wholes: visible or remembered, imagined or anticipated. That trait later became dominant in higher human types.

If one could sum up man's original constitution at the moment he became more than a mere animal, chained to the eternal round of feeding and sleeping and mating and rearing the young, one might do worse than describe him as Rousseau did, in his 'Discourse on the Origins of Inequality,' as an "animal weaker than some and less agile than others, but taking him all round, the most advantageously organized of any."

This general advantage may be summed up as his upright posture, his wide-ranging stereoscopic color vision, his ability to walk on two legs, with its freeing of the arms and hands for other purposes than locomotion and feeding. With this went a coordinate aptitude for persistent manipulation, rhythmic and repetitive bodily exercise, sound-making, and tool-shaping. Since, as Dr. Ernst Mayr has pointed out, very primitive hominids with brains hardly larger than those of apes had a capacity for tool-making, the last trait was probably only a minor component in the "selection pressure for increasing brain size." Later, I shall develop these points further and add one or two more traits in man's special mental equipment that have been strangely overlooked.

4: BRAIN AND MIND

The development of the central nervous system liberated man in large degree from the automatism of his instinctual patterns and his reflexes, and from confinement to the immediate environment in time and space. Instead of merely reacting to outer challenges or internal hormonal promptings, he had forethoughts and afterthoughts: more than that, he became a master of self-stimulation and self-direction, for his emergence from animalhood was marked by his ability to make proposals and plans other than those programmed in the genes for his species.

So far, purely for convenience, I have been describing man's special advantages solely in terms of his big brain and complex neural organization, as if these were the ultimate realities. But this is only a portion of the story, because the most radical step in man's evolution was not just the growth of the brain itself, a private organ with a limited span of life, but the emergence of mind, which superimposed upon purely electro-chemical changes a durable mode of symbolic organization. This created a sharable public world of organized sense impressions and supersensible meanings: and eventually a coherent domain of significance. These emergents from the brain's activities cannot be described in terms of mass and motion, electro-chemical changes, or DNA and RNA messages, for they exist on another plane.

If the big brain was an organ for maintaining a dynamic balance between the organism and the environment under unusual challenges and stresses, the mind became effective as an organizing center for bringing about counter-adaptations and reconstructions both in man's own self and in his habitat; for the mind found means of outlasting the brain that first brought it into existence. At the animal level, brain and mind are virtually one, and over a large portion of man's own existence they remain almost undistinguishable—though, be it noted, much was known about the mind, through its external activities and communal products, long before the brain was even identified as the prime organ of mind, rather than the pineal gland or the heart.

In speaking of man's nervous responses, I use brain and mind as closely connected but not interchangeable terms, whose full nature cannot be adequately described in terms of either aspect alone. But I purpose to avoid both the traditional error of making mind or soul into an impalpable entity unrelated to the brain, and the modern error of disregarding as subjective— that is, beyond trustworthy scientific investigation—all typical manifestations of mind: that is, the larger part of man's cultural history. Nothing that happens in the brain can be described except by means of symbols supplied by the mind, which is a cultural emergent, not by the brain, which is a biological organ.

The difference between brain and mind is surely as great as that between a phonograph and the music that comes forth from it. There is no hint of music in the disc's microgroove or the amplifier, except through the vibrations induced via the needle by the record's rotation: but these physical agents and events do not become music until a human ear hears the sounds and a human mind interprets them. For that final purposeful act, the whole apparatus, physical and neural, is indispensable: yet the most minute analysis of the brain tissue, along with the phonograph's mechanical paraphernalia, would still throw no light upon the emotional stimulation, the esthetic form, and the purpose and meaning of the music. An electro-encephalograph of the brain's response to music is void of anything that even slightly resembles musical sounds and phrases—as void as the physical disc that helps produce the sound.

When the reference is to meaning and the symbolic agents of meaning, I shall accordingly use the word mind. When the reference is to the cerebral organization that first receives and records and combines and conveys and stores up meanings, I shall refer to the brain. The mind could not come into existence without the active assistance of the brain, or indeed, without the whole organism and the environing world. Yet once the mind created, out of its overflow of images and sounds, a system of detachable and storable symbols, it gained a certain independence that other related animals possess only in a minor degree, and that most organisms, to judge by outward results, do not possess at all.

Evidence has accumulated to show that both sensory impressions and symbols leave imprints on the brain; and that without a constant flow of mental activity the nerves themselves shrink and deteriorate. This dynamic relation contrasts with the static impression of musical symbols on a phonograph record, which, rather, becomes worn out through use. But the interrelation of mind and brain is a two-way process: for direct electronic stimulation of certain areas of the brain can, as Dr. Wilder Penfield has shown, 'bring to mind' past experiences, in a way that suggests how similar electric currents self-induced may cause irrelevant images to pop up unexpectedly into consciousness, how new combinations of symbols may be effected without deliberate effort, or how breaks in the internal electric circuit may cause forgetfulness or oblivion.

The relations between psyche and soma, mind and brain, are peculiarly intimate: but, as in marriage, the partners are not inseparable: indeed their divorce was one of the conditions for the mind's independent history and its cumulative achievements.

But the human mind possesses a special advantage over the brain: for once it has created meaningful symbols and has stored significant memories, it can transfer its characteristic activities to materials like stone and paper that outlast the original brain's brief life-span. When the organism dies, the brain dies, too, with all its lifetime accumulations. But the mind reproduces itself by transmitting its symbols to other intermediaries, human and mechanical, than the particular brain that first assembled them. Thus in the very act of making life more meaningful, minds have learned to prolong their own existence, and influence other human beings remote in time and space, animating and vitalizing ever larger portions of experience. All living organisms die: through the mind alone man in some degree survives and continues to function.

As a physical organ, the brain is seemingly no bigger and little better today than it was when the first cave art appeared some thirty or forty thousand years ago—unless symbolic impressions have indeed been genetically enregistered and have made the brain more predisposed to mindfulness. But the human mind has enormously increased in size, extension, scope, and effectiveness; for it now has command of a vast and growing accumulation of symbolized experience, diffused through large populations. This experience was transmitted originally by impressive example and imitation and word-of-mouth from generation to generation. But during the last five thousand years, the mind has left its mark on buildings, monuments, books, paintings, towns, cultivated landscapes, and, of late, likewise upon photographs, phonograph records, and motion pictures. By these means, the human mind has in an increasing degree overcome the biological limitations of the brain: its frailty, its isolation, its privacy, its brief life-span.

This is by way of clarifying in advance the approach I shall soon make to the whole development of human culture. But one further point needs to be emphasized lest the reader overlook the basic assumption I have made: that brain and mind are non-comparable aspects of a single organic process. Though mind can exist and endure through many other vehicles than the brain, mind still needs to pass once more through a living brain to change from potential to actual expression or communication. In giving to the computer, for example, some of the functions of the brain, we do not dispense with the human brain or mind, but transfer their respective functions to the design of the computer, to its programming, and to the interpretation of the results. For the computer is a big brain in its most elementary state: a gigantic octopus, fed with symbols instead of crabs. No computer can make a new symbol out of its own resources.

5 : THE LIGHT OF CONSCIOUSNESS

At some stage, suddenly or gradually, man must have awakened from the complacent routines that characterize other species, escaping from the long night of instinctual groping and fumbling, with its slow, purely organic adaptations, its too well memorized 'messages,' to greet the faint dawn of consciousness. This brought an increasing awareness of past experience, along with fresh expectations of future possibility. Since evidence of fire has been found with the ancient bones of Peking Man, the first steps man took beyond animalhood may have been partly due to his courageous reaction to fire, which all other animals warily avoid or flee from.

This playing with fire was both a human and a technological turning point: all the more because fire has a threefold aspect—light, power, heat. The first artificially overcame the dark, in an environment filled with nocturnal predators; the second enabled man to change the face of nature, for the first time in a decisive way, by burning over the forest; while the third maintained his internal body temperature and transformed animal flesh and starchy plants into easily digestible food.

Let there be light! With those words, the story of man properly begins. All organic existence, not least man's, depends upon the sun and fluctuates with the sun's flares and spots, and with the earth's cyclical relations to the sun, with all the changes of the weather and the seasons that accompany these events. Without his timely command of fire, man could hardly have survived the vicissitudes of the Ice Ages. His ability to think under these trying conditions may have depended, like Descartes' first insights in philosophy, upon his ability to remain quiet for long periods in a warm, enclosed space. The cave was man's first cloister.

But it is not by the light of burning wood that one must seek ancestral man's source of power: the illumination that specifically identifies him came from within. The ant was a more industrious worker than early man, with a more articulate social organization. But no other creature has man's capacity for creating in his own image a symbolic world that both cloudily mirrors and yet transcends his immediate environment. Through his first awareness of himself man began the long process of enlarging the boundaries of the universe and giving to the dumb cosmic show the one attribute it lacked: a knowledge of what for billions of years had been going on.

The light of human consciousness is, so far, the ultimate wonder of life, and the main justification for all the suffering and misery that have accompanied human development. In the tending of that fire, in the building of that world, in the intensification of that light, in the widening of man's open-eyed and sympathetic fellowship with all created being, lies the meaning of human history.

Let us pause to consider how different the entire universe looks, once we take the light of human consciousness, rather than mass and energy, to be the central fact of existence.

When the theological concept of an eternity without beginning or end was translated into astronomical time, it became apparent that man was but a newcomer on the earth, and that the earth was only a particle in a solar system that had existed for many billions of years. As our telescopes probed further into space, it became plain, too, that our own sun was but a speck in the Milky Way, which in turn was part of far vaster galaxies and stellar clouds. With this extension of space and time, man, as a physical object, with his limited span of existence, seemed ridiculously insignificant. On first reading, this colossal magnification of space and time seemed to destroy, as mere empty brag and vanity, man's claim to being of central importance; and even his mightiest gods shriveled before this cosmic spectacle.

Yet this whole picture of cosmic evolution in terms of quantitative physical existence, with its immeasurable time and immeasurable space, reads quite differently if one returns to the center, where the scientific picture has been put together: the mind of man. When one observes cosmic evolution, not in terms of time and space, but in terms of mindful consciousness, with man in the central role as measurer and interpreter, the whole story reads quite differently.

Sentient creatures of any order, even the lowly amoeba, seem to be extremely rare and precious culminations of the whole cosmic process: so much so that the organism of a tiny ant, arrested in its development some sixty million years ago, still embodies in its mental organization and in its autonomous activities a higher mode of being than the whole earth afforded before life appeared. When we view organic change, not as mere motion, but as the increase of sentience and self-directed activity, as the lengthening of memory, the expansion of consciousness, and the exploration of organic potentialities in patterns of increasing significance, man's relation to the cosmos is reversed.

In the light of human consciousness, it is not man, but the whole universe of still 'lifeless' matter that turns out to be impotent and insignificant. That physical universe is unable to behold itself except through the eyes of man, unable to speak for itself, except through the human voice, unable to know itself, except through human intelligence: unable in fact to realize the potentialities of its own earlier development until man, or sentient creatures with similar mental capabilities, finally emerged from the utter darkness and dumbness of pre-organic existence.

Note that in the last paragraph I put 'lifeless' in quotation marks. What we call lifeless matter is an illusion, or rather, a now-obsolete description based on insufficient knowledge. For among the basic properties of 'matter,' we know now, is one that for long was ignored by the physicist: the propensity for forming more complex atoms out of the primordial hydrogen atom, and more complex molecules out of these atoms, until finally organized protoplasm, capable of growth, reproduction, memory, and purposeful behavior appeared: in other words, living organisms. At every meal, we transform 'lifeless' molecules into living tissue; and with that transformation come sensations, perceptions, feelings, emotions, dreams, bodily responses, proposals, self-directed activities: more abundant manifestations of life.

All these capacities were potentially present, as Leibnitz pointed out, in

the constitution of the primordial monad, along with the many other possibilities that have still to be plumbed. Man's own development and self-
discovery is part of a universal process: he may be described as that
minute, rare, but infinitely precious part of the universe which has, through
the invention of language, become aware of its own existence. Beside that
achievement of consciousness in a single being, the hugest star counts for
less than a cretinous dwarf.

Physicists now estimate the age of the earth as between four and five
billion years; and the earliest possible evidence of life comes about two
billion years later, though living or semi-living proto-organisms that were
not preserved surely must have come earlier. On that abstract time-scale
man's whole existence seems almost too brief and ephemeral to be noted.
But to accept this scale would betray a false humility. Time-scales are themselves human devices: the universe outside man neither constructs them nor
interprets them nor is governed by them.

In terms of the development of consciousness, those first three billion
years in all their repetitive blankness can be condensed into a brief moment
or two of preparation. With the evolution of lower organisms during the
next two billion years, those imperceptible seconds lengthened, psychologically speaking, into minutes: the first manifestation of organic sensitivity
and autonomous direction. Once the mobile explorations of the backboned
animals began, favored increasingly by their specialized nervous apparatus,
the brain made its first groping steps toward consciousness. After that, as
one species after another followed the same track, despite many branchings
out, arrests, and regressions, the seconds and minutes of mindfulness
lengthened into hours.

There is no need here to detail the anatomical changes and the constructive activities that accompanied the growth of consciousness in other
species, from the bees and the birds to the dolphins and the elephants, or
the ancestral species from which both apes and hominids evolved. But the
final break-through came with the appearance of the creature we now
identify as man, some five hundred thousand years ago, on our present
tentative estimates.

With man's extraordinary development of expressive feeling, impressive
sensitivity, and selective intelligence, which gave rise ultimately to language
and transmissible learning, the hours of consciousness were prolonged into
days. At first this change rested mainly on neural improvements; but as
man invented special devices for remembering the past, for recording new
experience, for teaching the young, for scanning the future, consciousness
lengthened into centuries and millennia: no longer confined to a single
lifetime.

In the late paleolithic period, certain 'Aurignacian' and 'Magdalenian' hunting peoples made another leap upward by transfixing conscious images in painted and sculptured objects. This left a trail that can now be followed into the later arts of architecture, painting, sculpture, and writing, the arts for intensifying and preserving consciousness in a sharable and communicable form. Finally, with the invention of writing some five thousand years or more ago, the domain of consciousness was further widened and lengthened.

When at last it emerges into recorded history, organic duration reverses the mechanical, externalized time that is measured by calendars and clocks. Not how long you live, but how much you have lived, how much meaning your life has absorbed and passed on, is what matters. The humblest human mind encompasses and transfigures more conscious experience in a single day than our entire solar system embraced in its first three billion years, before life appeared.

For man to feel belittled, as so many now do, by the vastness of the universe or the interminable corridors of time is precisely like his being frightened by his own shadow. It is only through the light of consciousness that the universe becomes visible, and should that light disappear, only nothingness would remain. Except on the lighted stage of human consciousness, the mighty cosmos is but a mindless nonentity. Only through human words and symbols, registering human thought, can the universe disclosed by astronomy be rescued from its everlasting vacuity. Without that lighted stage, without the human drama played upon it, the whole theater of the heavens, which so deeply moves the human soul, exalting and dismaying it, would dissolve again into its own existential nothingness, like Prospero's dream world.

The immensities of space and time which now daunt us when with the aid of science we confront them, are, it turns out, quite empty conceits except as related to man. The word 'year' is meaningless as applied to a physical system by itself: it is not the stars or the planets that experience years, still less measure them, but man. This very observation is the result of man's attention to recurrent movements, seasonable events, biological rhythms, measurable sequences. When the idea of a year is projected back upon the physical universe, it tells something further that is important to man: otherwise, it is a poetic fiction.

Every attempt to give objective reality to the billions of years the cosmos supposedly passed through before man appeared, secretly smuggles a human observer into the statement, for it is man's ability to think backwards and forwards that creates and counts and reckons with those years. Without man's time-keeping activities, the universe is yearless, as without his spatial conceptions, without his discovery of forms, patterns, rhythms, it is an insensate, formless, timeless, meaningless void. Meaning lives and dies with man, or rather, with the creative process that brought him into existence and gave him a mind.

Though human consciousness plays such a central part, and is the basis of all his creative and constructive activities, man is nevertheless no god: for his spiritual illumination and self-discovery only carry through and enlarge nature's creativity. Man's reason now informs him that even in his most inspired moments he is but a participating agent in a larger cosmic process he did not originate and can only in the most limited fashion control. Except through the expansion of his consciousness, his littleness and his loneliness remain real. Slowly, man has found out that, wonderful though his mind is, he must curb the egoistic elations and delusions it promotes; for his highest capacities are dependent upon the cooperation of a multitude of other forces and organisms, whose life-courses and life-needs must be respected.

The physical conditions that govern all life hem man in: his internal temperature must be kept within limits of a few degrees, and the acid-alkaline balance of his blood is even more delicate; while different hours of the day affect his ability to use his energies or rally against a disease, and the phases of the moon or changes in the weather willy-nilly have physiological or mental repercussions. In only one sense have man's powers become godlike: he has fabricated a symbolic universe of meaning that reveals his original nature and his slow cultural emergence; and up to a point this enables him to transcend in thought his many creaturely limitations. All his daily activities, feeding, working, mating, are necessary and therefore important: but only to the extent that they vivify his conscious participation in the creative process—that process which every religion recognizes as both immanent and transcendent and calls divine.

Theoretically the present conquest of time and space might make it possible for a few hardy astronauts to circumnavigate every planet in the solar system, or still more improbably, to travel to one of the nearest stars, four or five light years away. Let us grant both projects as within the realm of mechanical if not biological possibility. But even if miraculously successful, these feats would be nothing compared to the deepening of consciousness and the widening of purpose that the history of a single primitive tribe has brought into existence.

Comets travel as fast as man can probably ever hope to travel, and make longer journeys: but their endless space voyages alter nothing except the distribution of energy. Man's most valiant explorations in space would still be closer to a comet's restricted possibilities than to his own historic development; whereas his earliest attempts at self-exploration, which laid the foundation for symbolic interpretation of every kind, above all language, are still far from exhausted. What is more, it is these inner explorations, which date from man's first emergence from animalhood, that have made it possible to enlarge all the dimensions of being and crown mere existence with meaning. In this definite sense human history in its entirety, man's own voyage of self-discovery, is so far the climactic outcome of cosmic evolution.

We now have reason to suspect that the achievement of consciousness may have taken place at more than one point in the universe, even at many points, through creatures that perhaps exploited still other potentialities, or escaped better than man from the arrests and perversions and irrationalities that have checkered human history and which now, as man's powers vastly increase, seriously threaten his future. Yet, though organic life and sentient creatures may exist elsewhere, they are still infrequent enough to make man's achievement of his mind-molded culture infinitely more important than his present conquest of natural forces or his conceivable voyages through space. The technical feat of escaping from the field of gravitation is trivial compared to man's escape from the brute unconsciousness of matter and the closed cycle of organic life.

In short, without man's cumulative capacity to give symbolic form to experience, to reflect upon it and re-fashion it and project it, the physical universe would be as empty of meaning as a handless clock: its ticking would tell nothing. The mindfulness of man makes the difference.

6: MAN'S UNCOMMITTED CREATIVITY

Since man comes at the end of a long, widely ramifying evolutionary development, his singular capacities have, as their underpinning, the cumulative organic experience of many other species that preceded him. Though the old notion that 'man climbs his family tree' must not be taken too literally, the data that indicate the persistence of this rich inheritance in man, from the unicellular blastula through the fishlike gills of the embryo, and onward to the monkeylike fell of the seven months' embryo, cannot be tossed away as so much rubbish. Every organ in man's body, beginning with the blood, has a history that carries back into the earliest manifestations of life, for the salt content of his blood reproduces that of the sea from which the earliest organisms first emerged, whilst his backbone goes back to the early fishes, and the pattern of his belly muscles is already visible in the frog.

Man's own nature has been constantly fed and formed by the complex activities and interchanges and self-transformations that go on within all organisms; and neither his nature nor his culture can be abstracted from the great diversity of habitats he has explored, with their different geological formations, their different mantles of vegetation, their diverse groupings of animals, birds, fish, insects, bacteria, amid constantly changing climatic conditions. Man's life would be profoundly different if mammals and plants had not evolved together, if trees and grasses had not taken possession of the surface of the earth, if flowering plants and plumed birds, tumbling clouds and vivid sunsets, towering mountains, boundless oceans, starry skies had not captivated his imagination and awakened his mind. Neither the moon nor a rocket capsule bears the slightest resemblance to the environment in which man actually thought and throve. Would man have ever dreamed of flight in a world destitute of flying creatures?

Long before any richness of culture had been achieved, nature had provided man with its own master model of inexhaustible creativity, whereby randomness gave way to organization, and organization gradually embodied purpose and significance. This creativity is its own reason for existence and its own reward. To widen the sphere of significant creativity and prolong its period of development is man's only answer to his consciousness of his own death.

Unfortunately these ideas are foreign to our present machine-dominated culture. A contemporary geographer, already living in his imagination on an artificial asteroid, has observed: "There is no inherent merit in a tree, a blade of grass, a flowing stream, or a good soil profile; if in a million years our descendants inhabit a planet of rock, air, ocean, and space ships, it would still be a world of nature." In the light of natural history no statement could be more preposterous. The merit of all the original natural components that this geographer so cavalierly dismisses is precisely that, in their immensely varied totality, they have helped to create man.

As Lawrence Henderson brilliantly demonstrated, in 'The Fitness of the Environment,' even the physical properties of air and water and the carbon compounds were favorable to the appearance of life. If life had begun on the bald, sterile planet that the geographer quoted foresees as a possible future, man himself would have lacked all the necessary resources for his own development. And if our descendants reduce this planet to such a denatured state as the bulldozer, the chemical exterminators, and the nuclear bombs and reactors are already doing, then man himself will become equally denatured, that is to say, dehumanized.

Man's humanity is itself a special kind of efflorescence brought about by the favorable conditions under which countless other organisms have taken form and reproduced. Over six hundred thousand species of plants, over twelve hundred thousand species of animals, helped to compose the environment that man found at his disposal, to say nothing of countless varieties of other organisms: some two million species altogether. As human populations increased and became regionally differentiated and culturally identifiable they in turn introduced further variety. The maintenance of that variety has been one of the conditions of human prosperity; and though much of it is superfluous for man's mere survival, that very superfluity has been an incentive to his questing mind.

The student who asked Dr. Loren Eiseley why man, with his present capacity to create automatic machines and synthetic foods, should not do away with nature entirely, did not realize that, like the geographer I have quoted, he was fatuously cutting the ground from under his own feet. For the capacity to take in and make further use of nature's inexhaustible creativity is one of the underlying conditions for human development. Even lowly primitives seem to understand this basic relationship, though the 'post-historic' minds that are now being assembled and groomed in our multiversities, with their active hatred for whatever resists or threatens to escape the machine's control, evidently do not.

7 : THE SPECIALTY OF NON-SPECIALIZATION

The human race, we can now see retrospectively, had notable qualifications for making use of the earth's abundance; and perhaps one of the greatest of these was its disposition to break away from the restrictions imposed by specialized, single-purpose organs, adapted to a limited environment.

The complex assemblage that forms man's vocal organs began as highly specialized parts for tasting and biting and swallowing food, for inhaling air, for recording natural sounds; but while not ceasing to perform these functions man discovered a new use for them in making and modulating and responding to vocal expressions. Once properly assembled by the mind, lungs, larynx, palate, tongue, teeth, lips, cheek turned out to be a perfect orchestra of wind and percussion and stringed instruments. But even our closest surviving relatives never learned to compose an equivalent playable score. By accident, a few species of birds can mimic without effort the human voice: but it is only for man that the parrot's trick has significance.

But man's very freedom from stereotyped ancestral performances was accompanied by a loss of surety and readiness: for both walking and talking, those characteristic human acts, must be learned; and without doubt the greatest agent in freeing man from organic specialization was his highly developed brain. This concentration at the center both controlled and released every other activity. As his symbolically conditioned acts increased in number and complexity, it was only through the conscious mind that organic balance could be maintained.

The brain seems indeed to have begun as a limited single-purpose organ for receiving information and effecting appropriate motor responses. The oldest part is the olfactory bulb, devoted mainly to smell. Though the sense of smell has become progressively less essential as a guide to human behavior, it remains important in the enjoyment of food and in judging its edibility, or detecting unseen fire; and it is even useful in diagnosing bodily disorders like measles.

The next stage in brain growth increased the range of emotional responses; and before thought could be sufficiently symbolized to guide conduct, it ensured prompt and copious motor reactions—attacking, fleeing, cowering, shrinking, protecting, embracing, and copulating. But the great advance that separates man from his nearest probable relatives came through massive increases in the size and complexity of the forebrain, and therewith of the whole neural system. This mutation, or rather this succession of changes in the same direction, cannot yet be adequately accounted for by any biological theory; though C. H. Waddington, in 'The Nature of Life' has come closest to re-defining the organic changes that facilitate the formation and transmission of 'acquired characters.' The current cover-up phrase, 'selection pressures' explains the results, not the transformation itself.

But the facts themselves are reasonably plain. The size of the earliest cranium that can be identified as human is several hundred cubic centimeters larger than that of any ape; while the skull of latter-day man, as far back as Neanderthal man, is roughly three times that of the earliest Australopithecine hominid—now conjectured to be one of man's immediate forerunners—found in Africa. From this one may infer that there was, besides mere gain in mass, through the increase in the number of neurons and dendrites, a multiplication in the number of possible connections between them in more highly developed human specimens.

For the purposes of abstract thought alone, the brain contains ten thousand times as many components as the most complex computer today. That vast numerical superiority will doubtless diminish with miniaturization in electronics. But purely quantitative comparison does not begin to reveal the qualitative uniqueness of the brain's responses—the richness of odor, taste, color, tone, emotion, erotic feeling which underlies and suffuses both the reactions and the projections that take place in and through the human mind. Were these eliminated, the brain's creative capabilities would be reduced to the level of a computer, able to deal accurately and swiftly with pure abstractions, but paralyzed when faced with those organic concretions that are fatally lost by isolation or abstraction. While most of the 'emotional' responses to color, sound, odor, form, tactile values predate man's rich cortical development, they underlie and enrich his higher modes of thought.

Because of the extremely complex structure of man's large brain, uncertainty, unpredictability, counter-adaptability, and creativity (that is, purposeful novelty as distinguished from randomness) are constitutional functions, embedded in man's complex neural structure. In their readiness to meet unexpected challenges they surpass the surer instinctual patterns and the closer environmental adaptations of other species. But these very potentialities have made it necessary for man to invent an independent realm of stable, predictable order: internalized and under conscious control. The fact that order and creativity are complementary has been basic to man's cultural development; for he has to internalize order to be able to give external form to his creativity. Otherwise, as the painter Delacroix lamented in his diary, his tumultuous imagination would erupt in more images than he is able to hold together or utilize, as in fact it often does in nocturnal dreams.

But note: the oversized brain of *Homo sapiens* cannot be satisfactorily accounted for, at the beginning, as an adaptive mechanism that contributed to man's survival and his increasing domination of other species. Its adaptive contributions were valuable but only partial, for they were for long offset, as they still are, by maladaptations and perversions. For something like a hundred thousand years, the brain remained hugely disproportionate to the work it was called upon to do. As Alfred Russel Wallace pointed out long ago, the potential mental capacities of an Aristotle or a Galileo were already anatomically and physiologically present, waiting to be used, among people who had not yet learned to count on ten fingers. Much of this equipment is still unused, still waiting.

The 'overgrowth' of the brain may well, for a long period of prehistory, have been as much of an embarrassment to *Homo sapiens*' ancestors as a help; for it unfitted them in some degree for a purely instinctual animal role before they had developed any cultural apparatus capable of utilizing these powers. This neural efflorescence, like efflorescence in the botanical realm, is nevertheless typical of many other organic advances; for growth itself rests on the ability of the organism to produce a surplus of energy and organic capability well beyond that needed for bare survival.

Here again the arbitrary Victorian principle of parsimony has misled us: that principle does not do justice to the extravagances and exuberances of nature. Dr. Walter Cannon demonstrated the rationale of organic surpluses in his analysis of the paired organs of the body. The human kidneys have a reserve factor of four: even a quarter of a kidney will be sufficient to keep the organism alive. As concerns man's nervous system, Blake's aphorism holds: it was through the road of excess that man entered the palace of wisdom.

In an early essay, published in 'The Will-to-Believe' but never sufficiently followed up by him, William James put the case more clearly. "Man's chief difference from the brutes," he pointed out, "lies in the exuberant excess of his subjective propensities—his pre-eminence over them simply and solely in the number and in the fantastic and unnecessary character of his wants physical, moral, aesthetic, and intellectual. Had his whole life not been a quest for the superfluous, he would never have established himself as inexpugnably as he has done in the necessary. And from the consciousness of this he should draw the lesson that his wants are to be trusted; that even when their gratification seems furthest off, the uneasiness they occasion is still the best guide of his life, and will lead him to issues entirely beyond his present power of reckoning. Prune down his extravagance, sober him, and you undo him."

One might even, quite speculatively, go further. The gift of a rich neural structure so far exceeded man's original requirements that it may for long have endangered his survival. The very excess of 'braininess' set a problem for man not unlike that of finding a way of utilizing a high explosive through inventing a casing strong enough to hold the charge and deliver it: the limited usability of man's most powerful organ before its products could be stored in cultural containers perhaps accounts for the far from negligible manifestations of irrationality that underlie all recorded or observed human behavior. Either one must count this irrationality as an adaptive mechanism, too—which on the face seems absurd—or one must admit that the increase in 'braininess,' though partly adaptive, was repeatedly undermined by non-adaptive reactions from the same source. Without a large margin for misbehavior the human race could hardly have survived.

Through long, difficult, constructive effort man fabricated a cultural order that served as a container for his creativity, and reduced the danger of its many negative manifestations. But it was only by a multitude of experiments, discoveries, and inventions, lasting over hundreds of thousands of years and involving much more than tools and material equipment, that man created a culture sufficiently exhaustive to make use of even a part of the brain's immense potentialities. That development in turn brought its own dangers and disabilities. Sometimes, when the cultural complex became too elaborately structured, or too firmly fixed on retaining past acquisitions, as it repeatedly did in both early tribal and later civilized groups, it left no room for mental growth in new areas. But on the other hand when the cultural structure weakened and went to pieces, or when for some reason its components could not be internalized, then the unceasingly active, highly charged brain displayed hyperactivity of a manic and destructive kind, behaving like a racing motor that burns itself out for lack of a load. Today, despite the immense cultural assemblage at the disposal of Western man, we are all too intimately acquainted with both possibilities.

8: MIND IN THE MAKING

The size and neural complexity of the human brain brought about two familiar consequences. At birth, the head was big enough to increase the difficulties of childbirth, then, even more significantly, to demand extra care in handling during the period when the brain case was knitting together. This evoked a further display of normal mammalian tenderness. And since so much of man's behavior as was released from purely automatic internal controls had to be freshly learned, through imitation and conditioning, the period of infantile dependence was prolonged. The child's slow maturation demanded continuous parental solicitude and active intercourse, not observed in other less social species whose young can shift for themselves at a much earlier stage. Loving underlies effective learning: indeed, it is the basis of all cultural transference and interchange. No teaching machine can supply this.

The protracted phase of active mothering and minding was critical for the development of culture. A whole year must usually pass before an infant can walk: a still longer period before its babblings take form in recognizable human speech and effective communication. If speech is not acquired before the fourth year, it usually cannot, except in the crudest form, be acquired later, as we know from both deaf-mutes and a few attested examples of wild children; and without speech other forms of symbolism and abstraction remain defective, no matter how ample the physiological capacity of the brain.

The long period of emotional intimacy between parent and child remains essential, we know, to normal human growth: unless love is offered from the beginning, other necessary human qualities, including intelligence and emotional balance, will be deformed. Even with monkeys, experimenters at the University of Wisconsin have demonstrated, in the face of their barely concealed hopes of finding a cheap mechanical substitute for maternal care, the absence of a mother's affection and instruction, *including reproof for misbehavior,* leads to profound neurotic disturbances.

From the fact that the thalamus, the original seat of the emotions, is a far older portion of the vertebrate brain than the frontal cortex, one may conclude that man's emotional development became recognizably human, with a deepening and widening of earlier mammalian sensitivities, before his intelligence had risen sufficiently to produce an adequate means of expression or communication above the animal level. The earliest manifestation of culture that laid a basis for this growing intelligence, as I shall attempt to reconstruct it in the next chapter, was possibly a direct result of this emotional development.

Now, the activities of the brain ramify through all the organs of the body; and in turn, as Claude Bernard long ago proved in relation to the liver, the organs of the body affect the functioning of the brain, so that the slightest disequilibrium, as by a mild infection or muscular fatigue, may impair the working ability of the mind. In its peculiar vigilant way, the brain serves no single function or purpose: certainly—as I hope to show—it would be false even to say that it specializes in 'information' or 'communication,' though it would be correct to say that through the brain every internal activity, every act, every external impression, becomes related to a greater whole over which the mind presides.

Without contact with the greater whole—the domain of significance—man feels homeless and lost, or, as people now say, 'alienated.' Thus the human brain serves at once as a seat of government, a court of justice, a parliament, a marketplace, a police station, a telephone exchange, a temple, an art gallery, a library, a theater, an observatory, a central filing system, and a computer: or, to reverse Aristotle, it is nothing less than the whole polis, writ small.

The activity of the brain is as incessant as that of the lungs or the heart: the minding it sustains spreads over the better part of life. When needed, this activity is partly, though never entirely, under control, though the locus of that control may turn out to be another part of the brain. Even when the brain is not called upon for effort, the electro-encephalograph indicates that this organ is swept by electric impulses that suggest some underlying mental functioning. This predisposition holds, as the physiologist W. Grey Walter has pointed out, even at birth.

When the same scientist sought to make the simplest kind of two-element model of the brain, he noted that it must show in some measure the following attributes: "exploration, curiosity, free-will in the sense of unpredictability, goal-seeking, self-regulation, avoidance of dilemmas, foresight, memory, learning, forgetting, association of ideas, form recognition, and the elements of social accommodation." "Such," he wisely added, "is life!"

Instead of considering man's conventional tool-making as necessarily formative of the brain, would it not be more pertinent to ask what kind of tool could bring about this close relationship with the brain? The answer is almost implied in the question: namely, a kind of tool directly related to the mind, and fabricated out of its own special 'etherealized' resources: signs and symbols.

What concerns us in the present survey of the human past, in relation to technical history, is that there is a high probability that most of the present characteristics of the brain were at man's service, in an undeveloped state, before he uttered an intelligible sound or used a specialized tool. Further development doubtless came about with all of man's widened activities, with a progressive shifting of the higher functions from the 'old brain' to the 'new brain' where they came under conscious direction. The relationship between such increasing mental facility and the genetic enregistration by means of a larger brain with specialized areas and more complex neural patterns is still obscure, and probably cannot be uncovered without a radical change in the biologist's current approach. Until man fabricated a culture, his brain was undernourished and depleted.

What should be plain nevertheless is that man, at the outset of his development, had unusual gifts, far beyond his immediate capacity for using them. The fact that the human brain "is unique in being constantly speculative and expectant" shows that man's growth was not confined to problem-solving in an immediate situation or to adjusting himself to outside demands. He had 'a mind of his own,' as we say: an instrument for posing gratuitous problems, for making insurgent responses and counter-adaptive proposals, for seeking and fabricating patterns of significance. Therewith he showed a tendency to explore unknown territory and try alternative routes, never content for too long to follow a single way of life, no matter how perfect his 'adjustment' to it might be.

Despite the brain's capacity for absorbing information, man does not wait passively for instructions from the outside world. As Adelbert Ames put it, "It is within a context of expectancies that we perceive, judge, feel, act, and have our being."

Those who still take their biological models from physics fail to recognize this essential characteristic of organisms, as distinguished from unorganized matter. Unorganized matter neither records its past nor anticipates its future; whereas every organism has both its past and its potential future built into it, in terms of the life-cycle of its species; and the bodily structure of the higher organisms makes ample provisions for the future, as in the storage of fat and sugar to provide energy for future emergencies, or the progressive ripening of the sex organs, long before they are needed for reproduction.

In man, this pre-vision and pro-vision for the future become increasingly conscious and deliberate in dream images and playful anticipations, in the tentative trying out of imagined alternatives. So far from reacting only to the immediate sight or smell of food, like an animal confined in a laboratory, man goes about seeking it hours, days, or months ahead. Man is, one might say, a born prospector, though he often has only fool's gold to show for his pains. As an actor, he often projects himself in new roles before the play is written, the theater chosen, or the scenery is built.

Not the least contribution of man's extraordinary brain was this heightened concern for the future. Anxiety, prophetic apprehensiveness, imaginative anticipation, which came first perhaps with man's consciousness of seasonal changes, cosmic events, and death, have been man's chief incentives to creativity. As the instruments of culture become more adequate, the function of the mind becomes increasingly that of bringing larger areas of the past and future into coherent and meaningful patterns.

Now the delicacy and complexity of man's nervous organization make him unusually vulnerable: hence he has been constantly frustrated and disappointed, for his reach too often exceeds his grasp, and some of the most formidable obstacles to his development derive not from a harsh environment or from the menace of the carnivorous and poisonous creatures that share his living space, but from conflicts and contradictions within his own misguided or mismanaged self; indeed, often from that very oversensitiveness, that over-imaginativeness, that over-responsiveness, which set him apart from other species. Though all these traits have their basis in man's over-sized brain, their implications for the human condition have too often been forgotten.

Man's potentialities are still more important, infinitely more important, than all his present achievements. This was so at the beginning and it still holds. His greatest problem has been how to selectively organize and consciously direct both the internal and the external agents of the mind, so that they form more coherent and more intelligible wholes. Technics played a constructive part in solving this problem; but instruments of stone and wood and fiber could not be put to work on a sufficient scale until man had succeeded in inventing other impalpable tools wrought out of the very stuff of his own body, and not visible in any other form.

9: 'MAKER AND MOLDER'

If survival were all that mattered to primitive man, he could have survived with no better equipment than his immediate hominid ancestors had possessed. Some further unfocussed need, some inner striving, as difficult, indeed as impossible to explain by the outer pressures of the environment as the transformation of the crawling reptile into the flying bird, must have propelled man on his career; and something else besides scrabbling for food occupied his days. The favoring condition for this development was man's rich neural equipment; but by that very fact he was too open to subjective promptings to harden submissively in the mold of his species, sinking back into the repetitious animal round, and cooperating in the flowing process of organic change.

The critical moment, I suggest, was man's discovery of his own many-faceted mind, and his fascination with what he found there. Images that were independent of those that his eyes saw, rhythmic and repetitive body movements that served no immediate function but gratified him, remembered actions he could repeat more perfectly in fantasy and then after many rehearsals carry out—all these constituted so much raw material waiting to be shaped; and this material, given man's original deficiency of tools other than the organs of his own body, was more open to manipulation than the external environment. Or rather, man's own nature was the most plastic and responsive part of that environment; and his primary task was to fabricate a new self, mind-enriched, different in both appearance and behavior from his given anthropoid nature.

The establishment of human identity is no modern problem. Man had to learn to be human, just as he had to learn to talk; and the jump from animalhood to humanhood, definite but gradual and undatable, indeed still unfinished, came through man's endless efforts to shape and re-shape himself. For until he could establish an identifiable personality he was no longer an animal, but not yet a man. This self-transformation was, I take it, the first mission of human culture. Every cultural advance is in effect, if not in intention, an effort to remake the human personality. At the point where nature left off molding man, he undertook with all the audacity of ignorance to refashion himself.

If Julian Huxley is right, most of the physiological and anatomical possibilities of organic life had been exhausted some two million years ago: "size, power, speed, sensory and muscular efficiency, chemical combinations, temperature-regulation and the rest," and in addition an almost infinite number of changes, major and minor, had been tried out in color, texture, and form. Along purely organic lines radical innovations of any practical use or significance were hardly possible, though many improvements, as in the continued growth of the nervous system in the primates, actually occurred. Man opened a new way to evolution by self-experimentation: long before he attempted to master his physical environment he sought to transform himself.

This feat of self-transformation was accompanied by bodily changes, attested by fragments of surviving skeletons: but the cultural projection of man's selves was far more rapid, since man's prolonged biological infancy left him in a plastic and moldable state which encouraged experimentation with all the available organs of his body, no longer treated respectfully with a view to their purely functional offices, but fashioned for new purposes, as instruments of man's aspiring mind. The highly disciplined practices of Hindu yoga, with their conscious control of breathing, heart-beat, the bladder, and the rectum, for the sake of ultimate mental exaltation only carry to a sophisticated extreme primitive man's initial efforts either to control his bodily organs or put them to other than physiological uses.

Man might even be defined as a creature never found in a 'state of nature,' for as soon as he becomes recognizable *as* man he is already in a state of culture. In the rare exception of 'wild children,' who survived only through animal compassion, they not only lacked both the skill to walk erect and use words, but were closer in character to the animals they had associated with than to men, and in fact never learned to be fully human.

There have been many attempts during the past century to describe man's peculiar nature, but I am not sure that a better characterization has yet been made than that of the Renascence humanist, Pico della Mirandola, though couched in the now unfamiliar language of theology.

"God," observed Pico, "took man as a creature of indeterminate nature, and, assigning him a place in the middle of the world, addressed him thus: 'Neither a fixed body nor a form that is peculiar to thyself have we given thee, Adam; to the end that according to thy longing and according to thy judgment thou mayest have and possess what abode, what form, and what functions thought shalt desire. The nature of all other things is limited and constrained within the bounds of laws prescribed by us. Thou, constrained by no limits . . . shalt ordain for thyself the limits of thy nature. . . . As the maker and molder of thyself in whatever shape thou shalt prefer, thou shalt have the power to degenerate into lower forms of life, which are brutish. Thou shalt have the power, out of thy soul and judgment, to be reborn into the higher forms, which are divine.'" That choice recurs at every stage in man's development.

CHAPTER THREE

In the Dreamtime Long Ago

1: THE NEGLECTED FUNCTION

The exploration of the human psyche during the last half century suggests the need for a more complex, but necessarily more hazardous interpretation of man's early development than has so far been put together. To suppose that man's physical needs accounted for all his activities is precisely what must now be questioned. Doubtless early man had plenty to keep him busy, lest he starve: but sufficient evidence exists, for at least fifty thousand years back, that his mind was not entirely on his work. Was it perhaps rather on the queer things he found passing through his mind? He is the one creature whose external activities, we once more begin to see, cannot be fully accounted for without reference to a most peculiar kind of inner activity: the dream.

Before man emerged from unconsciousness, we must picture him as inarticulate and inexpressive to a degree one finds now only in idiots, for the symbolic instruments of consciousness, images and words, were lacking. We shall not go too far astray, I submit, if we picture this proto-human as a creature pestered and tantalized by dreams, too easily confusing the images of darkness and sleep with those of waking life, subject to misleading hallucinations, disordered memories, unaccountable impulses: but also perhaps animated occasionally by anticipatory images of joyous possibilities.

Now, in enumerating the traits that markedly differentiate man from all other animals, and from modern man's usual picture of himself as a sensible, matter-of-fact creature, man's dream life has usually been overlooked, as if beneath rational notice, simply because its most significant aspects lie outside direct scientific observation. The word 'dream' does not once occur in the index of an otherwise admirable three-volume symposium on Biological and Human Evolution. This seems a curious oversight, even among scientists who still shrink from accepting the methodologically illicit insights into human behavior achieved by psychoanalysis; for strict physiological observations on the brain, following more orthodox scientific procedures, indicate that the brain remains in a state of muted activity even when the rest of the body is quiescent, and the presence of dreams seems indicated, even if uninterpretable, by the rhythmic electrical patterns that accompany sleep.

Possibly many other animals share in some degree this propensity: such reactions as a dog's growling and twitching in sleep would indicate as much. But if so, man's way of dreaming has a feature peculiar to him: it spreads from his nocturnal to his daytime life: in waking moments the dream mingles increasingly with his vocalizing, manipulating, soliloquizing, playing; and at a very early date it leaves a mark on his whole behavior, for man's religious development with its significant 'other world' is inseparable from the dream.

From the beginning, one must infer, man was a dreaming animal; and possibly the richness of his dreams was what enabled him to depart from the restrictions of a purely animal career. Though dogs may dream, no dream ever taught a dog to imitate a bird or to behave like a God. Only in man do we have a plenitude of positive evidence that dream images constantly invade and quicken the waking life: only in him do they sometimes supplant reality, to his own peril or profit. If dreaming did not in fact leave visible imprints on human conduct, it would be only by each individual's own experience in dreaming that he would be able to accept, without incredulity, the reports that others give about their dreams.

Though the development of language and abstract intelligence to some extent replaces or suppresses the rich unconscious images of dream, these images still play a large part and often resume their sway with frightening compulsiveness; so that neurotics, in losing grip on reality, are thrown back on the disordered contents of their own minds. These very developments, good and bad as they may turn out, are but sublimations and extensions of the original functions of dreaming: an overflow of neural activity, the strangely liberating gift of the brain itself.

One grants, of course, that we have no proof that prehistoric man dreamed, in the sense that we have proof that he used fire or made tools. But the existence of dreams, visions, hallucinations, projections, is well attested in all peoples at all times; and since dreams, unlike other components of human culture, are involuntary reactions, over which the dreamer has little or no effective control, it would be absurd to assume that they are a late intrusion. If anything, it is likely that dreams were more abundant, more pervasive, more compelling until man learned to apply internal censorship and intelligent direction, along with the discipline of his practical activities, to narrow their role.

It seems reasonable, then, to suppose that dreams have always had some effect upon human behavior; and it seems likely, if scientifically undemonstrable, that, along with man's speech organs, they helped to make the whole structure of human culture possible. Creativity begins in the unconscious; and its first human manifestation is the dream.

The dream itself testifies to a more general organic exuberance that can hardly be accounted for on any purely adaptive principle, any more than one can account for the possession of 'absolute pitch' in music. Long before Freud, Emerson drew the proper conclusion from his own dream observations. "We know," he wrote in his Journal in March 1861, "vastly more than we digest. . . . I write this now in remembrance of some structural experience last night—a painful waking out of dreams as by violence and a rapid succession of quasi-optical shows following like a pyrotechnic exhibition of architectural or grotesque flourishes, which indicate magazines of talent and intention in our structure." Perhaps the first hint of this immense storehouse of sounds, images, patterns, in all their prodigality and superfluity, came to man in his dreams.

Through the dream, then, man became conscious of a haunting 'supernatural' environment: one that no other animal paid attention to. In this realm, the Ancestors lived on, mysteriously intervening at unexpected moments to give man the benefit of their wisdom or to punish him for departing from well-worn ancestral ways. These archetypal ancestor-images— ghosts, demons, spirits, gods—likewise issued from the same source, and may often have seemed closer to experienced reality than man's immediate surroundings, all the more because he himself had played a part in their creation. By his intercourse with this 'other world' man was perhaps prompted to free himself from his animal docility and fixation.

To ignore the immense psychic overflow from man's cerebral reservoir, to concentrate upon communicating and fabricating as the central human functions, is to pass over a basic clue to man's whole development: namely, the fact that it has always had a subjective, non-adaptive, sometimes irrational side, which frequently threatened his survival. Part of man's development may have taken place as an effort to control and counterbalance the inordinate pre-rational and irrational presentations of his unconscious. Like man's exuberant sexual life, with which the dream is closely connected, the dream holds at least part of the secret of human creativity: but likewise the secret of the blockages and breakdowns of that creativity, the monstrous destructions and debasements which the annals of history so constantly disclose.

With the gradual development of consciousness, civilized man has turned into a far more wakeful creature than any of his animal kindred: he learned to keep himself awake longer and to forget or disregard his dreams, as he suppressed the sloth that more primitive peoples, content to live less effortfully, may succumb to.

This brings us back to a paradoxical possibility, namely that consciousness may have been promoted by the strange disparity between man's inner environment, with its unexpected images and exciting, if disordered, events, and the outer scene to which he awakened. Did this breach between the inner and the outer world not merely cause wonderment but invite further comparison and demand interpretation? If so, it would lead to a greater paradox: that it was the dream that opened man's eyes to new possibilities in his waking life.

2: THE DANGER FROM WITHIN

Though the dream, if this interpretation is correct, was one of nature's most generous gifts to man, it required, before it could fully serve him, to be more firmly disciplined and controlled than any of his other aptitudes. In the raw, vivid state of sleep, the dream, through its very capacity to put together unrelated events or to reveal unacted desires and emotional eruptions, often suggested or incited demented behavior that animals in their wild state, with a few doubtful exceptions, seem altogether immune from.

All through history, man has been both instructed and frightened by his dreams. And he had good reason for both reactions: his inner world must often have been far more threatening and far less comprehensible than his outer world, as indeed it still is; and his first task was not to shape tools for controlling the environment, but to shape instruments even more powerful and compelling in order to control himself, above all, his unconscious. The invention and perfection of these instruments—rituals, symbols, words, images, standard modes of behavior (mores)—was, I hope to establish, the principal occupation of early man, more necessary to survival than tool-making, and far more essential to his later development.

Now, a realization that man's unconscious self often imperilled his life and brought his most sober plans to nothing, is far from a modern discovery, though it was first brought home to us through the bold, quasi-scientific explorations of Freud and Jung. That there is an antagonism between man's unconscious and his conscious selves, between his nocturnal and his daylight personality, has long been on record. Plato in 'The Republic' pointed out that "when the reasoning and humanizing and ruling power is altered . . . there is no conceivable folly or crime—not excepting incest or parricide or the eating of forbidden food—which at such time, when he has parted company with all shame and sense, a man may not be ready to commit. . . . Even in good men, there is a lawless wild-beast nature, which peers out in sleep." On our present hypothesis that large streak of irrationality which colors all human history becomes at least partly explainable. If man was originally a dreaming animal, it seems likely that he was also a disturbed one; and the source of his worst fears was his own hyper-active psyche. Man's early use of the opium poppy and other illusion-producing or tranquillizing plants may well indicate his underlying anxiety.

Modern psychologists, then, have only come abreast of Plato. And with the knowledge we now have of the unconscious—repellent and threatening as its contents often seem—we should have a better insight into the plight of early man. He was obviously, to a degree we can hardly picture today, culturally naked and therefore highly defenseless against internal assaults. Until he had laid down a solid floor of culture above his unformed 'id,' his inner life just to the degree that it had parted from its safe animal lethargy must have been teeming with archaic reptiles and blind monsters of the deep. Would this perhaps explain early man's long identification with the familiar animals around him: did their presence give him a reassuring sense of security that he now, in the very act of further development, had forfeited? They had a stability and poise he had reason to envy.

As soon as one begins to interpret man's pre-cultural state in terms of our present knowledge of the psyche, one realizes that his emergence from animalhood was beset by difficulties that were connected with the extraordinary qualities that made this passage possible and even, once well begun, imperative. Certainly it would be much easier to picture this transition could we still think of man as little more than an exceptionally intelligent and handy ape, at home in an increasingly intelligible and controllable world.

Unfortunately this rational picture does not correspond to either the surviving evidence or the necessary inferences one must make once one wipes out in one's mind all the cultural institutions that have become second nature to us. Before man achieved speech, his own unconscious alone must have been the only impelling voice he recognized, speaking to him in its own teasingly contradictory and confused images. Only a kind of dull doggedness can perhaps account for man's ability to get the better of these treacherous gifts and make something of them.

One of the most revealing hints about this state we derive from the Australian aborigines, who in equipment and habit of life were, when first encountered, almost as close as we can get in the flesh to early man. Deeply conscious of the continued presence of their ancestral spirits, carefully following in their tracks and respecting their injunctions, they still refer to the Alcheringa, 'the dreamtime of long ago,' from which all their valuable knowledge was derived. In various Australian languages, Roheim observes, we find identical words for dream, the mythical past, and ancestors.

What I am suggesting here is that this is no mere figure of speech: it is a reference to an actual period of human development when the inner eye of dream sometimes suppressed the open eye of observation, and thus helped to release man from his natural bondage to the immediate environment and the present moment. In that wordless period there were only two languages: the concrete language of associated things and events, and the ghostly language of dream. Until the dream finally helped to create culture it may have served as an impalpable substitute: tricky, delusive, misleading, but mind-stirring.

Our highly mechanized Western civilization has many devices for limiting the province of the dream: we even canalize the subjective life into collective mechanisms like the radio and television, and let a machine do our dreaming for us. But in childhood and adolescence, the dream still often dominates us, flowing into waking life so actively that the self-absorbed adolescent for hours at a time is 'not there,' 'lost to the world.' Even some of his seemingly awake behavior is little more than an activated dream. In that phase of growth day-dreaming may pervade the individual's whole existence, presenting him with an autodrama hardly different in content from that of sleep, though perhaps more directly related to wishes, like those for sexual fulfillment, that have come close to the surface of consciousness. In 'the dreamtime of long ago,' this may have been the normal state of man, still unable to project the dream into either collective acts or objects.

Do not dismiss this attempt to penetrate man's incommunicable wordless past as wholly empty speculation. Plentiful testimony from earlier cultures points to the central role persistently played by the dream. As A. I. Hallowell has pointed out about a surviving American hunting people: "The Ojibways are a dream-conscious people. . . . Although there is no lack of discrimination between the experience of the self when awake and when dreaming, both sets of experiences are equally self-related. Dream experiences function integrally with other recalled memory images. . . . And far from being of subordinate importance, such experiences are for them often of more vital importance than the events of daily life." The ancient peoples who shaped civilization, the Egyptians, the Babylonians, the Persians, the Romans, were no less heedful of the dream, even though they had a rich repository of culture to draw upon.

3: MAN'S TERRIBLE FREEDOM

In the dream world, space and time dissolve: near and far, past and future, normal and monstrous, possible and impossible merge into a hopelessly disordered conglomeration: what is exceptional is order, regularity, predictability, without which the dream and the world 'outside' are all sound and fury, signifying nothing. Yet from the dream man got his first hint that there is more to his experience than meets his eye: that there exists an unseen world, veiled from his senses and his daily experiences, as real as the food he eats or the hand he grasps.

What we know now by scientific demonstration, through microscopes, telescopes, and X-rays, early man seems to have stumbled upon through the dream: that a large part of our environment is in fact supersensible and only a small part of existence is open to direct observation. If man had not encountered dragons and hippogriffs in dream, he might never have conceived of the atom.

Eventually the primitives who learned to heed the messages of their unconscious, who thus lived by an ancestral wisdom other than the instincts, had at their disposal a source of further development. But if no sufficient outlet could be found, these same demonic powers might lead only to destructive activities.

All through history there is plenty of evidence that the destructive course too often was what followed, sometimes at the very moment when the group's collective energies were enhanced by a superior command of their physical resources. A. L. Kroeber has pointed out that one of man's distant cousins, the chimpanzees, when left to themselves, are studiously destructive: "they love to demolish; like small children who have grown up uncontrolled, they derive immediate satisfaction from prying, ripping, biting, and deliberately smashing. Once they begin, they rarely desist until an object has been reduced to its components."

Kroeber believed that this propensity might explain one phenomenon in human culture, the long precedence in time of chipping over grinding techniques in working stone. But I would urge a complementary interpretation: if the destructive impulse were sufficient unto itself, it should always have ended, like the Yankee whittler's shavings, in useless slivers. But the fact that tools, not just slivers, were produced shows that there is a countertendency in man, equally innate, and even more deeply, or at least more permanently, satisfying: the arts of creation and constructive organization, the deliberate forming of patterns, the putting together of ordered wholes. This principle lies at the base of all organic development, in defiance of the law of entropy; and it is fundamental both to human culture and purposeful development.

Even at the earliest moment of growth, this constructive bent is visible. The infant, too young for language, left completely to himself with a few blocks, will spontaneously place one on top of the other, as Arnold Gesell has experimentally shown, no less surely than he may, at another moment, fling them with a wild gesture to the floor. Thus we have reason to impute to our earliest ancestors the same qualities that Erich Fromm finds in the dream today: "the expression both of the lowest and most irrational and of the highest and most valuable functions of our minds."

Yet, when we attempt to contemplate man during the ages when his cultural acquisitions were few and erratic, one must allow for the possibility that his destructive tendencies may have been more easy to express than his constructive impulses. Precisely for lack of outlets, he may have overcome his blockages and frustrations in outbreaks of rage, in panics of fear, as devoid of rational support as the behavior of contemporary juvenile delinquents who are equally innocent of the disciplines and restraints of a living culture. Going berserk or running amok may have had a long history before history itself began. But fortunately for our remote ancestors these reactions were limited by their feebleness: unarmed man, using only his hands, feet, teeth, can do little damage to other men, still less to the environment: even with a stone or a club, his scope is limited, except in attacking helpless creatures. Real orgies of destruction, vast collective eruptions of hate, became possible only when civilization provided the technical means of accomplishing them. If the dream opened both outlets, circumstances at first probably favored the more benign outcome.

Still, the demonic promptings of the unconscious must be allowed for, in accounting for man's prehistoric development: are they not still with us? "Sleep," as Emerson noted, "takes off the costume of circumstance, arms us with terrible freedom, so that every will rushes to a head." Before man had achieved more than a glimmer of self-consciousness and moral discipline, this terrible freedom may from time to time have turned against himself. Bronislaw Malinowski, it is true, was inclined to belittle this underlayer of savage pathology, for he felt called upon to rectify the gross overemphasis on this disability of those 'civilized' observers who patronizingly underestimated the capacities of contemporary primitive peoples for even logical thought.

But in correcting one mistake Malinowski made another; for he curiously overlooked the formidable irrational components that remain in civilized man's own code and conduct. Possibly the domain of ignoble irrationality has widened in historic times, like the increase of collective destructiveness. But it would be strange if this irrational area had not existed from the very beginning, now increasing, now diminishing, never wholly suppressed, never wholly under control, but always there to be reckoned with, becoming embedded in the very culture that was in part created to cope with it.

Happily this side of our argument is open to demonstration. Let us consider a well-authenticated case of primitive irrationality from South Africa; for it illustrates the main functional aspects of dreaming: illusion, projection, wish-fulfillment, insulation from rational appraisal, and finally the possibility of turning into insane malevolence and destructiveness.

"One May morning in 1856," we are told, "a Xosa girl went to the stream to draw water and there met with members of the spirit world. Later her uncle went to the same place, and spoke with the strangers. . . . The spirits announced that they had come to assist the Xosa in driving the English from the land. After numerous cattle had been killed in accordance with the supernatural instructions as offerings to the spirits, an order went through Umhulakaza, the uncle, that every animal in the herds and every grain of corn in the granaries was to be destroyed. If this were done an earthly paradise would be created: myriads of beautiful cattle would issue forth from the earth and fill the pastures, and great fields of millet would spring forth ready to harvest. Trouble and sickness would be no more; youth and beauty would be restored to old people. The order was carried out: two hundred thousand cattle were slain; and as a result for a time the Xosa almost ceast to exist." (G. M. Theal, 'South Africa.')

Here the natural resentments of a people whose territory has been occupied by presumptuous white foreigners gave rise, since they had no more effective means of evacuating the intruders, to dream images of total deliverance, accompanied by a gigantic sacrifice and copious redemption. Such propulsive archetypal dreams have occurred often in recorded history: a whole series of similar visions of salvation, expressed in our own time by the so-called 'Cargo Cults' of the South Seas, have been sympathetically described by Margaret Mead; and these correspond with still other American Indian cults, like that of the Ghost Dance in the eighteen-nineties, with its promise that the "ancestors would return, game be replenished, white man driven away."

But in view of the palpable failure of such escapist dreams to cope with realities, indeed, their mischievous tendency often to make a bad situation worse, one must ask: How is it that this propensity for dreaming, which gives so many false clues and delusive promptings, which so often proposes disastrous actions or abortive goals, can nevertheless have persisted, without seriously lessening man's prospects for survival? Plainly the dream was a two-faced gift. Had there not been, in the long run, some compensating strength on the creative side—slight, no doubt, but decisive— it must have multiplied the recorded perversities of human conduct beyond any possibility of recovery.

The dangers from man's seething and bubbling unconscious were lessened in time, it would seem, by his special feats of intelligent insight, when finally he was able to utilize language: for he found that the dream must be interpreted skillfully before it could be safely acted upon, and long before we come upon historic evidence for shamans, priests, soothsayers, or oracles, every group probably had its wise old man, who could interpret the dream, blending its own suggestions with those of accredited and well-hoarded ancestral experience. But before that could have happened, early man had a long way to go. Until he had learned to inhibit his instinctual drives, to arrest their immediate translation into action, and to divert his autistic impulses from inappropriate goals, his behavior may sometimes have been as suicidal as that exhibited by the Xosa. But on this hypothesis those who erred too greatly would have been wiped out: so that human culture was favored by the emergence of those whose impulses were sufficiently moderate or under strict enough control to remain close to the animal norm of 'sanity.'

Until a firm basis for order was laid down, we can now see, it was almost as necessary to curb man's creativity as his destructiveness: that is perhaps why the whole weight of culture, down to modern times, has centered on its ties with the past, so that even fresh departures would be disguised as a replenishing of old sources. With good reason, archaic societies distrusted innovators and inventors as heartily as Philip II of Spain, who classed them, not without reason, as heretics. Even today that danger is still with us; for ungoverned creativity in science and invention has re-enforced unconscious demonic drives that have placed our whole civilization in a state of perilous unbalance: all the more because we have cast away at this critical moment, as an affront to our rationality, man's earliest forms of moral discipline and self-control.

The 'instructions' received by our military and political leaders for contriving atomic, bacterial, and chemical means of total human extermination have the same psychological status as the messages recorded by the Xosa girl: they are self-induced hallucinations that wantonly defy all the historic precepts of human experience. The fact that these dreams have been put forward under the pseudo-rational garb of advanced theoretic science and justified as a measure for national 'survival' does not disguise their bottomless malignity and irrationality, with its complete divorce from even an animal's instinct for self-preservation. But unlike the pitiable mistake of the Xosa, the colossal kind of error, or 'accident,' that the Pentagon and the Kremlin have already neatly set the fuse for, would be beyond redemption.

4: THE PRIMAL ARTS OF ORDER

We have at last to trace a path across the broken causeway of ordered activity, now sunken and almost invisible, that led to human culture. We must bridge in our own minds the distance between the immediately apprehended world of the animal, with its limited range and forced choices, and the first liberating glimmers of human intelligence, confused, partly enveloped in the mists of unconsciousness, broken erratically by the dream. We must follow early man across a swampy terrain where, for hundreds of centuries, only a few slippery hummocks of tested knowledge gave him a footing and led him to persevere till he reached the still-narrow shelf of firm earth beyond.

By what art did man build that causeway? The piling of those first stones across the bottomless morass of the unconscious was surely a greater feat than the later building of stone bridges, or even nuclear reactors. I have sought to show that though man's early quick-wittedness may have given him an edge over other animal rivals, freeing him in some degree from more fixed instinctual drives, it did not serve equally well in mastering the vagrant promptings of his hyperactive psyche: or so at least his later record gives us some reason to infer. His inner disorder could hardly have been offset by his hit-or-miss life in pursuit of food, grubbing, foraging, dependent for sustenance upon luck rather than his own steady application, feasting one week, starving the next.

Purely organic functions, indeed, produce their own kind of order and internal balance: the animal instincts are inherently functional and purposeful, and so, within their own context, rational: that is, appropriate to the situation and tending to promote survival, organic fulfillment, and reproduction of the species. But man had to re-form and re-instate these impulses at a higher level, illuminated by consciousness; and to make this transformation possible, he was driven to give some sort of orderly sequence and pattern to his daily activities, learning to associate what was immediately visible with something that had gone before or would come after—at first doubtless with purely bodily functions, as in the perception that green fruit today 'meant' stomach-ache tomorrow.

There must have been a long period when early man's burgeoning capacities pushed him to the verge of conscious ideated intelligence, even as his aimless babblings pushed him to the verge of speech—only to leave him baffled by an inability to express what was still inexpressible. We have all had some experience of this painful state, when a name or a word has dropped from memory, or at a higher stage of incipient thought, when we find that a dawning intuition cannot be translated into communicable speech for lack of a fresh vocabulary. In early man this impotence, this frustration, must have been intensified by the absence of any well-defined gestures that might have served as clumsy substitutes or analogues. Long before words had entered man's mind, he was forced to find another mode of expression.

In that state, what could our primeval ancestors do? They must have been driven to the only approach to language then available: the use of the whole body. No single part could yet suffice; for the organs of language and art had still to be mobilized and drilled. On that low level, both expression and the rudiments of communication exist among a wide range of animals. But we have a late example in literature of this primitive solution of an otherwise unbearable speechless frustration in Herman Melville's story of the British sailor, Billy Budd. Officially accused of treason by a villainous informer, Budd has no words sufficiently strong to record his horror or declare his innocence. Tongue-tied, he answers his accuser by the only form of speech open to him: by striking Claggart down with a deadly blow.

So early man must have first overcome his wordless paralysis by gestures and actions, supplemented by uncouth noises: his bodily movements, to the extent that they were deliberately made, would call for an audience and demand some answering response, as in the little child's insistent "Look at me!" when it has mastered a new trick. The enactment and establishment of meanings was not an individual discovery but a communal feat, performed in concert until gesture and sound were sufficiently shaped to be detached and passed on.

Now it would not be surprising if these first efforts at expression—unlike direct signals—were for no practical purpose whatever, but, as also in other animals, were in response to some seasonal event, prompted by the hormones: probably man was sky-conscious or season-conscious or earth-conscious and sex-conscious long before he was self-conscious. When such acts were performed by a whole group, under a strong emotional reaction, they would tend to be rhythmic and in unison; and since rhythm itself brings organic satisfaction, they would demand repetition, which in turn would bring a further reward in increased skill.

Such formative movements and gestures, if repeated often enough, in the same place, or in the same context of events—the rising sun, the new moon, the growth of vegetation—would begin to acquire meaning, though these pantomimic rituals might have to be performed endless times before that meaning would acquire sufficient definiteness by association to be usable outside the immediate milieu of the shared experience. Even today, Jung has reminded us, men act out ideas long before they understand them; and below the level of consciousness an illness may express a psychic conflict that has not yet found its way to the surface.

In the beginning was the word? No: in the beginning, as Goethe saw, was the act: meaningful behavior anticipated meaningful speech, and made it possible. But the only kind of act that could acquire a fresh meaning was one that was performed in company, shared with other members of the group, constantly repeated and thus perfected by repetition: in other words, the performance of a ritual.

In the course of time, such acts would be set apart and their exact performance would become inviolable: this apartness and this inviolability gave them the new property of being 'sacred.' Before anything that could be called connected discourse came into being, it seems likely that early man had produced sequences of connected actions that had many of the characteristics of verbal language, accompanied by shared feelings that would later be called religious. The proto-language of ritual laid down a strict pattern of order that would eventually be carried into many other expressions of human culture.

In all its many manifestations, ritual seems accompanied by a group of traits that may be innate, for one finds them in the untaught behavior of infants and little children as well as among primitive tribal groups: a need for repetition, a tendency for forming groups whose members respond to each other and imitate each other, and a delight in playful impersonation or make-believe. Sympathy, empathy, imitation, identification—these are the terms an anthropologist like Margaret Mead properly uses for the transmission of all culture: their existence in mammals, their conspicuous existence in other primates, led to their still fuller exercise in man. Within the field of ritual, these traits produced patterns and orderly sequences that could be memorized, repeated, transmitted to younger age-groups. It is surely here that sharable meanings have their beginnings; for naming, describing, relating, commanding, rationally communicating came as relatively late manifestations. Face-to-face communal expression through bodily movements almost surely comes first.

This interpretation of early man's behavior does not, I must point out, rest on pure conjecture. For one sees primordial human ritual against an older background of animal habits: the courting rituals of many animals and birds, the emotional cries uttered in the midst of sexual excitement, the howling of wolf-packs at the moon, the singing of gibbons, which impressed Darwin, the nocturnal dances of elephants, all support the notion that ritual is older than language in man's development and played an indispensable part.

Before they could utter an identifiable word, primitive hominids may have grunted or intoned in chorus: before man learned to sing, he probably engaged in dance and dramatic pantomime. Basic to all these performances was the strict order of ritual: the group's doing the same thing, in the same place, in the same way, without a hairbreadth's deviation. The meanings that emerged from such ritual had a different status—for they implied a higher degree of abstraction—from the visible signals by sight and sound with which animals communicate and learn; and that higher level of abstraction freed meaning, in time, from the here and now.

Long ago, in writing to me about my book, 'Technics and Civilization,' a friend, who is also a distinguished psychologist, observed: "I have always been puzzled by the widespread and spontaneous appearance of regular repetitive acts—touching things a certain number of times, counting steps, repeating words, etc.—in children, usually boys. In adults, it appears as a symptom associated with an unconscious sense of guilt. It is related to magic and religious ritual but is more fundamental than any of them. You find it in the infant who wants a story repeated with exactly the same words— it is the most elementary form of mechanization and is in contrast to the whimsies of impulse."

That clue waited thirty years before I dared to follow it up. Now my only amendment is to suggest that group ritual was historically more basic than any of the later acts he cites. If the sense of guilt was by any chance derived originally from the nature of man's frequently 'criminal' dreams, the mechanical order of ritual may have been a benign alternative to a compulsion neurosis.

By means of ritual, I suggest, early man first confronted and overcame his own strangeness, identified himself with cosmic events outside the animal pale, and allayed the uneasiness created by his huge but still largely unusable cerebral capacities. At a much later stage these inchoate impulses would come together under the rubric of religion. Actions still 'speak louder than words,' and the movements and gestures of ritual were the earliest foreshadowings of human speech. What could not yet be said in words or shaped in clay or stone, early man first danced or mimed; if he flapped his arms he was a bird: if the group formed a circle and revolved in measured steps they might be the moon. In short, what André Varagnac happily identified as the "technology of the body," expressed in dance and mimetic movements, was both the earliest form of any kind of technical order and the earliest manifestation of expressive and communicable meaning.

Once the inviolate pattern of ritual was established, it provided the security of reliable order, an order that primitive man did not at first find in his pressing immediate environment, not even in the starry sky. There were times, over a long period that lasted into the emergence of the ancient civilizations, when even the waxing and waning of the moon, or the return of the sun after the winter solstice, seemed attended by hazards and awakened collective anxiety. Before man could discover and project order outside himself he had first, by constant repetition, to establish it within. In this, the part played by ritual exactitude can hardly be overestimated. The original purpose of ritual was to create order and meaning where none existed; to affirm them when they had been achieved; to restore them when they were lost. What an old-fashioned rationalist would regard as 'meaningless ritual' was rather, on this interpretation, the ancient foundation layer of all modes of order and significance.

By close and ancient association we now tend to tie ritual to religion and even refer to ritual as the special language of religion, since the ultimate mysteries, cosmic and divine, that the high religions confront can only be obscured and mocked by words. But ritual pervades life: any act that lends itself to formalized repetition—even if no more than lunching once a week with a friend or wearing formal evening clothes at a public ceremony—has in it the basic traits of ritual: both the original creation of meaning by a repetitive act, and the later erosion or displacement of meaning by mechanical repetition when the original occasion ceases, or the original impulse has died. Since ritual order has now largely passed into mechanical order, the present revolt of the younger generation against the machine has made a practice of promoting disorder and randomness: but that, too, has turned into a ritual, just as compulsive and as 'meaningless' as the routine it seeks to assault.

5 : THE ART OF MAKE-BELIEVE

Once the human mind had begun to transcend its animal limitations, like-mindedness became an indispensable condition for mutual aid. Ritual promoted a social solidarity that might otherwise have been lost through the uneven development of human talents and the premature achievement of individual differences. Here the ritual act established the common emotional response that made man more ready for conscious cooperation and systematic ideation.

In the setting of shared communal experience, meaning in symbolic form first detached itself from the daily activities of identifying edible plants or inimical animals. Once achieved in pantomime and dance, some of this meaning would pass over to the spontaneous cries that accompanied the common action; and they in turn would take more definite and deliberate form through repetition.

From contemporary accounts of simple peoples, one can picture these aboriginal groups assembling, facing each other, repeating the same gestures, responding to the same facial expressions, moving to the same beat, uttering the same spontaneous sounds—sounds of joy, sounds of mourning, sounds of ecstasy: the members of the group at one with each other. This may have been one of the most profitable paths that led into the realm of speech, long before the necessities of hunting big game made speech an aid to cooperative attack.

Doubtless the development of ritual took countless years before anything like definite, associated, and abstractable meanings rose even dimly into consciousness. But what is striking, what indeed gives color to the notion that ritual is ancestral to all other forms of culture, is something noted by the distinguished philosopher of language, Edward Sapir, in relation to the Australian natives: however destitute a culture may be of clothing or shelter or tools, it will nevertheless show a richly developed ceremonialism. It is not sheer guesswork, but a highly probable inference, to suggest that it was through the social activities of ritual and language, rather than through command of tools alone, that early man flourished; and that tool-making and tool-using long remained backward arts, in comparison with ceremonial expression and speech-making. Man's most important tools at the beginning were those he extracted from his own body: formalized sounds and images and movements. And his efforts to share these goods promoted social solidarity.

On this matter the late Lili Peller's shrewd observation of the play of infants gives a special insight into the office of ritual in the life of early man. Bald, insistent repetition, which would be extremely irksome to an adult, she pointed out, is nevertheless utterly delightful to an infant—as many a weary parent has found out when compelled to repeat the same game or the same story without deviation a score of times.

"Early play," Mrs. Peller notes, "is repetitive because it yields pleasure of high intensity." Did not early man know that elemental infantile pleasure and make the most of it? Both wild spontaneity and monotonous repetition are equally native and equally pleasurable to the very young; and because this native capacity for forms that can be fixed and repeated was so deeply rooted and so subjectively rewarding it seems probable that it provided the groundwork of man's whole development.

In brief, the demand for ritual exactitude, the subjective satisfaction in the repeated rite, the security of seeking and receiving the expected answer, all these counterbalanced man's intense sensitivity and his psychal 'openness' and instability, and thus made possible the higher developments of his mind. But the conditions for ritual belong to the childhood of the race, and to return to a mechanical ritual, in which repetition, empty of potential meaning or purpose, is the sole source of satisfaction, is to regress to an infantile level.

Which, then, is the greater error—the failure to appreciate the basic importance of ritual before man's life had acquired any other mode of significant expression, or the failure to understand the threat to human development in contemporary mechanical mass rituals? For in the latter, order has been completely transferred to the machine, and no mode is internalized or acceptable unless it serves the machine. The specious idea that Marshall McLuhan has put forth in praise of mass communication— that the means are in fact the meaning—indicates a return to ritual at the most infantile, pre-human level.

The primal need for order and the achievement of order by increasingly formalized repetitive acts, I take to be basic to the whole development of human culture. Where this order became sufficiently solid and reliable, man had some control over his own irrational promptings, some security against the disruptive accidents of nature, and not least, some power of prediction over the otherwise often unaccountable behavior of his fellows: finally some ability to translate this order into the natural environment and to discover massive evidence of order in the movements of the planets and in the basic organization of the entire cosmos. But where this pattern of order collapses in the mind, as it does in some brain injuries, the commonest events, as Kurt Goldstein has demonstrated, become unintelligible, and awaken anxiety.

One must not, however, exaggerate the social benefits of the primordial rituals out of which so many other modes of human activity grew, great and far-reaching though these in fact were. For ritual has constantly borne within itself some of the very irrationality it existed to overcome. Susanne Langer, in her spirited exposition of symbolic ritual as an essential agent in human development, properly pointed out that the mind of early man was doing something more than recording and sorting out sense impressions, or labelling and cataloguing the contents of his environment: he was creating a meaningful world, a whole cosmos, with whose formation and manipulation he achieved a success that for long was denied him in the reconstruction of his natural habitat.

But there was a negative side to this whole effort that one must not overlook: an aspect that is still visible today. The author of 'Philosophy in a New Key' failed to account for the residue of gross absurdity, of magical hocus-pocus, of childish self-deception and paranoid inflation, that remains embedded in so much valued ceremonial throughout history. Though ritual provided an orderly channel for man's unconscious impulses, it has often obstructed the application of intelligence and hindered the development of consciousness. Rituals by their very success too often fall back into the automatism of unconscious existence, and so arrest human development.

Let me recall a familiar archeological example. What are we to make of the two hundred imprints of hands in the paleolithic cave at Gargas, a large number of which are severely mutilated, with two, three, or even four fingers missing? Such imprints are found all over the world, in America, India, Egypt, Australia; and they suggest a cult ritual, like the knocking out of the teeth still practiced in some tribes, which must often have handicapped the living. Though order and meaning may well have first taken form in ritual, disorder and delusion, one must allow, were likewise embedded there, and have long kept their hold in magical acts from which even well-disciplined minds are not altogether free. The amputated fingers, like other forms of ritual surgery, such as castration, involve a trait that has no animal parallel: self-induced sacrifice. Though the forms of sacrifice have often been described in detail, they have still not been satisfactorily explained; nor has the sense of guilt, with which both sacrifice and ritualistic repetition have been so often associated. Into that dark corner of the human psyche, the light of consciousness has yet to penetrate.

If anything were needed to establish the antiquity of ritual observances, the difficulty of throwing off ritual formalism even at an advanced stage of civilization would make it more than plausible. Long after languages of great grammatical complexity and metaphysical subtlety had been created, the practice of formal repetition, once so necessary, on this hypothesis, for creating meaning, clung to verbal expression. Even late documents, like the Egyptian burial texts or the Sumerian and Akkadian epic poems, disclose the same facile magic that first created abstract meanings: set phrases, sometimes quite lengthy ones, will be repeated again and again, to the point where the bored modern translator is driven to delete them, with a mere indication by asterisks of their existence. But we need not go back five thousand years for this evidence. In traditional ballads and songs, the repetition of the chorus still gives us some of the same satisfaction; and the fact that such choruses are often composed of nonsense syllables only links them more closely with the practices of primitive man; for words acquire meaning through use and association; and man's first sentences were probably more nonsensical than anything in Edward Lear.

Here again, speculation is permissible only because positive evidence will always be lacking. Geneticists may now breed back to a beast close to the ancestral form of the ox; but there is no prospect of breeding back to primitive man, and still less to the moment when meaning first emerged from ceremonial performances. Though the passage from gesture and bodily exercise to dance and song, from song to speech, seems a plausible one, the three perhaps developed together and the sequence may only indicate a difference in their rates of development.

Maurice Bowra is nevertheless probably justified in holding that the large number of primitive dances that are without words and are entirely self-sufficient and self-explanatory, indicates that this was the original order of development: all the more because surviving songs, prayers, and rituals are extremely limited in range, and often use archaic, no longer intelligible, expressions, without destroying their efficacy.

In short, the superstitious beliefs and rites that seemed to earlier interpreters of ritual, like James Frazer, to be the result of erroneous reasoning, were not unfortunate accidents that retarded human culture but the basis of stable social order and any rational system of interpretation. The act itself was rational, and the purpose valid, even though the contents proved otherwise. What Huizinga says of play is basic to man's early expression in ritual: ritual creates order and *is* order: indeed, it is probably the aboriginal form of that make-believe which is inseparable from human culture: the game, the drama, the ceremony, the contest, in fact the whole range of symbolic performances. Giambattista Vico's aphorism that one truly understands only what one can create applies particularly to this earliest form of creation. Ritual opened the path to the intelligible and the meaningful, and finally to constructive effort of every kind.

This essential function of ritual was long ago perceived by Friedrich Schiller, though in following the romantic revolt against tradition and convention of any kind he described it in disparaging words: "that which always was and ever more returns, which serves tomorrow because it served today: *und die Gewohnheit nennt er seine Amme.*' " Habituation was indeed man's nurse. Well before chipping and grinding stones had bound hand and eye in a firm chain of reflexes, ritual must have established order, preserved the past, and held the new-found world together. But to make ritual prevail, man paid a price: the tendency to overvalue the goods of the past, fearing to disturb them by further innovations, however slight. So far Schiller was right. Habit itself is, to speak paradoxically, the most habit-forming of drugs; and ritual is habit with group re-enforcement. Once ritual had been established as a basis for other forms of order, then, apart from the growth of language, the next step was to project a large part of its compulsive mechanism outside the human personality; and that process may well have taken as long as the original translation of actions into meanings.

So far I have discussed ritual, for the sake of clarity, as if it could be understood as a detached series of collective acts: but these acts had from the beginning, one must infer, a special quality: they had to do with that which was sacred. By 'sacred' I mean set apart from pressures of mere self-maintenance and self-preservation by reason of an important connection between the living and the dead. If ritual was the earliest form of work, it was sacred work; and the place where it was performed was a sacred place—identified by a spring, a great tree or stone, a cave, or a grotto. Those skilled in the performance of such sacred works developed into shamans, magicians, wizards, finally kings and priests: specialists set apart from the rest of the tribe by their superior talents, their gift for dreaming or interpreting dreams, for knowing the order of the ritual and interpreting natural signs.

The creation of this realm of the sacred, 'a realm apart' serving as a connecting link between the seen and the unseen, the temporal and the eternal, was one of the decisive steps in the transformation of man. From the beginning, one must infer, these three aspects of ritual, the sacred place, the sacred acts, and the sacred cult leaders, developed together for religious use at the appointed moment. Yet all the components changed so slowly that they preserved a thread of continuity in the midst of many later changes in the environment or the social order. And we shall not sufficiently understand the concentration of forces that made possible the technological civilizations that emerged in the Fourth Millennium B.C. unless we see that colossal change against the millennial background of the sacred rituals.

6: RITUAL, TABOO, MORALS

From what I have just set forth, it follows that, though the discipline of ritual played an important, indeed indispensable, part in human development, there is little doubt that it succeeded only with a certain loss of creativity. The prevalence of ritual and all its derivative institutional manifestations accounts therefore for both the facts of early human development and its extreme slowness. For long the brakes were far more powerful than the engine they controlled.

Wherever we find archaic man, we find no lawless creature, free to do what he pleases, when he pleases, how he pleases: we find rather one who at every moment of his life must walk warily and circumspectly, guided by the customs of his own kind, doing reverence to superhuman powers, be they the creator Gods of all being, the ghosts and demons associated with his remembered ancestors, or the sacred beasts, plants, insects, or stones of his totem. We can hardly doubt—though this, too, is an inference —that early man punctuated each phase of his development with appropriate rites of passage, those all but universal ceremonies that civilized man lately abandoned, only to manufacture hasty paper substitutes on The Care and Feeding of Infants, or The Sex Problems of Adolescents.

By inhibitions and severe abstentions, no less than by faithful acts of compliance, early man attempted to relate his activities to the unseen powers that surrounded him, seeking to capture some of their potency, forfend their malice, cajole—sometimes magically command—their cooperation. Nowhere is this circumspect attitude more fully revealed than in the two ancient institutions that Freud regarded with such suspicion and naive hostility: totem and taboo.

Now the concept of the totem, as Radcliffe-Brown and Lévi-Strauss have pointed out, displays many ambiguities and contradictions when one examines its varied applications. But so, for that matter, does the equally indispensable concept, the city, which covers a multitude of different urban functions and social needs, under an assemblage of only roughly similar structures. The binding element between all forms of totem is a special relationship of allegiance to a sacred object or force that must be piously respected. On a superficial rational reading, the affiliation of a group to a totem ancestor would be an effort to avoid the disruptive effects of incest in a small community. Hence marriage within the totem was prohibited, and sexual intercourse with a member of the same totem might be punished by death.

Unfortunately, this explanation does not hold. The fact is that a formalized sexual relationship under the banner of the totem grew up along with the maintenance of a normal family pattern, one practiced equally by many other species, even birds. This indicates a peculiarly human ambivalence—or should one call it complementarity?—between the biological and the cultural aspects of life. The complicated kinship regulations of 'primitive' peoples, no less than their taboos, reveal man's early preoccupation with making over his brutish biological drives and giving them a specifically human form, under the strict and deliberate control of his higher brain centers.

The pattern of totem affiliation was re-enforced by the taboo: a Polynesian word which means, simply, "That which is forbidden." Many aspects of life besides sexual intercourse come under this head: certain foods, especially those derived from totem animals, a corpse, a menstruating woman, a chief's game, like surf-riding, or a particular territory. Almost any part of the environment, indeed, may by some accidental association with good fortune or with evil become tabooed.

These prohibitions bear so little relation to common sense practices that one may easily be overwhelmed, as Freud was, by their unfathomable caprice, their willful unreason, their savage censorship of innocent acts; and it would seem, as it did to Freud, that such progress as man has made in achieving rational conduct has been in proportion to his capacity to defy or throw off taboos. This would be a grave error, and it has had even graver consequences. As with Freud's dismissal of religion, it rested on the strange assumption that a practice that did nothing whatever to aid human development, in some cases demonstrably worked counter to it, could nevertheless have persisted for ages with undiminished vigor. What Freud overlooked was something that a better-equipped observer, Radcliffe-Brown, has reminded us of with regard to all forms of ritual: the necessity to differentiate between the method itself and the social end. By invoking sacred powers, by prescribing awful penalties for breaking a taboo, early man built up habits of absolute control over his own conduct. The gain in group solidarity and predictable order for long offset the loss of freedom.

The ostensible purpose of the taboo might be infantile, perverse, unjust, like the denial to women of special privileges enjoyed by the men, and vice-versa at childbirth. But the habit of strictly obeying such commands and prohibitions was essential to man's achieving order and cooperation in other spheres.

Against the lawless absolutism of his unconscious, man needed a lawful counter-force equally absolute. At the beginning the taboo alone provided this necessary balance: it was man's earliest 'categorical imperative.' Along with ritual, with which it is closely connected, the taboo was man's most effective means of ensuring the practice of self-control. Such moral discipline, formalized as habit before it could be justified as a rational human necessity, was fundamental to human development.

Here again, the practice of a surviving primitive people, the Eualayi of Australia, provides an exemplary pattern, in a custom that Bowra reports. As soon as a child begins to crawl, the mother finds a centipede, half cooks it, and then catching the child's hands, beats them with it to the accompaniment of a song:

> Be kind,
> Do not steal,
> Do not touch what belongs to another,
> Leave all such alone.
> Be kind.

Not merely does the human mother exercise authority, but she associates this with a potentially poisonous creature, coupling her positive injunction with the symbolic imprinting of implied punishment. This is positive guidance, neither arbitrary command nor permissiveness. Moral order and mental order thus developed together.

So far has Western society departed from the ancient taboos against murder, theft, and rape that we are now faced with juvenile delinquents who have no inner check against wantonly assaulting other human beings at random 'for kicks' while we have adult delinquents capable of deliberately planning the extermination of tens of millions of human beings, in carrying out, also doubtless for kicks, a mathematical theory of games. Today our civilization is relapsing into a state far more primitive, far more irrational, than any taboo-ridden society now known—for lack of any effective taboos. If Western man could establish an inviolate taboo against random extermination, our society would enjoy a far more effective safeguard against both private violence and still impending collective nuclear horrors than the United Nations or the fallible mechanisms of Fail-Safe.

Just as ritual, if I have correctly interpreted it, was the first step toward effective expression and communication through language, so taboo was the first step toward moral discipline. Without both, man's career might have ended long ago, as so many powerful rulers and nations have ended their lives, in psychotic outbreaks and life-depressing perversions.

Human development at every point rests upon the ability to sustain tensions and control their release. At the lowest level this involves the control of the bladder and the bowels; and above that, the deliberate canalization of bodily appetites and genital urges into socially acceptable channels. What I am suggesting here, finally, is that the strict discipline of ritual, and the severe moral schooling of the taboo, were essential to man's self-control and in turn to his cultural creativity in every sphere. Only those who obey the rules are capable of playing the game; and up to a point, the strictness of the rules and the difficulty of winning without upsetting them increases the enjoyability of playing.

In short, the whole sphere of early man's existence which the modern scientific mind, in its consciousness of intellectual superiority, rejects, was the original source of man's self-transformation from an animal into a human person. Ritual, dance, totem, taboo, religion, magic—these provided the groundwork for man's later higher development. Even the first great division of labor, as A. M. Hocart pointed out, may have been established in ritual, with its fixed offices and functions, before it was carried over into technology. And this all began 'in the dreamtime long ago.'

CHAPTER FOUR

The Gift of Tongues

1: FROM ANIMAL SIGNALS
TO HUMAN SYMBOLS

If we are to take in the full sweep of early man's technical development, we must recognize it as flowing from sources deep in the whole human organism, building on the capacities of his primate ancestors and adding many items they lacked. Manual dexterity played a vital part in this development, but mental dexterity, the capacity to remember, to learn, to anticipate, played an even greater role; and that part of man's achievement which centered in symbols counted for more than the shaping of tools.

Man's greatest need in emerging from animalhood was to transform himself; and the chief instruments of his awakening consciousness were his own gestures and sounds reflected and imitated in those of his fellow creatures. The understanding of this original condition has been largely suppressed because our own culture over-stresses more practical interests.

This emphasis has embarrassed our interpretation of both language and technics; for as the Victorian observer, John Morley, recognized, modern man has prided himself upon being 'thing-minded' rather than 'word-minded,' and by the same token he has gratuitously read 'tools,' 'work,' 'the struggle for existence,' and 'survival' into creative exuberance, spontaneous motor activity, idle play, esthetic elaboration. And worse, he has neglected the early and persistent search for an underlying pattern of meaning that would give value to all his separate, inevitably transient, activities. Primitive man, in contrast to his present-day successors, could not boast of his 'know-how,' but he was far more engrossed in knowing why; and if his precipitate answers turned out too often to be a magic play of words, the fact that he produced these words made even his most commonplace activities significant.

In evaluating the function and purpose of language, our generation tends to begin at the wrong end: we take its most precious specialized characteristics, the property of forming abstract concepts, translating exact observations, and communicating definite messages as if they supplied the original motive for using words. But language was a life-reflecting, life-enhancing instrument long before it could be shaped for the restricted purposes of intelligent communication. The very qualities in language that offend the logical positivists—its vagueness, its indeterminateness, its ambiguity, its emotional coloring, its reference to unseen objects or unverifiable events, in short its 'subjectivity'—only indicate that from the beginning it was an instrument for embracing the living body of human experience, not just the bleached articulated skeleton of definable ideas. Voluminous oral expression must have preceded continent, intelligible speech by untold years.

Fortunately, in the long process of forming the complex structure of language, man did not, as many now do in the name of science, turn aside from life's irrationalities, its contradictions, its unexplorable and unexplainable cosmic mysteries. The abundance of archaic mythological lore indicates, even more definitely than ritual, one of the earliest human concerns. As for the effort to do away with emotionality, presupposing that respect for emotional values necessarily brings about a betrayal of truth, this view overlooks the fact that the very 'dryness' of so-called objective description may in itself be an indication of an unfortunate negative state, with equal dangers of distortion: except for the limited purposes of exact observation, not necessarily a desirable one. Carried into sexual intercourse, for instance, this austerity brings about impotence and frigidity; while in general human relations it produces the characteristic vices of bureaucracy and academicism.

Given the original condition of man, the acquisition of articulate speech, once it was sufficiently detached from animal signals and repetitive, ritualized acts, proved his greatest leap upward into a fully human state, though one must conceive that leap in terms of a slow-motion picture, which probably required more time and effort in the doing than any other phase of human culture. With the aid of vocal expression, man first increased the range of social communion and mutual sympathy. And when he finally reached the stage of intelligible speech, he created a proliferating symbolic world, partly independent of the flux of daily experience, detachable from any specific environment or occasion, and under constant human control as no other part of the world could become for many ages. The domain of significance. Here and here alone man reigned supreme.

This symbolic world paralleled, yet sometimes transcended, that taken in by the senses, because it could be held together in mind and recalled after the source of the sensations had disappeared and the visual memory of them had become dim. If words had crystallized as they were spoken, and left deposits like shells or shards, the paleontologist would hardly have paid any attention to early man's tool-making: the brittle deposits of words, in all their formative stages, would have commanded his attention, though the sheer mass of these verbal midden heaps would have overwhelmed him, and he would have been as baffled over interpreting the living structure of meaning as linguists still are by the Etruscan remains.

As it turned out, the most impalpable and evanescent of man's creations before writing was invented, the mere breath of his mind, has turned out to be the most formative human achievement: every other subsequent advance in human culture, even tool-making, depended upon it. Language not merely opened the doors of the mind to consciousness, but partly closed the cellar door to the unconscious and restricted the access of the ghosts and demons of that underworld to the increasingly well-ventilated and lighted chambers of the upper stories. That this vast inner transformation could ever have been neglected, and the radical changes it effected could have been attributed to tool-making, seems now an incredible oversight.

As Leslie White has well put it, "The ability to symbol, primarily in its expression in articulate speech, is the basis and substance of all human behavior. It was the means by which culture was brought into existence and the means of its perpetuation since the origin of man." That 'universe of discourse' was man's earliest model of the universe itself.

2: THINGS MAKE SENSE

Inevitably, it is only by hints and imperfect analogies that one may approach, even in the imagination, the critical moment in man's development when the highly abstract but fixed signals that animals use were replaced by significant gestures of much wider range, and at last by complex, ordered speech. In discussing ritual, I have sought to picture what 'thought' was like before man could put it into words; but it is only people who have recovered from brain injuries who have had even a shadowy equivalent of this experience—yet as soon as they find words for it they are no longer in the pre-linguistic, pre-human world.

Not that the animal mind is empty of rational associations and funded knowledge, re-enforced by definite signals and appropriate responses. On encountering suddenly a formidable male gorilla, George Schaller reports that he probably saved his life by remembering the meaning of a gorilla gesture—the slow shaking of the head from side to side—as a signal for abating hostility or breaking off contact. When Schaller nodded his head in the same fashion the gorilla turned away.

Nature had indeed paved the way for man's earliest attempts at significance; for there is an aboriginal semantics that precedes all special signals and signs. The 'semantics of concrete existence,' as one may call it, is basic to all languages and modes of interpretation.

Every being, whether it be a star or a rock, a flea or a fish, speaks for itself: its own shape, magnitude, character, identify it and concretely 'symbolize' it: by association, that shape and character constitute its meaning to other higher organisms that encounter it. A lion says 'lion' by its own presence far more emphatically than the word 'lion' even if shouted: and the lion's roar, an abstraction, calls up by association the predatory threat of the animal himself. No words are needed to make an antelope flee. Free-moving animals live in a meaningful environment; and their correct interpretation of these manifold concrete meanings is essential to their physical survival. By an elementary system of signals—cries, barks, gestures—they convey meanings to their own kind: Eat! Flee! Fight! Follow me!

In Swift's Academy of Lagado, the School of Languages proposed to abolish words altogether: in the new 'Pop' speech invented by its professors, things took the place of words. As so often happens in intended satire, this points to the significant fact that the concrete experience of every animal including man 'makes sense' without the intervention of symbols, if the creature is alert and responsive. This Swiftian 'symbolism of things' has in fact left a deep mark on speech, which only a specially contrived language, like that of mathematics, has been able to shake off: for this is essentially the language of myth, metaphor and graphic art, and eventually that of early hieroglyphic writing. However tenuous the final abstractions of the painter or the sculptor, the arts have always revelled in the symbolism of the concrete.

Symbolic figures are first of all living figures. A king is represented by a bull, because a bull is endowed in its own being with the prior meanings of physical power, sexual potency, dominance. This method of representation even lent itself to partial abstraction, as noted in an observation by Backhouse in 1843, quoted by Sollas in 'Ancient Hunters.' "One day we noticed a woman arranging stones; they were flat, oval, about two inches wide, and marked in various directions with red and blue lines. These we learned represented absent friends, and one larger than the rest was a corpulent woman on Flinders Island."

That mode of concrete representation has not entirely disappeared. On my own desk, as I write, an array of stone paperweights still conveys the same primitive messages from distant places and dead people. The male penguin who announces his desire to mate by rolling a pebble toward a prospective female has gone so far in symbolism. But if human communication had kept within that concrete frame of reference, 'speech' would have been like playing a game of chess with real bishops and knights, in which the moving of a castle would have required a whole army of pawns. Only when the semantics of things treated as abstractions yielded to symbolic sounds did the mind have an effective means for re-presenting experience.

From this point of view, it was extremely important for his mental development that man, once he had left his original animal niche, had the run of a far larger territory than any other animal: not merely was he equipped to take in more of the concrete world of arresting patterns and identifiable objects, mineral, vegetable, animal, and human; but these existed in an overwhelming abundance and variety. If man had originally inhabited a world as blankly uniform as a 'high-rise' housing development, as featureless as a parking lot, as destitute of life as an automated factory, it is doubtful if he would have had a sufficiently varied sensory experience to retain images, mold language, or acquire ideas.

The valuable work that has been done recently on communication among social animals and birds establishes what a distance exists between the coded instructions of these creatures and even the simplest usages of human language. The breaking of one particular code by von Frisch, that of the socially advanced honey bees, does not lift their dance ritual to the level of language, though it is true symbolic communication. Animal signals become meaningless when detached from the situation that evokes them. What is more, these signals derive mainly from the past experience of the species: they neither anticipate fresh experience nor open a path to it. To make up for this meager vocabulary and these fixed messages, Konrad Lorenz points out, animals have learned to look more closely at other animals and 'read' their intentions from the faintest physiological indications—an involuntary tremor, or the whiff of a secretion.

Man himself must have been in much the same state before he learned to increase his repertory of expression: indeed these bodily clues to meaning are still useful in human encounters, especially when it is a private emotional state that can be read from a wince, a frown, or a blush. But as a newcomer among the primates, man had largely thrown away the frozen vocabulary of instinct: indeed, his very lack of pre-formed responses put him under the necessity of inventing new gestures and sounds that could be applied in unfamiliar associations and understood by his own kind.

Here again, man's recalcitrance to docile adaptation, his insurgence, as Patrick Geddes used to call it, was an incentive to invention. But endless trouble was involved in this effort: for even if he was a more constant chatterer than any of the apes, the muscular control that turned this infantile flow of sounds into articulate speech could not have been easily achieved.

Before phonetic symbols were formed the images of dream may have served as a sort of transitional make-believe language: the only symbolic language man originally possessed, yet one that has stayed with him, only lightly modified by actual experiences and memories, to this day. But now that the psychoanalysts have given us a key to dream symbols and have taught us some of the odd, deliberately deceptive ways in which this language functions, we realize what a baffling mode of expression it is, and how impossible an instrument of thought. For the dream can present ideas only in disguised story form, a wild masquerade. The dream was perhaps the first glimmer of a significance that transcended the senses: but only when it was harnessed to conscious experience, through words and images, could it be put to constructive use.

So in accounting for man's success with language I am driven back to two points already made. First: man's greatest need was to form and establish his own liberated human personality; and language proved to be better than any kind of cosmetic or surgery for identifying and defining that new, non-animal self, and giving it a more genial social character. Second: the intense satisfaction, indeed infantile delight, in repetition, which was one of early man's outstanding biological traits, laid the basis for language no less than for ritual, while ritual by itself continued useful at a lower level as a universal social cement.

Language, one must suppose, was put together out of many disparate experiments and efforts, in the face of many collapses into incoherence and misunderstanding: so the chances are all against its having developed in one place or at one time, through any single agency or method, or to fulfill any single function. From time to time a sudden outbreak of phonetic invention or semantic insight may have occurred—such as the separation, as Jespersen suggested, of long holophrases into discrete words.

There is historic confirmation for such a hypothetical display of singular linguistic genius in the work of an illiterate Cherokee chief, named Sequoyah, who actually invented a composite syllabic alphabet, with many new signs, in order to put the language of his tribe into writing. But the best proof of early man's inventive ability lies in the product itself. No complex machine that has yet been made approaches the uniformity, the variety, the adaptability, the efficiency of language: to say nothing of its unique capacity, derived from the human organism itself, for orderly growth.

At the beginning, then, ritual and language were the chief means of maintaining order and establishing human identity: an increase of cultural continuity and predictability, the basis for further creativity, were the proofs of their success. Later these offices were partly taken over by the graphic arts, by building, by social organization, moral rules, and codified law. And as these other arts improved, language became better able to fulfill its own special function, that of summarizing experience in concepts and ideal structures of increasing complexity. By means of language, each group progressively organized its immediate impressions, its memories, its anticipations into a highly individualized and articulated design, which continued to embrace and absorb fresh experiences while giving them its own idiomatic stamp. It was mainly by creating these elaborate structures of significance that man eventually mastered—though still imperfectly—the art of becoming human.

Once language mediated in every activity, man could not obey Whitman's urge to go and live with the animals without losing his connections with the real world: that which he had reconstituted in his own mind. The world that was symbolically organized, mainly in language, became more significant, more essential to all specifically human activities, than the raw 'outer' world mutely taken in by the senses, or the private inner world beheld in dream. Thus the transmission of speech, from generation to generation, became the chief task of parental nurture; and the acquisition of the language of the group became more essential to kinship organization than physical consanguinity. Language, far more than tools, established human identity.

Even if the rest of man's history were lost, the vocabularies, the grammars, and the literature of all his present languages would testify to a mind infinitely above the level of any other living creature's. And if some sudden mutation afflicting the progeny of the entire human race resulted in the birth of only deaf-mutes, the outcome would be almost as fatal to human existence as that of a nuclear chain reaction.

Most working tools that were fashioned as late as five thousand years ago were still extremely primitive by present-day standards. But, as I have already pointed out, there was nothing primitive then in the structure of the Egyptian or Sumerian languages, nor yet is there anything primitive in the language of the lowliest tribes now known. Early Victorian observers of 'savage' peoples were scarcely willing to credit them with speech: even Darwin was at fault here. Listening to the natives of the Tierra del Fuego, small bands almost destitute of any material culture except fire, with a single sealskin to keep off the harsh weather, he thought their speech was barely human. But an English clergyman, Thomas Bridges, who lived among a single one of these tribes, the Yahgans, between 1861 and 1879, recorded a vocabulary of thirty thousand Yahgan words.

If identified by their technical equipment, the Yahgans could hardly be said to have reached the level of beavers: it is their language that demonstrated that they had grown to human stature. Though another people, the Arunta of Australia, developed four hundred and fifty signs made by hands and arms, only their spoken words suffice to show that they were more than exceptionally intelligent and expressive animals.

Many non-civilized languages show a grammatical complexity and a metaphysical subtlety, as Benjamin Whorf demonstrated, that in themselves testify to the overwhelming concern their speakers must have had to transform the raw materials of experience into an intelligible, richly patterned whole, related comprehensively to reality, both seen and unseen. These vast symbolic structures were built and transmitted by sounds: a feat of abstracting, associating, memorizing, recognizing, recalling, that at first must have demanded strenuous collective effort. This effort continued in popular speech long after writing was invented, and it still goes on in every living language.

The period during which the aboriginal languages were formed would seem, on this reading, to have been the period of mankind's most intense mental activity—not perhaps duplicated again at such a high level of abstraction until modern times. Without such steadfast application to the formation of mind and the intensification of consciousness, all the tools in the world would not have raised man above the level of the ants and termites. The invention and perfection of language was the work of countless lifetimes, lived at a meager material level, because man's mind was busy day in and day out with more important matters. Until he had learned to speak, his mind had no direct organ of expression. So most of the other components of culture might wait.

The emphasis I have placed here upon dream, ritual, and myth as basic to both the formation of language and the social function it performs, does not overlook the part that the more practical concerns of primitive man also played in its development. Once the initial experiments in verbal symbolization showed some success, one would naturally expect that this new performance would invade many other departments of human activity, to their immense advantage. But on my hypothesis these secondary applications were relatively late ones: the primary effort may have required more than a million years, if we are to account for the specialization and localization of speech functions and the coordinate motor controls in the brain: a point demonstrated by Dr. Wilder Penfield and others during the last generation.

In reconstructing hypothetically the development of language there is a natural temptation to associate it with a specific need or a specific mode of life. One linguist, Révész, goes so far as to say that "speech evolved when it was necessary and not before, and that it evolved for purposes for which it was necessary." But except in the sense that all organic acts, even when unconscious, are purposeful, this is far from obvious. Those who hold to the utilitarian explanation couple the wider use of language with the more extensive employment of tools and with the great increase of brain capacity that occurred something like a hundred thousand years ago. Doubtless they interacted: but tool-making as such calls forth little verbal facility, and most of the knowledge needed can be handed on without verbal instruction, by example, as one best learns to tie a knot.

Some anthropologists and biologists are now inclined to associate the formation of language with the cooperative hunting techniques that must have come into existence when, during the successive glacial periods, man in Europe and Asia became dependent upon hunting big game—the hairy mammoth, the bison, the rhinoceros, and the horse. This hypothesis is all the more attractive, since an increase in brain size seems to have occurred at about the same time. On the surface, hunting seems a better explanation than simple tool-making; there is reason, indeed, to suspect that one extremely primitive mode of communication then came into existence: verbs of command. This form has come down to us in the imperative mood: a useful vocabulary, not a language.

But the hypothetical vocabulary of the hunt, like the later similar vocabulary of military organization, can be encompassed in a few limited sounds: the demand for a prompt response removes all possibility of elaboration or shading. Nothing more is needed for concerted action in surrounding or killing game than a few intelligible words and tones. That in itself was, without a doubt, a valuable contribution to communication, particularly for language meant to control conduct by urgent direction, warning, exhortation, or prohibition; and it still remains useful in situations of danger, when the need for prompt action demands the imperative mood, brevity, and—obedience! Yet even for the purposes of organized hunting and foraging, something more than a hoard of common words, applicable to the immediate task in hand, was necessary: for a hunt has to be planned, particularly if it involves ambushing the game or driving it into a trap. There is scattered evidence in cave art of the re-enactment of the hunt in ritual, probably both as a rehearsal in preparation and as an aftermath of celebration.

Here again I venture to press the point that formal communal ceremonies were essential for creating and refining the vocabulary and grammar of paleolithic man, at least from the Aurignacian culture onward. For while the 'language of the concrete' and words of command might often suffice for present purposes, only a comprehensive language structure can recall the past, anticipate the future, or embrace the invisible and the distant. A general improvement in symbolic thinking was what possibly gave *Homo sapiens* the edge over the earlier Neanderthal types.

Though words are the bricks that go into the architectural structure of language, the entire contents of a dictionary would not constitute a language; and we can hardly call the terse vocabulary of command used in hunting more than an approach to language. Divorced from the act of hunting that gave these words meaning, they would cease to be more useful than a bird's mating call out of season. The same objection holds, incidentally, with even greater force to those theories of language that have sought to trace language back to interjections, expletives, or imitative sounds.

Even more than the hunter's words of command it is likely that a more notable contribution was made by the food-gatherer, for he may, even before the Ice Ages, have contributed one of the earliest and most useful functions of language, that of identification by naming. This phase of identification is one of the earliest traits one finds in the infant's achievement of language. Pattern identification and re-cognition are the first steps to cognition itself.

What we need, to reveal the final achievement of a full-blown language, is a type of explanation similar to that which has lately rounded out the Darwinian doctrine of evolution: a composite model which will include and unify the many different factors that contribute to language, at various stages of man's development, and which will relate what we know of language acquisition—and language loss—in present-day man to what we can surmise, at many removes, in the families and tribes that, at a very early date, pushed this particular aspect of culture to the highest degree of perfection. I have not the linguistic qualifications for this job; possibly no one as yet possesses them. But even an inadequate attempt to present the whole picture, in a blurred sketch, may be more satisfactory than an accurately drawn fragment, unrelated to the prehistoric social background.

3: THE BIRTH OF HUMAN LANGUAGE

The biography of Helen Keller, the blind and deaf mute whose early development was frustrated almost to the point of neurotic breakdown, throws a partial light on the origin of language. Though it has often been cited, it remains too important to be passed over. For almost seven years she lived in darkness and mental isolation, not merely without clues for identifying the world about her, but often full of savage rage because unable to articulate or communicate her own feelings. Intelligible messages between her and the outside world neither came nor went. (Much the same state has recently been explored experimentally under laboratory conditions; and even a short period of such blind, soundless confinement, neither giving out nor receiving messages or sensory clues, may cause a swift breakdown of the personality.)

Then for Helen Keller came the famous moment when she was suddenly able to couple the sensation of water with the symbolic taps made by her teacher on the palm of her hand. With that, the meaning of a word dawned on her: she found a way of coupling symbols with things, sensations, actions, events. The over-used term 'break-through' surely applies to that moment.

One dare not even guess where or when or how such a break-through occurred in man's development: or how many times the opened passage may have led into a blind alley, and thrown him back again, baffled. True: before speech developed, man was in a better state than Helen Keller; for his ears and eyes were open; and things around him had meaning before words gave them wings.

But primordial man, on the other hand, did not have little Helen's great advantage, the intelligent companionship of other human beings in full possession of the art of using signs and symbols, by gesture and touch as well as sound. This makes man's aboriginal state roughly comparable to hers, and gives one license to suppose that at a similar moment—by 'moment' one means many repeated moments over perhaps thousands and thousands of years—he experienced a similar illumination, and, like her, was dazzled by the new possibilities words opened up. Once mere animal signals could be translated into complex human messages, the entire horizon of existence widened.

In evaluating this final step which made possible an extensive and continuous dialogue between man and the universe he inhabited, as well as between the members of a group, one must not forget the persistence of the earlier stage I have dwelt on. I mean the fundamental need for autistic expression: an outer manifestation of the human organism's insistent neural activity, and its heightened responsiveness. This oral need, as Edward Tylor long ago shrewdly noted, was admirably illustrated in the case of Laura Bridgman, because, "being blind as well as deaf and dumb, she could not even have imitated words by seeing them made. Yet she would utter sounds, as 'Ho-o-ph-ph' for wonder, and a sort of chuckling or grunting as a sign of satisfaction. When she did not like to be touched, she would say 'F!' Her teachers used to restrain her from making inarticulate sounds; but she felt a great desire to make them, and would sometimes shut herself up and 'indulge herself in a surfeit of sounds.' "

One final step remained to be taken: yet this came about so slowly that the results were in operation long before they impinged on consciousness. This was the passage from the symbolic translation of immediate things and events to creating new entities and situations within the mind, purely by the manipulation of symbols. In this latter change, it was not the individual words or phrases that conveyed meaning but the patterns formed by combinations of words, each varying with the speaker, the occasion, and the contents.

That abstract sounds could bring to mind actual people, concrete places and objects, was the fundamental magic property of speech: but there was an even more potent magic in the fact that these same or similar sounds, differently organized, might bring into the mind events that had ceased, or project entirely new experiences. This was the passage from the closed codes of the animal's world to the open languages of man: one full of endless potentialities, that at last matched the unfathomable potentialities of man's own brain. When language had reached this point, both past and future became a living part of the present.

With the development of language, its main components, autistic expression, social communion, group identification, and intelligent communication persisted and interacted: in living speech, they are still almost inseparable, though for the practical purposes of conveying information, the first three are minimized or ruled out. The original expressive aspect of language, which still lingers in the color, tone, rhythm, and stress of words, cannot be presented except in oral intercourse; and something essential of man's own nature would disappear, if, with one-way communication and a pragmatic over-emphasis on abstract thought, he lost contact with those parts of his own nature which cannot be so processed.

And how important this intimate range of expression was in the development of man! In first forming human character, in establishing group identity, and creating a conscious solidarity that did not depend solely upon kinship or cohabitation in a particular area, conceptual thought played little part. The formative function of language, in establishing a fully human self, is lost in any reduction of speech to a mere communication system. Languages, for all their wealth of abstract terms, still show the marks of their primeval office: the disciplining of the unconscious, the establishment of a coherent and stable social order, the perfection of the social bond.

Mark how the subtlest gradations of tone and pronunciation, suffusing all words and sentences, identifies the 'in-group,' whether it be that of a tribe, a caste, a village, a region, or a nation; while the possession of a particular vocabulary quickly reveals status and vocation without further evidence. No other art rivals speech in inviting a contribution from each member of the group: no other art expresses individuality so definitely, so economically.

Though birds use vocal warnings to exclude others from their territory, language for long served man as a unifying agent to keep his separate communal organizations within bounds. Linguistically, each group is surrounded by an invisible wall of silence, in the form of a different language group. The multiplicity of existing languages and dialects (some four thousand in all), despite the unifying processes of trade, transportation, and travel, suggests that the expressive and emotive functions of language remained as important in the formation of a culture as the function of communication: if nothing else, they prevented a flattening out of human potentiality through mechanization. Hence one of the first efforts of a political conqueror is to suppress the popular language of the conquered; and the most effective means of defence against such suppression, first suggested by Rousseau, is the revival of the national language and its literature.

4: THE NURTURE OF LANGUAGE

Our speculations upon the beginnings of language would be worthless if they did not have for support a certain amount of contemporary observation—though doubtless the last hundred thousand years of language development have left genetic changes that are recorded in the very facial expressions and babblings of an infant, before he utters a word.

Imitation, 'consciousness of kind,' identification, ritual order—where and when do these begin? No one can say. Jespersen's tracing back of the origin of language to the play of lovers may be accepted, like the commands of hunters, as one of a hundred different sources: but the archetypal situation for language training lies, as he also saw, in the relation of mother and child. Early on, the baby has the bodily beginnings of symbolic expression—reaching and grasping, gurgling and smiling, howling and bawling.

By bodily movements, voice, facial expression, the baby brings about a response from the part of the environment most necessary to him, his mother; and there the basic human dialogue begins. At first, mother and milk are one. But at the point where 'mama' brings mother, not milk, and where 'milk' brings milk, not mother, there is a slowly grasped but often repeated situation that corresponds to Helen Keller's sudden breakthrough: special sounds stand for things, relations, acts, feelings, desires. At that moment the crude earlier bow-wow and dingdong theories of the origin of language completely collapse, for the true symbol, a marriage of internal need with external experience, at last comes forth.

This intimate family situation may seem remote from the communal ritual expressions that both Susanne Langer and I take to be basic; but not if one remembers that the education of the infant, even before it is out of the mother's arms, takes place within the ambiance of a larger adult group. Margaret Mead has taken due note of this larger environment in which motherhood itself functions. She observes that "when a Manus child learns from an adult or an older child to say a word . . . the teacher sets up an imitative sing-song: the child says pa piven, the adult says pa piven . . . as many as sixty times. Here the learning can be said to proceed by imitation of a specific act. . . . Imitation of this sort begins a few seconds after birth, when one of the officiating midwives . . . imitates the cry of the newborn child." Here is the first 'imprinting' of order, moral authority, meaning.

Not the first nor yet the hundredth association between word and movement and gesture and dream-ridden inner state would have produced the first faint glimmer of coherent meaning. Probably years and centuries of such effort, sustained only by acts which for long were enjoyed only for their own sake, went into the formation of language. Without the foundation in a fixed, indeed compulsive, ritual, the unexpected result, meaningful sounds, could never have been achieved: a whole world of meaning that revealed an increasingly meaningful world. Whatever its many imaginable sources, the fabrication of language was no happy series of accidents, still less a vacation hobby, or after-work recreation: it was rather the sustained and purposeful occupation of early man from the moment of his emergence.

Without these laborious imitative, repetitive efforts, beginning, as I suppose, in an originally wordless but not soundless ritual, the delicate coordinations of the vocal organs would never have become sufficiently articulate to produce the stable phonetic elements of speech: speech would have remained an incoherent flux of sounds, inimitable and useless. Thus a certain amount of mechanical drill was essential to even the simplest speech; and such drill must have been far more constant than tool-making or hunting.

But we must not overlook the vital connection between *all* physical movement and the acquisition of speech, for this has now been established independently by psychologists. In the case of children whose speech has been retarded or has become disordered, they have found that the child's ability to handle words can be recovered by re-training his motor behavior through inducing him to resume the earlier posture of crawling, the stage that usually accompanies, or slightly precedes, the first efforts at speech.

The Australian aborigines seem to have discovered this important connection long before modern investigators—as might seem likely in view of our hypothesis about the primacy of ritual. When a child is a year old, just before he is ready for speech, the Berndts report, the grandparents will teach the baby the steps of a simple dance. The old people may thus be recapitulating the very association that first made connected speech possible: all the more likely if we remember that the primitive dance itself is a repetitive activity. Obviously, the child is ready for ritual and speech long before he is ready for work. Marxian linguists have stubbornly overlooked this obvious biological fact.

In following this passage from animal signals to coherent human speech, man could not possibly have guessed his destination until he finally reached it and beheld the results: only now, indeed, do we have sufficient information to take in that whole passage and fill in, with our imagination, the otherwise untraceable stages of this journey. Once man had created even the beginnings of language, there was no turning back: he had to hold on to speech for dear life, for he had lost forever many of his pre-lingual animal reactions.

Significantly even chimpanzees lack certain vital instinctive responses: they do not know how to copulate or nurse their babies unless they are reared in the presence of older animals and have acquired the proper knack by imitation. In cases of brain damage affecting the speech centers, the rest of the personality is affected, until sometimes other parts of the brain take over this specialized function. Without speech associations, the world one sees is no longer as meaningful as it is to other animals. In one case I observed, the loss of coherent speech in senility even produced an illusion of blindness: what the eye beheld had become 'invisible': it no longer 'made sense.' Lacking words, the modes of meaning that other animals utilize vanish.

What distinguishes language from isolated gestures and signs, no matter how numerous, is that it forms a complex ramifying structure, which in its conceptual entirety presents a *Weltbild* or comprehensive symbolic framework capable of embracing many aspects of reality: not a static representation like a picture or a sculpture, but a moving picture of things, events, processes, ideas, purposes, in which every word is surrounded by a rich penumbra of original concrete experiences, and every sentence brings with it some degree of novelty, if only because time and place, intention and recipient, change its meaning. Contrary to Bergson, language is the least geometric, the least static, of all the arts.

Among many primitive peoples, anthropologists have discovered, the tribe feels that it has a heavy responsibility for ensuring, by ritual and verbal spells performed punctually from day to day, that the sun shall rise and the universe shall not fall apart. This is a far more intelligent reading of the actual office of language than the contemporary belief that man's command of words makes no difference whatever, that consciousness is an illusion, that all human behavior is capable of being translated, by means of suitable mechanical apparatus and symbolic abstractions, into a quantitative system, free from subjectivity, that does not need further human interpretation. About this assumption only one question remains to be asked: How much meaning will be left in the world when the scientific observer eliminates his own subjective contribution? No mechanical system knows the meaning of meaning.

One more point deserves perhaps to be emphasized. We have good reason to think that only to the extent that sounds and words could be standardized and fixed could individualized meanings be derived from their different combinations and sequences. To manipulate the infinite number of variables opened up by language, words themselves must remain relatively constant, just as to achieve the complexities of the protein molecule, carbon, oxygen, hydrogen, and nitrogen atoms have under normal conditions to remain stable. Obviously it was not the words themselves—as discrete containers of meaning—but their power of combination that gave language its ability to enter into every function of man's life, every aspect of his habitat, every impulse of his nature.

This suggests a different interpretation of the relation of exact verbal formulae to magic than the common one: namely, that words originally were not merely a means to the performance of magic, but were in themselves the archetypal form of magic. The right use of words created for the

first time a new world seemingly under human control: any departure
from meaningful order, any confusion of tongues, was fatal to this magic.
The passion for mechanical precision which man now pours into science
and technics stems originally, if I guess correctly, from the primordial magic
of words. Only if the right word were used in the right order did the spell
work.

Robert Braidwood notes that a similar standardization can be detected
in fairly early paleolithic culture in the shaping of tools. Once a good
functional shape of hand-axe was achieved, it was repeated and not wantonly modified. While the two modes of standardization doubtless, in time,
re-enforced each other, the standardization of language was more essential,
and would seem, on the basis of their comparative rates of improvement
and development, to have come first.

Without that strictness of standardization, without that emphasis on
magical correctness, man's earliest words might have dissolved into thin
air, leaving not a trace, long before writing could be invented. Awe and
reverence for the word, as for the magic spell, was probably needed to keep
language from being eroded or mutilated when passing from mouth to
mouth. In the formative stage of language, that compulsive orderliness was
essential; language was, in its own right, 'sacred,' inviolable.

If meanings were not standardized and stabilized in words, so that
changes would take generations or even centuries before they were generally accepted, each person would soon speak a private language to which,
as with an infant, only those in most intimate contact would have a clue:
baby talk. And if words changed as rapidly as the events they describe, we
would be back again in a pre-lingual state, unable to hold experience in
our minds. For individual words are containers; and as I observed in 'The
City in History' containers can serve their function only if they change more
slowly than their contents.

What Révész observes at a later stage of language is implied almost from
the beginning: "Without the verbal formulation of subjective experience
and ethical standards, self-consciousness is incomplete and self-knowledge
and self-control equally so." The subjective ordering of experience reached
a higher stage in language, in its intensification of consciousness and
rationality, than was possible by ritual or taboo.

In our day, unfortunately, the reverse process has become evident. The
present failure to use the words 'good' and 'bad,' 'higher' and 'lower,' in
judging conduct, as if such differences were unreal, and such words nonsensical, has brought on a total de-moralization of behavior. Yet so important is
the directive and formative function of language that the essential human
values now secretly re-assert themselves in topsy-turvy form: for intellectual

confusion, crime, perversion, debasement, torture, random murder have in the language of many of our contemporaries become 'good' while rational thought, continence, personal probity, and loving-kindness have become 'bad' and hateful. This negation and corruption of language is a plunge into a murkier darkness than that from which man emerged when he first achieved speech.

We can now perhaps understand why one of the greatest and most influential of moralists, Confucius, relied upon two instruments for re-establishing the social order of his time on a sound footing. One was the restoration of the ancient rituals, the other was the clarification of language. These were the two most ancient instruments of social cooperation and control, basic to any further advances in humanization.

If the building up of the complex structure of language was the central feat of human culture, that effort itself, as most linguists now hold, must have been begun with the emergence of the first hominids. But the difficulty of creating, not a few score words, but a highly organized structure comparable in its purposeful self-direction to a living organism, capable of encompassing almost every aspect of experience, not merely identifying things, but interpreting processes, functions, relationships, mechanisms, goals, must have demanded unflagging effort.

In that effort language itself, by its very successes, happily furnished the extra incentive needed. This linguistic concentration may well explain the relative slowness in fabricating other necessary instruments of culture for the greater part of half a million years. And now that the reverse process is swiftly taking place in all the arts—the relapse of articulate speech into sloppy grammar, inarticulate mumble, and willful written gibberish, we can perhaps understand the immense effort needed to create the complex structures of significance that made it possible for primitive groups to become human.

No modern technological device surpasses in the articulation of its parts or its functional fitness the qualities of the least important language. Lévy-Bruhl has pointed out that, in the language of the small Ngeumba Tribe in New South Wales, "there are endings which vary for the purpose of indicating that the act has been accomplished in the immediate, the recent, or the far-off past, or that it will take place shortly, or within a more or less distant future, that there has been or will be a repetition or a continuation of action." These subtle discriminations are far from primitive: such analysis, if applied to the making of tools, might have produced elaborate artifacts at a much earlier stage than the shapely Solutrean laurel-leaf points.

Yet, once language had evolved beyond a certain point, it may have engrossed man as a game, even at the expense of his putting it to more practical social uses—though certainly primitive man's elaborate kinship organizations would imply a complex linguistic structure. In all likelihood, conversation became early man's principal amusement, apart from sexual intercourse. Primitive peoples excel at conversation and delight in it; and among peasant populations, as in Ireland, it still ranks high as *the* social occupation.

5: MYTH AS 'LINGUISTIC DISEASE'

This generalized description of the origins of language has sought to bring out those pre-logical and pre-utilitarian functions of language that the conventional definitions of language as primarily an instrument of conceptual thinking and organized intelligence ignore. Languages, even at the dawn of civilization, had achieved a high degree of terminological exactitude and grammatical differentiation, without yet being efficient instruments of thought; and though accurate symbolic description was essential to effective communication and cooperation, that achievement was long delayed. The most notable contribution of industrial and agricultural technology, from the neolithic phase onward, was to rescue thought from its helpless immersion in dream and myth.

Perhaps the only systematic linguist who would readily have sympathized with the interpretation I have put forward was Max Müller, though we were both in fact anticipated by the original intuitions of the great Neapolitan philosopher, Giambattista Vico. Müller intuitively grasped the important part that metaphor and myth had played in the early formation of language, when the purpose of speech was not to convey specific information but to enable primitive man to infuse every part of his experience with significance and to cope with the mystery of his own existence.

When Müller made the startling statement that mythology was the "disease of language" he came close to hitting upon the role that the dream had played in the first formation of verbal symbolisms. But he read the facts in reverse order: the 'disease'—dream symbols and dream myths— was, we must now infer, one of the sources from which more abstract forms of speech were derived. As a vehicle for intelligent discourse, then, rational language was the final seed in a prolonged cycle of human growth from unconsciousness to consciousness, from immediate concrete presentations and associations to organized mental patterns, in which 'mythology' was the first efflorescence. Coherent verbal discourse, rational speech, abstract symbolism, analytic dismemberment, were not possible till the flowers had faded and the petals fallen.

In 'The Science of Thought' Müller put his basic intuition in these words: "It was absolutely impossible to grasp and hold, to know and understand, to conceive and name the world without us, except through this Fundamental Metaphor, this universal mythology, this flowing of our spirit into objective chaos and recreating it in our own image." No one else has better described the original feat of language or has given a more adequate reason for the fact that the rational uses of language were so long delayed, and its application to numbering, sorting, ordering, defining, exactly describing took such a tediously roundabout route. The terms 'metaphor' and 'mythology' describe the original nature of language and apply to the preliminary stages of marking the passage from the disordered symbolic contents of dream, the ceremonial associations of 'festal play' and religious rite into the richly structured world of definable meanings and conscious purposes. In the formation of language, thought, so to say, was an afterthought.

In this whole transformation 'mythology,' with its early and constant association with ritual, promoted the first flowering of language. Matter-of-fact prose appears in early texts only in temple accounts or military instructions: indeed, even in the latter it is far from pure. When the purpose speech served was a practical one, the intended core of abstract meaning would still be buried in metaphor. Archaic language must, to judge by later written languages, have been largely double-talk: allegoric meaning, lush in imagery, mingled with instrumental intentions that were long hidden under this efflorescence.

Nothing could better illustrate this aboriginal characteristic than Malinowski's interpretation of the magic formula among the Trobriand Islanders for ensuring a heavy crop of *taytu:* an incantation that invokes the dolphin. "We know that the dolphin is big and long as tubers should become, that its weaving in and out of the rising and falling waves is associated with the winding and interweaving of the luxuriant vines whose rich foliage means a plentiful *taytu* harvest." Though superficially no organisms could be more unlike than a sea-animal and a vegetable, the first implies by its imagery the *abstract* qualities of the other. It is just his inability to hold such images that, according to Wolfgang Köhler, has kept the chimpanzee from developing articulate speech.

In suggesting that language was at the beginning overrun by myth and metaphor Max Müller was, I think, indicating an important clue to most of early man's mental activities. In all that we know of his linguistic expression, we must allow for fanciful excess and speculative superabundance, as well as for a serious lack of interest in those many practical concerns that now often take a heavy toll of human vitality. Whitehead did well to remind us, in his 'Symbolism,' that "no account of symbolism is complete without recognition that the symbolic elements in life have a tendency to run wild, like vegetation in a tropical forest."

But magic itself long preserved an even more primitive characteristic of speech derived from ritual: a large part of all magic formulae consists of a precise series of nonsense syllables repeated *ad nauseam*. This is perhaps the buried bedrock foundation of all language, retained unaltered in magic, as a secret for the initiated, long after the more public uses of intelligible speech had begun. So far has language passed beyond this state of magic and mystery that Malinowski, in setting down the magic spells, confessed his bafflement in "translating the untranslatable" or finding the "meaning of meaningless words." Do these phrases not take us back to the beginnings of language? Yet in its efforts to achieve exactitude does not current scientific language approach closely the style of a secret ritual formula, jealously guarded from the uninitiated? This magical component has never been expelled from language; and like ritual itself it may long have been a retarding factor.

As with existing 'primitives,' early man must have delighted in the symptoms of Müller's imputed disease: myth and word-magic for long flourished at the expense of more definite meanings attached to the operational commonplaces of daily experience; for among most primitive peoples, matters of fact and matters of magic are equally real. Even today, as Schuyler Cammann tells us, the Mongol tribesman regards the hemispherical shape of his tent as the dome of the sky, and the circular smoke vent at the top as the sun gate or sky door, while the column of rising smoke is the world pillar or the world tree, the *axis mundi*. Only by sloughing off these mythopoetic attributes did a tent finally become a mere tent, a hole a hole, a column of smoke just smoke.

By sedulously cultivating metaphor, I suggest, primitive man first developed playfully and dramatically the art of language, well before he learned to put it to effective use for accurate description and record, and eventually for directed, organized thought. The very words I unintentionally used to characterize this transformation—seed, efflorescence, vehicle, bedrock—show how much metaphor may still cling to what are nevertheless relatively commonplace statements that seek only to convey information, not to produce emotional arousal.

Those who seek an exact scientific transcription of abstracted events rightly choose to use the translucent symbols of mathematics. But those who would use language to deal with cosmic processes, organic functions, and human relations, as operative interacting wholes, must realize that they can only be represented loosely in the language of myth: in their dynamic complexity and wholeness, they evade other modes of abstraction and representation.

The closer language comes to the dense concretions of all related being, the less abstract and exact it can be. The final word about human experience is the human experience itself, unmediated by words; and every creature knows in the act of living something about life that will forever escape scientific analysis, even after the scientist has reduced to a chemical formula or an electric charge every observable manifestation of the living organism. So, too, only in silent face-to-face encounters can the last word be said.

When Vico characterized the earliest state of human development as the Age of Poetry, he anticipated Jespersen's description of it as an Age of Song; but rather it was an age when dance, song, poesy, and prose, myth, ceremonial, magic, and brute fact intermingled and were almost indistinguishable. By the very nature of this mythological affluence, it held a charm for man's still unformed mind. And it was finally by reaction against unbearable subjective confusion that the peculiar mythology of our own age has been built up: one that gives exclusively to quantitative measures and logical abstractions the same magical properties that the primitive mind gave to colorful figures of speech.

But we do ill to read back our own highly specialized 'disease of abstraction,' which reached an apex in Wittgenstein's analysis of language, into the origins of human language. This rejection of myth and metaphor produces an equally great distortion. The effort to dissect human experience with entirely sterile instruments, aimed to prevent the slightest invasion of the germs of language's original metaphorical 'disease,' transfers the danger to the surgeon's knife which, in the act of removing a pocket of infection, also impatiently removes other organs needed for maintaining the patient's life. Something essential to man's creativity, even in science, may disappear when the defiantly metaphoric language of poetry gives way completely to the denatured language of the computer.

That there is some relation between the dynamic structure of language and the nature of the cosmos, as Benjamin Whorf thought, is highly probable, though no single language can fully reveal that nature. For man, the creator of language, is himself a representative sample of the cosmos and embodies its emergent characteristics at their highest point of organization and self-awareness. But the larger structure man discovers is that which he has necessarily helped to create. Those who would turn man into a passive instrument for registering sensations, a mere recording and communicating device, would rob even their own philosophy of meaning.

6: THE LINGUISTIC ECONOMY
OF ABUNDANCE

Yet is it any wonder that man was intoxicated with the miracle of language? Did it not give him powers that no other animal possessed? With language, the light of consciousness soon spread over man's whole sky. So remarkable is the actual power of words that man often succumbed to the temptation, similar to that exhibited in the first exultation of discovering 'wonder drugs,' to apply verbal incantations or exhortations in situations where they could have no efficacy: to influence not only the spirits of men, but the behavior of natural processes and objects. As with our boasted antibiotics, the side effects often proved disastrous.

Even in historic times, to utter a secret name was a means of getting hold of power. Egyptian myth tells how once the goddess, Inanna, by a trick found the 'true' name of the 'all-powerful' Atum, so that she had him at her mercy. In the same spirit, a present-day illiterate tribesman may swallow a written medical prescription instead of taking the medicine, though in this case the power of suggestion may counterbalance the therapeutic uselessness of paper. These lingering misapplications of verbal magic only bear testimony to the heady original potency of the word itself.

So effectual and compelling was the operative magic of words that even after many other inventions had extended man's control over the physical environment and increased his prospects of survival, the word itself still took precedence as the chief source of human creativity. In the 'Instructions for King Merikere,' written in the inter-regnum between the Old and the Middle Kingdoms of Egypt, we read: "Be a craftsman in speech, so that thou mayest prevail, for the power [of a man] is the tongue, and speech is mightier than fighting."

In an earlier text we read that the creativity of Ptah, the god who created all the other gods, "is the teeth and lips of his mouth, which pronounce the name of everything . . . all the divine order came into being through what the heart thought and the tongue commanded." As James Breasted observed, "the extraordinary basis of this early system [of hieroglyphics] is the fundamental assumption that mind or thought is the source of everything." By the same token, a group of Northwest Indians whom Kroeber studied, believed that "the supreme god of the Wiyot needed no sand, earth, clay or sticks for the creation of man. God merely thought, and man was there." In these observations lies a truth essential once more for modern man to grasp: 'minding' is still more important than making.

Now the immediate efficacy of speech in affecting human behavior contrasts with all the more laborious processes of shaping and controlling the environment; and this very fact may, to man's own disadvantage, have diverted him from all the effort he might have put forth to provide a more commodious home for himself. Craftsmen, as the author of Ecclesiasticus smugly pointed out, become unfitted by their drudging industry in the field, the forge, or the potter's workshop for the higher exercises of the mind. Thus the God of Genesis, like Ptah, performed no actual labor in creating the universe. He merely *said:* "Let there be light!" and there was light.

I cite these relatively late examples of the potent role of speech because we shall see further on that the immense technological achievements of civilization would have been long delayed if unqualified reverence for the magic of the word, as "spoken from above," had not been established as the firm base for the efficient collective organization of work. The 'myth of the machine' would have been inconceivable, and its operations impracticable, without the magic of language and the formidable increase in its power and scope through the invention of writing.

In acknowledging the critical contribution of language to technology itself, one need not deny that this may nevertheless have ultimately slowed up the whole process of invention. As Allier has suggested, the application of verbal magic to the work process may have caused technique to remain stationary. "The man who believes in magic uses the technical methods discovered before his time and handed down by tradition. . . . It appears to him that, were he to make any change, he would cause them to lose their efficiency." Perhaps this is still another factor that helps account for the slowness and imperfection of prehistoric technical development, as compared with that of language. Likewise it would explain the rapidity of invention in recent centuries, with their proper disdain of verbal magic and their improper de-sacralization of the word.

All these blockages and handicaps must be freely granted. The invention of language certainly did not remove all other human infirmities; on the contrary, it all too easily inflated the ego and made people over-rate the efficiency of words in controlling the visible and invisible forces that surround man. Even after endless disillusions, these handicaps persisted in highly developed societies, as with the great Roman physician, Galen, who coupled magic formulae with rational medical prescriptions. Does not this misuse of repetitive spells still continue in the form of advertising and propaganda? Word magic is one of the chief means of attaining power and status in the 'affluent society.'

Since our own age has become acutely conscious of the possible misuses of words through semantic confusion and magical misapplication—perhaps because it has specialized in the debasement of language through unscrupulous political and commercial deceptions—let me stress rather the extraordinary nature of this invention, whose wonder should never be lost to us. If language for long diverted human energy from manipulation and fabrication and environmental transformation, it nevertheless in itself has all the attributes of an elaborate technology, including certain desirable features that have not yet been translated into the mechanical-electronic system now in the course of development.

What Freud mistakenly regarded as the infantile illusion of the omnipotence of thought is indeed an over-credulous surrender to the magic power of words. Yet to deny their omnipotence or their omnicompetence is not to belittle their actual function in influencing human conduct and in making possible the correct interpretation of natural events, beyond what is open to animal intelligence. The fact is that, right up to our own time, language has surpassed any other form of tool or machine as a technical instrument: in its ideal structure and its daily performance, it still stands as a model, though an unnoticed one, for all other kinds of effective prefabrication, standardization, and mass consumption.

This is not so absurd a claim as it may at first seem. Language, to begin with, is the most transportable and storable, the most easily diffusible, of all social artifacts: the most ethereal of cultural agents, and for that reason the only one capable of indefinite multiplication and storage of meanings without overcrowding the living spaces of the planet. Once well started, the production of words introduced the first real economy of abundance, which provided for continuous production, replacement, and ceaseless invention, yet incorporated built-in controls that prevented the present-day malpractices of automatic expansion, reckless inflation, and premature obsolescence. Language is the great container of culture. Because of the stability of every language, each generation has been able to carry over and pass on a significant portion of previous history, even when it has not been otherwise recorded. And no matter how much the outer scene changes, through language man retains an inner scene where he is at home with his own mind, among his own kind.

Though words are often described as tools, they may be more properly regarded as the cells of a complex living structure, units quickly mobilized in orderly formations to function on particular occasions for particular uses. Every member of the community has access to this linguistic organization and can use it up to the capacities of his experience and intelligence, his emotional responsiveness, and his insight. At no point, except by the invention of writing, has language ever been the monopoly of a dominant minority, despite class differentiations of usage; while the medium itself is so complex and so subtle that no centralized system of control was ever, even after the invention of writing, completely effective.

There is one further attribute of language that places it at a higher level than any existing technological organization or facility; and that is, to function at all, it demands a reciprocal relation between producer and consumer, between sayer and listener: an inequality of advantage destroys in some degree the integrity and common value of the product. Unlike any historic economic system, the demand for words may be limited without embarrassing the supply: the capital reserves (vocabulary) may become huger and the capacity for production (speech, literature, sharable meanings) continue to increase without imposing any collective duty to consume the surplus. This relationship, embedded in the special form of language, the dialogue, is at last being undermined by a new system of control and one-way communication that has now found an electronic mode of operation; and the grave issues that have thus been raised must now be faced.

But though the parts of language are standardized and in a sense mass-produced, they achieve the maximum of variety, individuality, and autonomy. No technology has yet approached this degree of refinement: the intricate mechanisms of the so-called Nuclear Age are extremely primitive in comparison, for they can utilize and express only a narrow segment of the human personality divorced from its total historic expression.

If one asks why early man took so long to improve his technical skills and his material facilities, the answer must be: he concentrated upon the greatest of all utilities first. By his command of words he increasingly embraced every aspect of life and gave it significance as part of a larger whole he retained in his mind. Only within that whole could technics itself have significance. The pursuit of significance crowns every other human achievement.

CHAPTER FIVE

Finders and Makers

1: THE ELEPHANT'S CHILD

In giving equal weight to dreams, ritual, speech, social intercourse, and social organization rather than to tools alone as the prime agents of early man's development, I am not suggesting that any of these were separated from the whole range of human activities. Still less do I suppose that early man withdrew to a sheltered spot and spent his days brooding over his impressions and re-enacting his dreams until he ultimately gave birth to significant pantomime and communicable verbal discourse. The present interpretation of the office of language only accepts with reverse emphasis something that Kenneth Oakley, an authority on prehistoric technics, noted when he pointed out that the slow improvement of 'Chellean' tools was probably the sign of a failure as yet to achieve speech.

This slowness, before language gave continuity and coherence to individual experience by making it sharable, was well accounted for by Leroi-Gourhan, who observed that "if the slightest gap had ever interrupted the slow acquisition of basic techniques, everything would have been to do over again." Before speech was achieved, those gaps may have occurred all too often; and the need to avoid such setbacks may well account for the anxiety exhibited by all cultures down to our own time not to lose what the Ancestors had achieved. Tradition was more precious than invention. To keep even the smallest gain was more important than to make new ones at the risk of forgetting or forfeiting the old. It was not nostalgia but the necessity for preserving the hard-won symbols of culture that made man treat the ancestral past as inviolable: at once too valuable and too vulnerable to be lightly altered.

At all events, even the improvement one notes in Acheulian tool-making, after the hundreds of thousands of years of fumbling 'Chellean' effort, leaves us with an assemblage of very primitive artifacts, which are usually treated in too facile a manner as weapons of the hunt; though as one museum label confesses, what is called a "weapon is sometimes called a pick and also a borer, but it probably served as a thrusting weapon, and as such may properly be designated as a poignard."

But were the earliest men chiefly hunters? That question demands to be considered if we would assign a proper value to early tool-making. Most of the 'weapons' in the primeval period assigned to hunting have a more plausible function as tools, if we connect them rather with food-gathering and trapping—activities that may have sufficed for survival in warmer southern climates even during the Ice Ages. What is called a hand-axe or a fist weapon would be useful in digging up tubers, or in finishing off an animal caught in a trap.

Those who stick to the view of early man as a specialist in hunting have not really taken into account his omnivorousness or explained how he could have acquired a strong liking for meat before he had learned either to shape weapons of bone, stone, or wood, or to kill big animals without their aid. Nor do they explain why the diet of mankind at all times has been predominantly vegetarian. Even Leakey's demonstration of what a present-day man can do with crude bone or stone tools and weapons, such as are attributed to the Australopithecines, does not prove that a smaller, weaker creature with a meager brain and inadequate teeth for chewing raw flesh could or did do the same thing.

Is not the answer to this that early man lived mainly by his wits? At the beginning 'braininess' stood him in better stead than either ferocity or dogged industry. Can there be any doubt that in the very earliest stages of his 'hunting' career man was forced to do what the Pygmies of Africa still do to achieve results that were otherwise far beyond his technological horizon—contrive ingenious traps and a daring strategy, such as the Pygmies use to capture and kill elephants, by hiding in pits where, once the elephant is caught, they can attack his soft underbelly with the weapons they command. Only at close quarters, with a far more vulnerable creature than an elephant, would a 'fist-weapon' be of any service beyond that which any unshaped stone would perform.

Snares and traps could be made with the naked hands out of reeds and creepers and young branches, long before man had an axe capable of chopping off the cave-man's legendary club—a weapon never found, nor shown in any ancient cave painting!—or of fashioning a wooden spear out of live, hard wood. When Columbus discovered the West Indies the natives still used "traps and snares of vines and other devices of nets" to capture deer.

The greatest use of the pointed hand-axe may have been as a digging tool for getting at succulent roots and hollowing out sufficiently deep pits. On this whole phase of trapping, as the fore-stage of hunting, Julius Lips has assembled much pertinent evidence. The aborigines of the Tierra del Fuego made bird snares; and as for larger traps, they must have preceded the stone-pointed spear and the bow-and-arrow, though of course they would leave no trace except by becoming part of a widespread human tradition.

But trapping, like nest-building, is an even more ancient art, practiced by organisms as different as the pitcher plant and the spider. "Many varieties of primitive stage weapons, such as traps or pounds," Daryll Forde points out, "have long been employed. . . . The main types of net [hunting and fishing] and the basic techniques of their manufacture is so widely distributed, that like the cordage of which they are made, they must be among man's oldest inventions." Even entangling devices for trapping an animal at a distance—the noose and the bola—seem likewise of extreme antiquity, for the principle of the running noose is found on every continent.

A. M. Hocart reports seeing a 'primitive' tear off a stick from a tree, sharpen the end with his teeth, and dig for a tuber. At the moment I write, report comes from Australia of a hitherto undiscovered tribe, the Bindibu, who "used their feet as clamps and anvils, and their tightly clenched jaws as a combined vise and knife during tool-making operations"—even flaking stones with their teeth. The hand long continued to serve as a cup or a shovel or a trowel, before specialized tools were 'handy.' At the dawn of civilization in the Near East the pickaxe broke the soil, but no shovel has been found—or shown in pictures—for digging earth or transferring it to a basket.

What I would emphasize here is the number of technical feats man can achieve solely with the use of his bodily organs: digging, scraping with his fingernails, pounding with his fists, fiber-twisting, thread-spinning, weaving, plaiting, knotting, shelter-building with twigs and leaves, basket-making, pot-making, clay-modelling, peeling fruits, opening nut shells, weight-lifting and carrying, cutting threads or fibers with his teeth, softening hides by chewing, wine-pressing with bare feet. While durable stone or bone tools would in time usefully assist many of these operations, they were not essential. Where suitable shells and gourds were available no comparable cutting edges or containers existed till middle paleolithic culture.

Reading back in this fashion from the residual practices of primitive peoples, particularly noting those traits that have a worldwide provenance, one sees that many advances in technology were both necessary and feasible before adequate, manufactured tools, utensils, and weapons were conceived and invented. In the primeval phase of technical development, ingenuity in utilizing the organs of the body, without turning any one part—not even the hand—into a limited specialized instrument, made a whole set of bodily appliances available hundreds of thousands of years before any comparable array of specialized stone tools can even be hinted at. In man's earliest career as finder and maker, as I noted before, his biggest find and his first shapable artifact was himself. No chipped pebble before *Homo sapiens* appeared shows any comparable proof of his technical ability.

2 : THE OLD EXPLORATION

In collecting food man was also incited to collect information. The two pursuits went together. Being imitative as well as curious, he may have learned trapping from the spider, basketry from the birds' nests, dam-building from beavers, burrowing from rabbits, and the art of using poisons from snakes. Unlike most species, man did not hesitate to learn from other creatures and copy their ways; by appropriating their dietary and methods of getting food he multiplied his own chances for survival. Though he did not at first set up beehives, a cave painting shows that he imitated the better-protected bear and dared to gather honey.

Human society was founded from the beginning, then, not on a hunting but a collecting economy—and for ninety-five per cent of man's existence, as Forde points out, man was dependent upon food-gathering for his daily nourishment. Under these conditions his exceptional curiosity, his ingenuity, his facility in learning, his retentive memory, were put to work and tested. Constantly picking and choosing, identifying, sampling, and exploring, watching over his young and caring for his own kind—all this did more to develop human intelligence than any intermittent chipping of tools could have done.

Again the over-emphasis on the surviving material evidence, the stone tool, has led to an under-estimation in most accounts of prehistoric equipment of the organic resources that probably contributed most heavily to early technology. In avoiding the dangers of wanton speculation, many sober scholars have surrounded themselves with a veritable stone wall that hides much that is essential to know, at least by inference, about the nature and habits of early man. The aboriginal creature they present as man, *Homo faber,* Man the Tool-Maker, is a late arrival. Before him, even were one to overlook or deny the special contribution of speech, stands Man the Finder who, before he engaged heavily in constructive activities, explored the planet, and, before he had begun to exhaust the earth's gifts, found and embellished himself.

Early man, in his own self-absorption, perhaps tended too often to be immersed in wishful dreams or haunted by nightmares; and quite possibly the latter increased alarmingly as his mind continued to develop. But from the outset he was rescued from any tendency to torpid adjustment to the conditions of his life by the fact that he was, pre-eminently, a 'nosey' animal: restlessly exploring every part of his environment, beginning with the most immediate part, his own body: smelling and tasting, seeking and sampling, comparing and selecting. These are the qualities that Kipling humorously made use of in his 'Just So' tale of 'The Elephant's Child'— man's insatiable curiosity.

Most of our current definitions of intelligence involve problem-solving and construction, more or less conditioned by acquired facility with abstractions, such as comes only with the use of language. But we overlook another kind of mental effort, common to all animals but probably greatly intensified in man: the capacity to recognize and identify the characteristic shapes and patterns in our environment: to detect rapidly the difference between a frog and a toad, between a poisonous mushroom and an edible one. Within the realm of science this is the grand labor of taxonomy: but early man must have been an acute taxonomist under sheer stress of daily existence. He must have made many intelligent identifications and associations long before he had formed words that helped keep this knowledge at his disposal in memory for future uses. Intimate contact with and appreciation of the environment, as Adolf Portmann has shown, brings quite different rewards than intelligent manipulation—but equally real ones. Pattern identification, as a necessary part of environmental exploration, stimulated man's active intelligence.

There is good reason to believe, indeed, that man used an immense variety of foods, far beyond the range of any other species, before he had invented any adequate tools. As long as the hunter-image of early man prevailed, the significance of man's omnivorousness was overlooked. The enriching of his botanical vocabulary extended in time to poisons and medicines, sometimes derived from sources like the venomous caterpillar the African Bushmen use, which no modern would think of testing.

The botanist, Oakes Ames, was surely right in suggesting that while early man already possessed much botanical knowledge acquired by kindred primates and hominids—the gorilla forages on more than two dozen plants—man added to it greatly not only by utilizing roots, stems, nuts, either repulsive to taste or toxic in their raw state, but also by experimenting with the properties of herbs that other animals seem by 'instinct' to avoid. Almost the first two phrases learned by the children of Australian aborigines are "good-to-eat" and "not-good-to-eat."

Unfortunately, we can hardly venture to guess how much of the knowledge accumulated by late paleolithic times had reached the point we discover in surviving primitives. Did Magdalenian hunters already follow the Bushmen's practice of cunningly tipping their arrows with stronger or weaker poisons, derived from the amaryllis, the scorpion, the spider, or the snake, according to the size and vitality of the intended victim? Quite possibly. But plainly, this kind of observation, which goes likewise into primitive medicine, is of the same order that makes science possible; and to explain all that followed one must perhaps assign an even longer period to its acquisition than to language itself.

What I would stress in all this dim but indubitable evidence is the amount of intelligent discrimination, evaluation, and ingenuity it reveals: equivalent to that displayed in the evolution of ritual and language, and far beyond what one finds, until late paleolithic culture, in the shaping of stone tools. At first probably the only animals included in the diet of early man were the small creatures, like rodents, turtles, frogs, insects, that could be caught by hand, as they are still caught in the Kalahari Desert or the Australian bush by small primitive groups surviving with meager paleolithic equipment—stones, throwing-sticks, bows—supplemented by possibly later blow-guns and boomerangs. In so far as primitive man killed bigger animals, as the collection of bones in widely scattered caves would indicate, it is safer to assume that they were ambushed or trapped rather than killed in the chase. Superior social coordination and cunning alone could offset the absence of effective weapons.

What early man's diet lacked in quantity, except perhaps in tropical habitats, it made up for in variety, thanks to his persistent experiments. But new foods brought about more than bodily nourishment: the constant practice of searching, tasting, selecting, identifying, and above all noting the results—which must sometimes have been cramps, illness, early death —was, I repeat, a more important contribution to man's mental development than ages of flint-knapping and big-game hunting could have been. Such searching and experimenting demanded plenty of motor activity; and this exploratory foraging, along with ritual and dance, must be given a fuller share of credit for man's development.

Let me give a concrete example of the way intelligence must have developed long before man had a large kit of tools or material equipment comparable to that of the Aurignacian hunters. One finds an excellent description of a truly primitive economy, from which almost every trace of later culture except language and tradition is absent, in Elizabeth Marshall's account of the Bushmen of the Kalahari Desert.

In the dry season, when there is a dire lack of water, the Bushmen gather a plant called *bi* for its watery fibrous root, and bring it back to the *werf,* the hollow that serves as a home, before the sun is hot; it is scraped, and the scrapings are squeezed dry. The people drink the juice they squeeze. They then dig shallow pits for themselves in the shade. They urinate on the *bi* scrapings and line the pits with the moist pulp, then lie in the pits and spend the day letting the moisture evaporating from the urine preserve the moisture of their bodies. Except in scraping, tools play no part in this process; but the kind of causal insight and seasoned observation disclosed by this life-preserving routine shows a high order of mind. Here the strategy of survival was worked out through close observation of such a far-from-obvious process as evaporation, by utilizing all the materials at hand, including the water from their own bodies, to counteract it.

In this we see at work three aspects of mind that were intertwined with the development of language as well as with adaptation to the environment: identification, discrimination, and causal insight. The last named, which Western man has too often assumed as his special triumph, and a very recent one, could never have been absent from primitive existence: if anything, early man's mistake was rather to overstress and misplace the role of causality, and to attribute both accidental events or autonomous organic processes, as in a disease, to the deliberate intervention of malicious men or demons.

Unlike the later hunting cultures that were based upon following wide-ranging herds of bison or reindeer, the early food-grubbing and foraging activities must have been relatively sedentary: for this literally hand-to-mouth existence demands an intimate knowledge of the habitat through all its seasonal changes, as well as firmly established knowledge of the properties of plants, insects, small animals, birds, such as can be achieved only by recurrent occupation for generations in an area small enough to be ransacked in every nook and corner. Thoreau, not Leatherstocking, is the contemporary exemplar of truly primitive man.

The kind of detailed knowledge coming from this sort of exploration must have been subject to heavy loss until language was developed. But long before even the crudest form of domestication can be suggested, man must have achieved an encyclopedic inventory of the contents of his environment: what plants had edible seeds or fruits, what others had nourishing roots or leaves, what nuts needed leaching or roasting, what insects tasted good, what fibers were tough enough to shred, and a thousand other discoveries upon which his life depended.

All these insights denote not merely habits of curiosity but powers of abstraction and qualitative appreciation. If we may judge by later evidence, some of this knowledge was intellectually quite detached and had nothing to do with ensuring physical survival. Lévi-Strauss cites an observer of the Penobscot Indians who found that they had the most exact knowledge of reptiles, but except on rare occasions when they wanted charms against sickness or sorcery, made no use of them.

When we persist in thinking of hunting as the primary source of food throughout man's early existence, and stone-shaping as his principal manual occupation, man's cultural progress seems unaccountably slow, for essentially the same process was used in chipping the fine Solutrean tools as in rough-hewing the Acheulian ones: stone striking stone.

This snail-like pace has been somewhat concealed by the practice of preserving paleolithic tools and weapons in museums, where they are telescoped in space and show marked improvement within a relatively short distance. If each foot represented a year, these improvements would have to be strung along a distance of roughly ninety miles, of which only the last five or ten miles would denote a period of rapid advance. But if one accepts the notion that tool-making began with the Australopithecines, the rate of advance is three times slower, and the effect of 'selection pressure' favoring the brain development supposedly derived from tool-making becomes even more questionable.

What is missing from the usual petrified model is all the knowledge and art and equipment passed on by example from man's early exploration of his environment. It was this foraging activity, with its limited need for tools, that possibly accounts for the slowness in their improvement. His best extraneous tools, for long his only ones, were, as Daryll Forde points out, staves—"for knocking down fruits, prizing shellfish from rocks, and digging for buried organisms."

Yet the continued occupation and intensive exploitation of a small territory must have been favorable not only to the increase of knowledge but to the stability of family life; and the better care of the young under such conditions would increase the prospect of transmitting learned behavior by imitation. Darwin was impressed by the exquisite mimicry of both words and bodily movements that primitive peoples displayed, along with unusual retentiveness. These traits would seem to indicate some environmental continuity. So there is reason to support Carl Sauer's contention that paleolithic man was in the main no roving nomad, but a territory-occupying, family-maintaining, child-nurturing, settled creature, who habitually accumulated and stored the necessities of life, and at most shifted with the season from open clearing or prairie to woods, from valley bottom to hillside.

Such a life would help account for early man's opportunity, if my original hypothesis is sound, for giving so much of his attention to ritual and language. "Historical tradition," the philosopher Whitehead observed, "is handed down by the direct experience of physical surroundings"— provided of course that these surroundings remain coherent and stable. Under such conditions, material accumulations would remain small: but the immaterial accumulations, which have left no visible record, might be considerable.

Looked at in one way, man's original method of getting a living, by foraging and food-gathering, seems a desperately shiftless, penurious, anxious, cultural vacuous existence. Yet it brought genuine rewards, and it left a deep mark on human life; for by the very terms of his existence, the food-gatherer must have ransacked the natural environment more thoroughly than it was combed again until the nineteenth century; and if he often had experience of the parsimonies and stringencies of nature, he also knew something of its variegated largesse when the means of life might be obtained without much forethought and often without any serious muscular effort.

Gathering, collecting, storing, went hand in hand; and some of the earliest cave finds bear witness to the fact that primitive man did not hoard merely foods or dead bodies. In the caves inhabited by Peking Man stones were found which had been transported from a distance, but for no visible purpose; while Leroi-Gourhan notes that in the Périgordian stage, on two different occasions, chunks of galena were found: collected, like later stones and jewels, because of their shining faces and cubic crystalline structure.

These first efforts of man to master his environment, blank though they seem when one looks for visible results, left their mark on every subsequent achievement in culture, even though the actual connection cannot be established. On this I may again quote Oakes Ames. "When the elaborate methods of preparation of some of the plants used to break down the monotony of life are studied, it becomes quite evident that primitive man must have possessed something other than chance to reveal to him the properties of food and drug plants. He must have been a keen observer of accidents to discover fermentation, the effect and localization of alkaloids and toxic resins, and the arts of roasting and burning a product to gain from it the desired narcotization or pleasing aromas (coffee). To fermentation and fire, civilization owes a tremendous debt." But before the knowledge could be transmitted by language, to say nothing of a written record, it might well take a thousand years to register a single gain.

This fore-stage of prospecting and collecting, then, was a prelude to the later arts of agriculture and metallurgy; and it still goes on today with every kind of object, from postage stamps and coins to weapons and bones, books and paintings: so that as the end product of this oldest manifestation of human culture we have had to create a specialized institution, the museum, to house such collections. From this it would seem that the foundations of an acquisitive society were laid before those of an affluent society. But if the vices of the collecting economy were hoarding and niggardliness, secrecy and avarice, it carried with it also on happier occasions a marvellous sense of release, when most human needs were met directly without all the circuitous preparations and the painful physical exertions that even hunting entails.

From that ancient collection economy, perhaps, mankind is still haunted by dreams of effortless superabundance: dreams that come back in swift realization to those who go berry-picking or mushroom-hunting or flower-gathering, when more fruit or blossoms are there for the asking than can ever be picked. Hours so sunnily spent possess an innocent enchantment that only the gold or the diamond prospector can rival, though perhaps not so innocently. Even at a more sophisticated level this ancient propensity shows through. The attraction that the huge supermarket holds for the present generation may be partly due to the fact that it is a mechanical replica of that earlier Eden—until one reaches the pay-counter.

But in giving first place to finding over making, to collecting over hunting, let us not make the error of exchanging the term food-collecting for hunting, to describe early man's means of getting a living. "Man is by nature omnivorous," Daryll Forde correctly reminds us, "and we shall seek in vain for pure gleaners, pure hunters, or pure fishers." Early man never committed himself to a single source of food, or a single mode of life: he spread over the whole planet and tested life under radically different circumstances, taking the bad with the good, the harsh with the temperate, glacial cold and tropical heat. His adaptability, his non-specialization, his readiness to come up with more than one answer to the same problem of animal existence—all this was his salvation.

3 : TECHNICAL NARCISSISM

Reading back from the preoccupations of our own immensely productive
—and prodigiously wasteful and destructive—era, we tend to bestow on
early man too generous a measure of our own greedy, aggressive attributes.
Too often we patronizingly picture the small scattered groups of the early
Stone Age as engaged in a desperate struggle for survival, in ferocious
competition with equally forlorn and savage beings. Observing that once-
flourishing Neanderthal man died out, even some well-trained anthropologists too quickly jumped to the conclusion that *Homo sapiens* must
have slaughtered him, though in the absence of evidence they might at
least have allowed for the possibility that a new disease, a volcanic up-
heaval, a reduced food supply, or some stubborn fixation and inadapta-
bility might have been responsible.

Until comparatively late paleolithic times, there is little evidence to
show that man was half as effective as a bee in re-shaping his domestic
environment, though perhaps he already had a symbolic home, like the
hollowed out *werf* of the African Bushman, or the crossed sticks of the
Somali tribesman, which may have actualized the idea before he made the
first dugout or stamped down the first clay hearth or built the gable-roofed
'houses' that may be skeletonized in the 'tectiform' drawings found in
Magdalenian caves.

But there was one sphere besides language where all the traits I have
been sorting out and evaluating were put to work: ancient cave finds
show that one of the phenomena man investigated most eagerly and
altered most ingeniously was his own physical body. As with the gift of
tongues, this was not only the most accessible part of his environment,
but the one that continually fascinated him, and upon which he was
readily able to effect radical if not always salutary changes. Though Greek
myth has anticipated the modern psychologist's discovery that man during
adolescence falls in love with his own image (narcissism) early man
strangely did not love that image in its own right: he rather treated it as
raw material for special 'improvements' of his own that would thereby
change his nature and give expression to another self. One might say he
sought to rectify his bodily appearance almost before he had identified his
original self.

This propensity may go back to the widespread animal practice of
grooming, notable among apes. Without this persistent cosmetic drive, in

which licking and picking are almost indistinguishable from fondling and petting, man's early social life would have been poorer: indeed, without attentive grooming, the long head and facial hair of many races would have turned into an embarrassing matted bundle, dirty, vermin-infected, sight-obscuring. Such a grotesque growth of hair made the Indian wild girl, Kamala, seem more of an animal when captured than the wolves that reared her.

From the universality of ornaments, cosmetics, body decorations, masks, costumes, wigs, tattooings, scarifications, in all peoples down to our own day, one must assume, as I noted before, that this character-transforming practice is an extremely ancient one; and that the naked, unpainted, undeformed, unadorned human body would be either an extremely early or an extremely late and rare cultural achievement.

If we are to judge by surviving primitives—our own little children or the few remaining groups that were still mainly in the Stone Age when discovered by Western man—there was no function of the body that did not at an early stage provoke curiosity and invite experiment. Early man looked with respect, often with awe, on the effluvia and excreta of the body: not merely on the blood whose unabated oozing might bring life to an end, but on the placenta or caul of the newborn, the urine, the feces, the semen, the menstrual flow. All these phenomena roused either wonder or fear and in some sense were sacred; while the very air he exhaled eventually became identified with the highest manifestation of life: the soul.

This kind of infantile interest, which is still exhibited by contemporary adults afflicted by various neurotic disorders, must have occupied some little part of primitive man's days, if we take full note of the many marks it has left on our culture. Some of this dabbling with his own waste products might in time serve a utilitarian turn, as with the urine that the Bushman still uses for tanning leather, just as the Roman metal founders mixed urine with fuller's earth. Kroeber notes that this whole group of traits characterizes "backward as against advanced cultures," though he had no premonition when he wrote that within a few years the novelists and painters of so-called 'advanced Western cultures' would express their own disintegration by wallowing once more in this infantile symbolism.

The structure of the human body, no less than its functions and its excreta, called forth early efforts at modification. The cutting or braiding or curling or pasting together of the hair, the removal of the male foreskin, the piercing of the penis, the extirpation of the testicles, even the trepanning of the skull were among the many ingenious experiments man first made on himself, prompted perhaps by magical illusions, long before he sheared wool from the backs of sheep or turned the fierce bull by castration into the docile ox, in a religious ceremony where the animal may well have served as a substitute for a human victim.

On the surface, most of these efforts seem to come under the head of 'brotlose Künste,' as my grandmother used to call such unrewarding practices. But they are not without resemblance to the displays of 'idle curiosity' that Thorstein Veblen looked upon as the surest mark of scientific inquiry, or to even more startling parallels in 'idle experimentation' that goes on in many biological laboratories today, as in the experiment of flaying live dogs to death in order to ascertain the bodily changes that take place under shock. Primitive man, less cultivated but perhaps more fully human, was content to visit the most fiendish tortures upon himself; and some of these mutilations turned out to be far from futile.

What incited man to operate on his own body is hard to fathom. Many of these primitive body transformations involved difficult and painful surgery, and were often, considering the likelihood of infection, highly dangerous. Yet tattooing, scarification, sexual alteration, were all in evidence, the Abbé Breuil reported, in the cave he explored at Abacete in Spain. What is more, many of these surgical operations not merely deformed the body but lowered its capacities: witness the late Pleistocene Negro skull, with its central upper incisors knocked out, which imposed a willful handicap in eating. That sort of self-injury has remained a savage practice in more than one tribe till today.

What all this seems to add up to is that primitive man's first attack upon his 'environment' was probably an 'attack' upon his own body; and that his first efforts at magical control were visited upon himself. As if life were not hard enough under these rude conditions, he toughened himself further in such grotesque ordeals of beautification. Whether decoration or surgery was involved, none of these practices had any direct contribution to make to physical survival. They count rather as the earliest evidence of an even deeper tendency in man: to impose his own conditions, however ill-conceived, upon nature. Yet what they point to even more significantly is a conscious effort toward self-mastery and self-actualization; and even— though often exhibited in perverse, irrational ways—at self-perfection.

But the technological implications of body alteration and decoration must not be overlooked. Possibly the passage from purely symbolic ritual to an effective technics was opened through surgery and body ornament. Scarification, tooth-extraction, skin-painting, to say nothing of later elaborations like tattooing, lip-extension, ear-enlarging, and skull-lengthening were the first steps in man's emancipation from the complacent animal self that nature had provided. Our own contemporaries should hardly be surprised, still less shocked, at this emphasis. Despite our wholehearted dedication to the machine, the amount of money spent in technically advanced countries on cosmetics, perfumes, hairdressing, and cosmetic surgery vies with that spent on education, and until only recently the barber and the surgeon were one.

Yet in some obscure way, still not quite explicable, the arts of body decoration may have been helpful to hominization. For a dawning sense of formal beauty, otherwise found only in the bower-bird's confections, accompanied this. Captain Cook noted of the inhabitants of the Tierra del Fuego, "Although they are content to be naked they are very ambitious to be fine. Their faces were painted in various forms; the region of the eye was in general white, and the rest of the face adorned in horizontal streaks of red and black; yet scarcely any two were exactly alike. . . . Both men and women wore bracelets of such beads as they could make themselves of shells or bones."

We are never so sure of the presence of a creature who thought and acted like ourselves as when we find, alongside his bones, even when tools are absent, the first necklaces of teeth or shells. If one looks for the first evidence of the wheel, one will discover the earliest form of it, not in the fire drill or the potter's wheel, but in the hollow ivory rings, carved out of an elephant's tusk, in Aurignacian finds. And it is not without significance that three of the most important components of modern technics—copper, iron, and glass—were first used as bead ornaments, perhaps with magical associations, thousands of years before they had an industrial use. Thus, while the Iron Age begins around 1400 B.C., iron beads from before 3000 B.C. have been found.

As with language and ritual, body decoration was an effort to establish a human identity, a human significance, a human purpose. Without that, all other acts and labors would be performed in vain.

4: STONE AND THE HUNTER

The glacial epoch, which geologists call Pleistocene, extended over the last million years and buried much of the northern hemisphere under ice. Four long periods of cold alternated with briefer periods of more temperate climate—warmer, wetter, cloudier. Under such formidable environmental pressures, early man made his appearance, perfecting the anatomical structure that made walking, talking, and fabricating possible—above all, learning to bring these traits into the service of a more fully socialized and humanized personality.

Man's survival at the margin of the ice sheet—"By the Skin of His Teeth"—testifies to both his stubborn fortitude and his adaptability. The evidences of fire and hunting go back five hundred thousand years: tool-making perhaps farther. Whatever man's ancestral deficiencies, he was at least well-seasoned. Many animals found these conditions difficult. Some surmounted them by acquiring heavy coverings of wool, like the mammoth and the rhinoceros; and man himself, once he had developed sufficient skill in the hunt, not merely protected himself with the fur of better-equipped animals, but even pieced them into more or less fitting, Eskimo-like clothes.

In the terminal period of glaciation, beginning roughly a hundred thousand years ago, the geographic horizon narrowed and the human horizon widened. Here A. J. Toynbee's belief that the challenge of hard conditions may draw forth resourceful human responses, which the easier existence of the tropics does not encourage, seems if anywhere to hold. Part way through this period a new mutant of the human species appeared, *Homo sapiens;* and he made greater advances in every department of culture than his predecessors had been able to make in ten times the span of his existence—if only because the last steps always prove the easiest.

The relative rapidity of man's advance, in a period when the physical conditions of existence, up to 10,000 B.C. were often quite formidable, would indicate two things: further genetic and social changes that favored intelligence, and a sufficient advance in the art of symbol-making in speech and images to permit the transmission of acquired habits and knowledge more effectively than ever before. We find plentiful testimony to both conditions in the paintings and artifacts first discovered in the caves of France and Spain little more than a century ago. These discoveries revolutionized the picture of early man: but so fixed are the ancient images of his brutish existence that even now the pat association-word for 'cave man' would be 'club.'

Up to this phase there is no evidence of vocational specialism, and therefore no craft incentive to make improvements in stone tools: such improvements as are noticeable must be measured in terms not of ten thousand but fifty thousand years. From middle Pleistocene times on, Braidwood notes, the standardization of at least chipped tools is assured. This shows that the "users had developed notions of an ideal standard form or two for some particular job (or jobs) and could reproduce it well." Braidwood properly treats this as implying both a sense of future occasions, when the tool could be used, and a capacity for symbol-making, in which a visible or audible *this* refers to an invisible *that*.

This is the most generous judgement that can be made about man's earliest technological accomplishment. The same rough shapes that characterized the Acheulian culture persisted for over two hundred thousand years, while the somewhat improved implements of the succeeding Levalloisian phase lasted almost equally long: forty times as long as the entire period of recorded history. Even Neanderthal man, who had a big brain case and buried his dead, some fifty thousand or more years ago, cannot be accused of making headlong advances.

But around thirty thousand years ago the time-scale changes. Though new finds may shift these provisional datings, one definable culture follows another at intervals of three to five thousand years: very short periods compared with the earlier phases. The cold of this last glacial period wrought severe changes in the plant and animal life of the Northern Hemisphere; for the growing season became as short as that around the Arctic Circle today, and groups mainly dependent upon food gathering had the alternative of retreating into a warmer zone, or changing their way of life and living off the big herd animals that also stood their ground.

Under these pressures, great and rapid advances were made in the fabrication of tools; quarrying and even mining began; and the marked increase of skill shown in shaping stone implies specialization and possibly lifetime employment.

So far from being daunted by the dour conditions of the glacial climate, paleolithic man was later keyed up by them, and in many ways seems even to have prospered under them, for once he had mastered the art of big-game hunting, he had a more plentiful supply of protein and fat than he had ever, in all likelihood, had before. The tall skeletons of Aurignacian man, like the tall frames of our own youth of today, testify to this rich diet. By exercising great ingenuity and cooperative effort in contriving pitfalls or deadfalls for big animals, by starting or taking advantage of forest fires that panicked great herds, by perfecting his stone weapons so that they would pierce tough hides impossible to penetrate with fire-charred spear-points, and doubtless by availing himself of the wintry coldness to preserve the meat he had accumulated, these new hunters mastered the environment as never before, and even, thanks to the extra supply of fat, were able to endure the long winters. Though this existence was strenuous, and the life span possibly short, there was time for reflection and invention, for ritual and art.

Here again, at the risk of tedious overemphasis, I must point out that the fixation on stone tools has turned attention away from the bigger technical equipment of leather, sinew, fiber, and wood, and in particular has failed to give sufficient weight to the one outstanding weapon produced under these conditions, a weapon that reveals a remarkable capacity for abstract thought. For between thirty thousand and fifteen thousand years ago, paleolithic man invented and perfected the bow-and-arrow. This turns out to be the first real machine.

Up to this point, tools and weapons had been mere extensions of the human body, like the throwing stick, or, like the boomerang, an imitation of some other creature's specialized organ. But the bow-and-arrow is like nothing whatever in nature: as strange, as peculiarly a product of the human mind as the square root of minus one. This weapon is a pure abstraction translated into physical form: but at the same time it drew upon the three major sources of primitive technics: wood, stone, and animal guts.

Now, a creature who was clever enough to use the potential energy of a drawn bowstring to propel a small spear (arrow) beyond ordinary hurling distance had reached a new level of thought. This was an advance upon an even simpler device, midway between a tool and a machine—the throwing stick. Yet that combination of slot and javelin was so effective that Captain James Cook noted it was more accurate and deadly at a hundred and fifty feet than his own eighteenth-century muskets.

These technical improvements were contemporary with equivalent advances in art, though the fore-stages in the latter case are obscure, for suddenly well-modelled figures appear, out of a 'nowhere' that cannot yet be adequately described. From the very nature of these advances one may legitimately infer an even more decisive improvement in the older art of language, with such finer differentiation of the meaning of events in time and space as later languages show. The first musical pipe, an instrument we associate with Pan, appears in a Magdalenian picture of a very Pan-like figure, either a masked sorcerer or an imaginary being, half man, half animal, like Pan himself. But who knows when a music-making reed was first invented?

"Ukwane took out his hunting-bow and, setting one tip on a dry melon shell, he began to tap the string with a reed, making a sound." This image, taken from the admirable book about the Bushmen that I have drawn on before, bears witness to an earlier interplay of art and technics: it takes us back to a point where Prometheus and Orpheus were twins, almost Siamese twins. Conceivably, if doubtfully, the first use of a bow was as a musical instrument, before the twanging gut suggested one of its many later uses, as a weapon of the hunt, or as an instrument for causing rotary motion in the fire-drill or the bow-drill. This hypothetical history of the bow would end, then, with its return to its point of origin in the last exquisite refinements of the Cremona violin.

The bow-and-arrow might serve as archetypal model for many later mechanical inventions, by the translation of human needs—but not necessarily organic aptitudes—into detachable, specialized, abstract forms. As with language, the key idea is 'detachable.' And yet the feathering of the arrow, which ensured the accuracy of the weapon, was possibly due to a purely magical identification of the arrow with the wings of a living bird. This is one of those cases in which magical thinking often misled men further by sometimes actually paying off. But between the bow and the next visible machines, such as the potter's wheel, something like ten or twenty thousand years seems to have elapsed.

Meanwhile, the improvement of tools and the fabrication of varied objects made with tools had served these paleolithic artisans in three ways. For one thing, the regularity of effort needed acted as a counter-discipline to the spasmodic life of the hunter: then, too, the intractability of the hard materials made the craftsman come to terms with his environment and realize the impotence of purely subjective desire or magic ritual unaccompanied by intelligent insight and effort: both were needed. Finally, paleolithic man's increasing skill heightened his self-confidence and brought an immediate reward: not merely the joy of work, but the finished object itself—his own creation.

Now at this point, since I have sufficiently offset the over-petrified image of man's earliest economy, it is time to do justice to the positive role stone actually played from an early moment in human development. Stone singled itself out from every other part of the environment by its own special character, its hardness and durability. Rivers might change their course, great trees, struck by lightning, might fall or be burned up: but stone outcrops and columns remained fixed landmarks in a changing landscape. Throughout human history stone has been the agent and symbol of continuity; and its very hardness, color, and texture seem to have fascinated early man and challenged him. Stone prospecting may have gone along with food collecting, well before stones like flint and obsidian, that were especially fit for making tools, were identified and skillfully put to use.

The mining of flints and the fabrication of stone tools gave man his first taste of systematic, unremitting work. That the digging of flint with reindeer picks required hard muscular effort I can personally bear witness, since one of my tasks, as a naval recruit in 1918, was to dig into a hummock of flint on our island base at Newport, Rhode Island. Even with a steel pickaxe that was no light task: so it may well have required the hopeful imaginative re-enforcement of magic to summon early man to this back-breaking effort, though it brought an extra reward in building up a certain masculine pride: a pride that long clung, before the days of automation, to the miner.

Partly through working stone, early man learned to respect the 'reality principle': the need for persistence and intense effort in order to achieve a distant reward, as opposed to the pleasure principle of obeying momentary impulse and expecting an immediate response, with little effort. If paleolithic man had been as indifferent to stone as civilized man has long proved indifferent to the organic environment, civilization itself would never have taken form, for, as we shall soon see, it was originally a Stone Age artifact, shaped by stone tools and stony-hearted men.

5: HUNTING, RITUAL, AND ART

Behind the fine craftsmanship and expressive art that characterized the last phases of paleolithic culture was the mode of life brought about by specialization in hunting big game. In this pursuit a more cooperative strategy, requiring larger numbers of trackers, beaters, and killers, was required; and that presupposes a tribal or clan organization. Single family groups of less than fifty people, only a minority being adult males, could hardly have done the job. That Ice Age hunting life was necessarily dependent upon the movement of the great herds in search of fresh grazing or browsing grounds: yet it developed fixed points of reference and return—water courses, springs, camping grounds, summer pastures, not least caves, and even, in late paleolithic times, hamlets of houses.

If curiosity, cunning, adaptability, inurement to repetition were—along with sociability—the prime virtues of early man, the later paleolithic hunter needed still other traits: courage, imagination, adroitness, readiness to face the unexpected. At a critical moment in the hunt, when an enraged buffalo, already wounded, turned upon the hunters closing in upon him, the ability to act in concert at the command of the most experienced and daring hunter was the price of avoiding injury and sudden death. There was no parallel to this situation in food-gathering, nor yet in the later modes of neolithic agriculture.

Probably the nearest modern equivalent to the slaying of big game in paleolithic times was the hunting of another big mammal, the sperm whale, a century and more ago. Without unduly stretching the imagination, one can find in the pages of Melville's 'Moby-Dick' the psychal and social parallels of the paleolithic hunt. In both cases, fortitude in pursuit, unflinching courage, and the ability on the part of the leader to give commands and exact obedience were necessary to the success of the venture; and here, too, youth was probably a greater qualification than age and experience. Leadership and loyalty, those keys to military success and large-scale organization, flourished in this milieu. Both were at a later period to have technological consequences.

It was from this cultural complex that a commanding personage, the hunting chief, finally walks onto the stage of civilized history, in the epic of Gilgamesh, and in the pre-dynastic Egyptian Hunter's Palette. As we shall shortly see, that combination of docile ritualistic conformity, an early and deeply engraved trait, with exhilarating self-confidence, venturesome command, and, not least, a certain savage readiness to take life, were the essential prerequisites for the greatest early achievement in technics: the collective human machine.

Unlike food-gathering, be it noted, hunting carried with it an insidious danger to man's tenderer, parental, life-fostering nature: the necessity to kill as a recurrent occupation. The stone-pointed javelin or arrow, with its capacity to strike home at a distance as well as at close quarters, enlarged the range of killing and appears at first to have awakened anxious misgivings as to its effects. Even toward the cave-bears he expelled from their shelters and ate for food, paleolithic man seems to have nourished a sacred fear, as with his later totemic animals. The skulls of these animals have been found arranged as if they were the objects of a cult. Like some hunting tribes to this day, paleolithic hunters possibly begged the slain creatures' forgiveness, pleading hunger as justification, and limiting the kill to such food as was actually needed. Millennia passed before man would take the life of his own kind in cold blood, without even the excuse, magical or otherwise, of needing to eat them.

Yet the very compulsion to overplay the more brutal masculine qualities may have resulted, on a Jungian interpretation, in the enlargement of the feminine component in the male unconscious. The so-called mother goddesses in paleolithic art may represent the hunter's instinctual attempt to counterbalance the occupational over-emphasis on killing by an increased sensitiveness to sexual enjoyment and protective tenderness. When in the Army, my son Geddes noticed that it was the ugliest and toughest characters in his unit who often showed the greatest gentleness toward children: a similar compensation.

The systematic killing of big game probably had still another effect upon paleolithic man: he was confronted by the fact of death, not at infrequent intervals, but as an everyday accompaniment to life. To the extent that he may have identified himself with his victim, he was forced to take into consciousness his own death, too, and that of his family, his kinsmen, his fellow tribesmen.

Here, under the further incitements of dream, may lie the beginnings of man's devious efforts to prolong his life in the imagination, by assuming that the dead, though physically removed from the scene, are still in some sense alive, watching, intervening, prompting: sometimes benignly, as a source of wisdom and comfort; but in no small number of instances the spirits of the departed, haunting the dream life, are full of malice and must be exorcised, or propitiated, lest they bring on disaster. Perhaps the memorial arts of sculpture and painting which flourished now for the first time, were deliberate attempts to outwit death. Life departs, but the image remains and continues to enhance other lives.

The greater part of paleolithic art was preserved in caves; and in the case of some of the painted images and sculpture found there—about ten per cent of the total number—we have reason to associate the art with magic rituals to invoke success in hunting. But the artists who painted these images, under the most difficult conditions, undeterred by the rough surfaces, sometimes taking advantage of their contours, must have acquired their skill by long practice somewhere else than on the walls of the cave. The corroborating evidence for this comes from Leo Frobenius' experience with a group of African Pygmies. Once when he suggested going on an elephant hunt, they professed to find the conditions unfavorable and refused. But next morning he discovered that the hunters had gathered at a secret spot. Drawing the outline of an elephant on the cleared surface of the earth, they then thrust a spear into it, while going through a magic formula. Only then were they ready for the hunt.

This lucky bit of evidence throws a light on certain aspects of both paleolithic ritual and art. Hunting under paleolithic conditions was not a hit-or-miss scramble for food: it required forethought, a studious, carefully rehearsed strategy, intimate knowledge, graphically conveyable, of the anatomy of the hunted creature: such a knowledge as that which, embodied in the drawings illustrating Vesalius, preceded the advances of surgery and medicine in our own age. One finds a similar magic ritual, Sollas pointed out, among the Ojibway Indians of North America, where the shaman made a drawing on the ground of the animal to be hunted, painted the heart in vermilion, as European cave animals were often painted, and drew a line from it to the creature's mouth, along which the magic would flow to ensure death. In the same vein, the Mandan Indians welcomed George Catlin as a medicine man because his paintings "brought the bisons."

"Lately," Fernand Windels reports in his study of the Lascaux caves, "ethnologists have spent months with a desert tribe in Australia and brought back films. In one of these an Australian, the head of his tribe, is seen decorating the wall of his cave. . . . The sight is quite astounding. The film shows not an artist working, but a priest, or a sorcerer, officiating. Each gesture is accompanied by songs and ritual dances, and these take a much more important place in the whole ceremony than the decoration itself."

If dance, song, and language detached themselves from ritual, as I have argued, painting may likewise have done so: originally all the arts were sacred, since it was only to achieve communion with sacred forces that man would make the necessary efforts and sacrifices for esthetic perfection. With these associated acts of dance, ritual, and graphic motion we perhaps have a clue to the mysterious macaroni-like tracings on the walls of various caves: those abstract images may have been a by-product of ritual gestures—an act recorded on the walls as it can now be recorded on film.

If magic ritual was invoked by the hunter, it was because in the very performance of it he acquired both the insight and the skill necessary to carry out his task successfully. The kind of graphic line achieved in the paintings of the bisons of Altamira or the deer of Lascaux implies fine sensory-muscular coordination, along with the sharpest kind of eye for subtle detail. Hunting, as everyone who has hunted even in the most desultory way knows, requires a high degree of visual and aural alertness to the least quiver of movement in leaves or grass, along with hair-trigger readiness to react promptly. That the Magdalenian hunter had attained this state of sensory vividness and esthetic tension is shown, not merely by the evocative realism of his highly abstract representations, but by the fact that many of his animals are depicted in motion. This was a higher achievement than static symbolization.

One of the purposes of creating a realistic image of an animal was to 'capture' it, and what greater triumph was possible than to capture it in motion, the highest test of the hunter's skill with assegai or bow-and-arrow? In English we still speak in portraiture of 'capturing a likeness.' But this art was not just an agent of practical magic: it was likewise a higher mode of magic in its own right, as miraculous as the magic of words, yet even more secret and sacrosanct. Like the interior of the cave itself, walled and vaulted by natural forces, which gave man his first insight into the possibilities of symbolic architecture, these images opened up a world of color and form that transcended the esthetic range of natural objects because it likewise included, as an unavoidable ingredient, man's own personality.

Was this art not merely magical and sacred, but even more, a cult-secret not open to all members of the tribe? The difficult physical passage to the painted interior walls, which often necessitated dangerous crawls, may indicate more than an initiation ordeal. Was this perhaps a device deliberately chosen by the élite, in their selection of the almost closed cave, to keep for their own use and instruction the art of making images: an early anticipation and equivalent of the esoteric language and the inviolable temple sanctum of later priesthoods? Was some memory of the layout of the cave carried over into the secret passage to the interior of an Egyptian pyramid? These will always remain unanswerable questions: yet it is important that we should keep on asking them, lest we close our eyes too quickly to related positive evidence we may yet find.

Something of this cavernous secrecy and mystery normally attended all the great moments of life, up to our own de-sacralized modern culture: birth, sexual intercourse, initiation into the stages of life, and death. And if to capture a likeness was to have power over the soul, as many primitives still believe, this may account for the fact that the human face is so conspicuously absent from cave paintings, though attenuated bodies, with masks or birdlike features, may be depicted. This failure would not be due to any lack of skill, but was rather because of the inherent magical danger to the person so represented. The threatening scowl and protesting gesture that met me when I once photographed from a decent distance a Hawaiian native at a Honolulu market, reminds me of how deep the fear of the transferred image remains.

One does not exhaust the meanings of paleolithic art by referring some, but not all, of the cave paintings to magic ritual. In his comprehensive examination of cave art, as rich in evidence as it is fertile in fresh hypotheses and discriminating in judgement, André Leroi-Gourhan shows reason to deduce from the nature and the position of the images and signs that the cave artists were endeavoring to formulate their new religious perspectives, based on the polarity of male and female principles. Surely, these images transcended any practical effort to promote the reproduction of game animals and ensure good hunting. What can hardly be doubted, in an art whose practice was attended by so many difficulties, is that beliefs of supreme importance, which seemed more essential to man's development than mere food and bodily security, must have been the source of this cave painting. It is only in the pursuit of a more significant life that man has exhibited such devotion or been ready without a sense of grievance to make such sacrifices.

With sculpture in the round another interest came into play and another function was possibly served. Here the unabashed handling of the human body, including feminine nudes unequalled till the Egyptians, may indicate a pre-magical culture. Even in the cave paintings, I would not be so sure as some interpreters have been that the paintings of allegedly pregnant animals are, inevitably, only attempts to ensure by some sympathetic magic large yields of food. This explanation hardly fits in with the evidence of a teeming abundance of these animals, well beyond the capacity of a small hunting population to decimate. But sculpture shows a quite different range of interests and feelings: the carved ibexes confronting each other found at Le Roc-de-Sers seem scarcely symbols of anything but themselves: and the 'Venus' of Laussel is a woman in every dimension, even her head. Did sculpture represent the plane of daily experience, while painting kept closer to dream, magic, religion?

All that we can say with any surety about this phase of human development is that hunting was a propitious medium for imaginative art. And at last man's overcharged nervous system had material worthy of its potentialities to work on. The dangers of the big game hunt brought forth vigorous self-confident human breeds, with swift emotional responses, a ready supply of adrenalin, stimulated by fear, excitement, and rage, and above all fine coordinations that would serve in painting and carving as well as in slaying animals. Both kinds of skill, both kinds of sensitiveness, were brought into play.

So while hunting in the grand style required daring muscular exploits and promoted a surgical hardness about inflicting pain and taking life, it was also accompanied by an increase in esthetic sensitiveness and emotional richness—preludes to further symbolic expression. This combination of traits is not unusual. That murderous cruelty and extreme esthetic refinement are not incompatible we know from a long succession of historic examples, stretching from China to Aztec Mexico, from the Rome of Nero to the Florence of the Medicis, not forgetting our own times, with the exhibition of nicely planted flower beds at the entrance to the Nazi extermination camps.

Whatever the hardships and gambles of the hunting mode of life, it unleashed the imagination and turned it to the arts: above all, if one may judge from the frail evidence, it seems marked by an exuberant display of sexuality, presenting us with widely distributed images of the naked female form, with the interest centered in the vulva, the breasts, the buttocks, all magnified and swollen in many figures besides the famous 'Venus' of Willendorf.

These figures are usually called Mother Goddesses; and many ethnologists assume that they were the center of a religious cult. But this is imputing to the earliest cultures the same meaning such figures would have had in a much later one. All we can strictly infer is a heightened consciousness of sex, and a deliberate effort, by means of symbolic images, to hold it and prolong its effects in the mind, instead of letting it be dissipated in immediate copulation. Sexual intercourse, the earliest mode of social communion and cooperation, was now directed and enriched by mind.

Since representations of the male phallus are sometimes found in the same cave as female forms with open vulvas—an association that has continued to this day in Hindu temples—one has indeed reason to suspect ritual performances to awaken, enhance, and intensify interest in sex: perhaps even some definite group initiation and instruction, such as is widespread among almost all primitive peoples. An extra encouragement to sexual activity may have been imperative in a harsh climate whose long winters and forced hibernation, sometimes accompanied by a meager diet, had the usual effects of extreme cold and fasting: it depleted sexual interest and lowered sexual performance. Because both male and female figurines have been found close together one is tempted to challenge the Mother Goddess explanation, for the fact that the figurines are small shows that they were meant to be carried along easily, as personal equipment, almost as if they were domestic mementos or keepsakes, rather than objects of group worship.

We are faced here with the contradiction of an intensely masculine society whose main occupations excluded woman, except in the secondary capacity of butcher, cook, and tanner of skins, nevertheless elevating woman's peculiar functions and aptitudes, her capacity for sexual play, reproduction, and child-rearing, to a point where sex seized the imagination as never before. Both sculpture and the many surviving forms of ornament from shells to reindeer necklaces point to a considerable effort to enhance feminine bodily beauty and increase sexual appeal. Here was a gift that did not come to full fruition until another set of technical inventions, those of domestication, had pushed hunting into the background.

This view of the imaginative transformation of art and sex that accompanied the improvement of weapons and collective hunting techniques gets support from the very distribution of the female figurines found in this period. As Grahame Clark points out, the same type of sexually over-emphasized figures is found from France and Italy over to the South Russian plain, "most commonly made from mammoth ivory or of various kinds of stone," but also, in Czechoslovakia, from fired clay. "The fact that all those with a definite provenance," Clark notes, "came from settlements, whether from caves or from artificial dwellings, argues for their domestic rather than public or ceremonial significance." But domesticity, I would add from the historic evidence from Ur to Rome, of priestly functions performed by the paterfamilias, surely does not preclude ceremony. Even today in orthodox Jewish households, the housefather performs this role.

Along with this symbolic concentration on sex, the first evidences of hearth and home as central to a settled life appear: a mutant in hunting culture that became a dominant in the succeeding phases of neolithic culture, and has remained a persistent since. And not without further technical significance is the fact that the first use of clay, apart from the hearth itself, is its employment as a material for art, as in the bisons in the Tuc d'Audoubert cave, thousands of years before there is evidence of pottery. The suggestion here is plain: paleolithic man began to domesticate himself before he domesticated either plants or animals. This was the first step, beyond ritual, language, and cosmetics, in transforming the human personality.

And here, precisely at the point where the symbolic arts join together and supplement each other, *Homo sapiens*—Man the Knower and Interpreter—appears in the character that marks all his later history: not just doggedly scratching for a living, grubbing and picking, tool-making and hunting, but partly detached from those animal necessities, dancing, singing, playing, painting, modelling, gesturing, mimicking, dramatizing, conversing —certainly conversing!—and perhaps for the first time laughing. That laughter would identify him and certify his mastery better than tools.

Like Lazarus, late paleolithic man had at length fully emerged from the grave of pre-conscious existence, and there was reason for his laughter. His mind, increasingly released from brute necessity, from anxiety, dreamy confusion, panicky fear, had become fully alive. With command over words and images, no part of his world, inner or outer, animate or inanimate, lay entirely beyond his reach or his mental grasp. Man had at last perfected the kind of artifact, the symbol, upon which his highly organized brain could work directly, without any further tools than those the body itself provided. As for Magdalenian cave paintings they are the proof of a still more general and many-sided achievement in building a symbolic world.

These gifts were scattered and unevenly distributed: and they continued to develop unevenly, so that no generalization one makes about 'Man' applies to the entire human race, at all times and all places: far from it. Yet every symbolic advance has proved as transferable and as communicable as the genetic inheritance that binds the human race together: and man's pre-eminently social nature ensured that in time no population, however small, remote, or isolated, would be cut off completely from this common cultural heritage.

6: AROUND THE FIRE

One cannot do justice to the achievements of paleolithic man without coming back at the end to the capital discovery that ensured his survival after he had lost his own hairy fell: the utilization and perpetuation of fire. Language apart, this counts as man's unique technological achievement: unparalleled in any other species. Other creatures use tools, construct dwellings, dams, bridges, and tunnels, swim, fly, perform rituals, cooperate as families in raising the young; and even, among the social ants, wage war with soldiers, domesticate other species, and plant gardens. But man alone dared play with fire: so he learned to court danger and to discipline his own fears; and both practices must have enormously increased his self-confidence and effective mastery.

Many conditions existed to retard mental activity in the Ice Age: recurrent threats of starvation, fatigue from excessive physical activity, and torpor under the cold, which produces mental sluggishness and sleep. But fire saved man, awakened him, further helped to socialize him. What is more, the command of fire released this naked creature from his original tropical habitat. The hearth lay at the center of his life; and once the beginnings of language were made, it was surely in endless conversations and story-tellings around the fire that he perfected the vehicle of speech. This ancient art charmingly surprised that sympathetic proto-anthropologist, Schoolcraft, when he encountered it around the campfire of Indian tribes he had once supposed sullen, brutish, fierce—and silent. Is it merely a coincidence that the cultural hearths where one now discovers the first evidences of 'neolithic' domestication are those highland areas in Palestine and Asia Minor where ample supplies of firewood were once available?

Starting with fire, most of the equipment needed to ensure man's further development—except domesticated animals and plants—was already in existence before the last Ice Age came to an end around, say, 10,000 B.C. Let us sum up these mainly paleolithic contributions before neolithic domestication widened their scope and supplemented their deficiencies.

Taking material equipment alone, one finds: cordage, traps, nets, skin-containers, lamps, possibly baskets, along with hearths, huts, hamlets: likewise specialized tools, including surgical instruments, varied weapons, paints, masks, painted images, graphic signs. But even more important than this array of material inventions was the steady accretion of the agents of significance, the social heritage or tradition, expressed in every

mode of ritual, custom, religion, social organization, art—and above all
language. By Magdalenian times, minds of a superior order had not merely
come into existence but had produced a culture through which potentialities
hitherto undiscovered could be expressed and used.

In surveying paleolithic technics I have sought to counterbalance the
over-emphasis of tools and weapons as such by concentrating rather upon
the way of life they helped to bring into existence. The harshness of the
conditions under which paleolithic man operated—at least in the Northern
Hemisphere—seems to have intensified the human reaction and widened
the distance from his animal point of origin: he was fortified, not destroyed,
by these ordeals.

Under these circumstances, the fears and anxieties and eruptive fantasies that I have posited as an attribute of 'the dreamtime long ago' may
have been reduced to manageable quantities, much as the neuroses of
many Londoners were wiped out, psychiatrists found, during the Blitz.
Men have frequently risen to their highest potentials under conditions of
stress and physical danger: a tempest, an earthquake, a battle may summon forth energies and displays of selfless devotion and sacrifice not
elicited by a more prosperous way of life. It would be strange if some of
the qualities selectively preserved by paleolithic man were not still part of
our biological inheritance.

CHAPTER SIX

Fore-stages of Domestication

1: THE 'AGRICULTURAL REVOLUTION'
—REVISED

When the Stone Age was first seen as falling into two broad periods, there seemed a sharp dividing line between the earlier flaked tools and a later assortment which were ground and polished. The first belonged to the supposedly nomad food-gatherers and hunters, while the latter belonged to herdsmen and settled farmers, who within a period of about five thousand years brought about the domestication of plants and animals. But the changes in tools and weapons and utensils were much easier for the archeologist to read than the much more significant changes in breeding: hence until recently the neolithic phase was mainly identified with polished stone tools and—mistakenly—with clay pots.

For a time this picture seemed plausible; but almost every feature of it has, during the past generation, been revised. Tools and utensils constitute only a small part of the total equipment needed for bodily survival, to say nothing of cultural development. Even a purely technical story of the material improvements that took place is far from self-explanatory. To know how and why and when an invention became important one must know more than the materials and processes and previous inventions that went into it. One must likewise seek to understand the needs, the desires, the hopes, the opportunities, the magical or religious conceptions with which they have, from the beginning, been connected.

In order to clarify the immense changes ultimately brought about by domestication, I shall use the terms Paleolithic, Mesolithic, and Neolithic only to describe time-sequences, without necessarily tying them to any particular cultural or technical contents. Roughly, from 30,000 to 15,000 B.C. would be Late Paleolithic: from 15,000 to 8000 B.C. Mesolithic, and from then on to around 3500 B.C. Neolithic—provided one uses these dates only to describe the areas where significant changes first took place and where they reached their particular climax. The technical facilities and habits of life brought in by each of these phases are still with us.

The domestication of plants itself appeared as a mutant well before the end of the last glacial period. To associate this process with the moment when we behold the final results, or to attribute the change to improvements in tool-making draws our attention away from the real problems. Clay sickles in Palestine show that grain was being systematically garnered before it was deliberately planted, and stone mortars were used for grinding mineral paints thousands of years before they were used for grain. There are, nevertheless, profound cultural differences between the two epochs, despite all the evidence of continuous cultural fibers running through the successive layers that the archeologist finds in his diggings.

Partly because living conditions were so formidable during the Ice Age, paleolithic man, apart from playing with fire, largely took his habitat as given, and bowed to its demands, even becoming a specialist in one particular mode of adaptation—hunting. Such shaping as was open to him, I have sought to show, had mainly to do with his own body and mind. But the neolithic cultivator made numerous constructive changes in his environment, aided by the warming of the climate, and—after the great melting and flooding that followed—by the drying up of the swamps. With the aid of the axe he opened up the heavy forest, built dams and reservoirs and irrigation ditches, erected stockades, terraced hills, staked out permanent fields, drove piles, built clay or wooden dwellings. What neither the miner nor the hunter had been able to accomplish, the woodman and the farmer, able to support larger numbers in a small area, actually achieved: an increasingly humanized habitat.

Later civilization would be inconceivable without this mighty neolithic contribution: for only in their relatively large communities could work on this new scale be accomplished. Where the paleolithic artist, intent on the image, was satisfied to leave the walls of his caves rough and uneven, now wooden planks were split or stone was ground and polished, or clay and plaster were applied to form smooth surfaces for house walls or painted images.

Viewing this work as a whole, one must admit that there is little in either mesolithic or neolithic art, till we reach the threshold of urban life, that competes esthetically with either the early carved or modelled figures of the caves, or the paintings of Altamira and Lascaux. But a new trait appears in neolithic culture: 'industriousness,' the capacity for assiduous application to a single task, sometimes carried over years and generations. The intermittent technical activities of paleolithic man no longer sufficed: it was by prolonged, dogged, unremitting effort that all the typical neolithic achievements, from breeding to building, were accomplished. Paleolithic males, if we may judge by most surviving hunting peoples, had an aristocratic contempt for work in any form: they left such drudgery to their womenfolk. So when neolithic peoples turned to work, one need hardly wonder that woman, with her patient, inexorable ways, took command.

In this transformation, from a predominantly hunting economy to an agricultural one, much was gained; but also something was lost. And the contrast between the two cultures underlies a large part of human history; it can still be seen in more primitive communities even today. A modern observer in Africa, innocent of any of my present preoccupations, noticed a difference between the Batwa hunters, "gay with an uncomplicated cheerfulness" and the "rather sullen demeanor of the average Bantu" in his employ. And he asks himself: "Is it possible that the hard but unfettered life of the hunter brings a freedom of the spirit that the sedentary agriculturalists have lost?" Looking only at the surviving art and artifacts, one is driven to answer: Very possibly indeed—for reasons that will soon come out.

2: THE BREEDER'S EYE

Under the watchful breeder's eye of neolithic man, and even more neolithic woman, almost every part of the environment became plastic and responsive to the human touch. In a sense, the wider use of clay as against stone symbolizes this new trait in technics. Certain animals, now more important for food, became gentle and docile under man's tutelage: wild plants that once yielded only a modicum of nutriment, under continued selection and special care in prepared plots, swelled at the roots or burgeoned with edible beans, aromatic seeds, juicy flesh, colorful flowers. With the efficient stone axe it was possible to create open glades in the forest, where, around charred stumps and roots, herbaceous annuals long in use could be planted; and under such open, protected cultivation, plants hybridized rapidly, while at the edge of the woods, edible berry bushes, whose seeds were spread by cardinals and finches, multiplied.

For the first time, under neolithic cultivation and building, man began deliberately to change the face of the earth. In the open landscape, the signs of man's year-round occupation began to multiply: little hamlets and villages made their appearance in every part of the world. Instead of the random richness and variety of nature, one finds in the neolithic economy the beginning of a well-defined order; and this orderliness, this industriousness, transposes into physical structures much that had so long been confined to ritual and oral traditions.

If it is unwise to characterize this new period in terms of polished tools alone, it is equally misleading to regard the process of domestication as a sudden one: an agricultural 'revolution.' The implications of 'revolution,' which derives from the hopes and fantasies of the eighteenth century, are deceptive: for revolution implies a peremptory rejection of the past, a break in its ways; and in that sense no agricultural revolution took place until our own day. Archeologists have been slow to recognize what Oakes Ames called the "carry-over period" of uninterrupted knowledge of food plants, from primate times on, which led in the mesolithic phase to the deliberate selection and improvement of food plants, particularly the tropical fruit and nut trees valued by food-gathering groups, before systematic planting of annuals began.

The significance of this long prelude has been stressed by Ames, the botanist whose work on cultivated plants carried further the original researches of de Candolle. "The more important annuals," he pointed out, "are unknown in their wild state. They first appear in association with man. They are as much a part of his history as is the worship of the gods to whose beneficence he attributed the origin of wheat and barley. Therefore, their appearance simultaneously on the historic record indicates a greater age for agriculture than the archaeologists and anthropologists have allowed"—or, I might add, as yet allow.

Though there is still a tendency to date the great agricultural advance from 9000 to 7000 B.C., we now have reason to see that it was a much more gradual process that took place over a much longer period in four, possibly five, stages. First, the early knowledge of plants and their properties that paleolithic foragers had acquired and retained: knowledge that might have been partly lost in the northern zones but probably was continuous in the tropical and sub-tropical areas. Some of this early plant usage goes back so far that even the opium poppy, the first of the pain-killers, is no longer found wild. At this early period, familiarity with the feeding and breeding habits of wild animals must likewise be assumed, to account for the first domestication of animals.

This began, it would seem, with the dog, but if Eduard Hahn was correct, it included such barnyard animals as the pig and the duck. The third stage, then, would be mesolithic plant cultivation, which would include the care and eventual planting of various nutritive starchy tropical roots, like the yam and the taro plants. At length came the twofold process of domestication, plants and animals together, that ushered in the neolithic phase, and created in most of the Old World, though unfortunately not in the New, the soil-regenerative practices of mixed farming. The domestication of the ox, the sheep, and the goat went along with the introduction into the garden of beans, squashes, cabbages, and onions, and the wider selective cultivation, probably initiated long before, of the fruit-bearing trees: the apple, the olive, the orange, the fig, and the date. With the pressing of olives and grapes and the fermentation of grain for beer, fired pottery containers became indispensable.

Now, on the eve of civilization, the last stage in this complex, long-pursued process was taken, with the domestication of the cereals (einkorn, barley, wheat) and the beginnings of large-scale open-field, clean-crop agriculture. This brought about an immense increase of food in the rich lands of Mesopotamia and Egypt, for the very dryness of the cereal seeds makes these grains storable at ordinary temperatures for a longer period than most other foods except nuts; and the richness in proteins and minerals gives them exceptional food value. Stored grain was potential energy: also the oldest form of capital—witness pre-monetary commercial transactions in measures of grain.

But to signalize this last step as 'the' agricultural revolution is to overlook all the earlier steps that made it possible; for while still in their wild state many of the domesticates already served for tools, food, utensils, lashings, dyes, and medicines. Even after this phase matured, the impetus of domestication continued for a few thousand years, with the training of the llama, the vicuña, the onager, the camel, the elephant, and above all the horse, for traction and transport.

The most striking events of the agricultural transformation did indeed fall within the neolithic phase. Soon after it reached a climax, its original impulse toward domestication was exhausted. While some of the oldest domesticates, like the amaranth grains, fell out of cultivation, hardly any new species were added: instead, both in nature and on the farm, there was an unceasing proliferation of new varieties of the existing species, strikingly exemplified in the oldest of domesticates, the dog. In various parts of the world the neolithic technology was only partly carried out by the indigenous inhabitants, often content to stop at a half-way house.

But even where the changeover was fully consummated, food-gatherers still abounded, and nearer at hand, the hunter still performed a necessary function, for nowhere can crops be successfully cultivated or domestic cattle be kept safe, without trappers and hunters keeping down both the predators and the mischievous crop eaters, like the deer and the monkeys. In my own long-inhabited area of Dutchess County, the multiplying raccoons, no longer hunted for their fur, now often ruin whole fields of maize.

Not merely did the 'paleolithic' hunter remain at hand: but his special characteristics, as a wielder of weapons and a leader of men, played an essential part in the transition to the highly organized urban civilization that neolithic agriculture made possible. As with the fable of the flowers and the weeds, what one finds depends upon what one looks for. If one seeks evidence only of changes in a culture, one may pass by equally significant evidence of continuity. For culture is a compost in which many traits temporarily disappear or become unidentifiable, but few are ever completely lost.

In 'The Culture of Cities,' let me add parenthetically, I pointed out that every culture can be separated into four main components, which I then called dominants and recessives, mutations and survivals. Today, to get rid of this inappropriate genetic metaphor, I would change these terms to dominants and persistents, emergents—or mutants—and remnants. The dominants are what give each historic phase its style and color: but without the substratum of active persistents, and without the vast underlayer of remnants, whose existence remains as unnoticed as the foundations of a house until it sags or crumbles, no fresh invention in culture could achieve dominance. If one bears this in mind, it is legitimate to characterize a cultural phase by its largest visible new feature: but in the total body of a culture the persistents and the remnants, however hidden, necessarily occupy a far larger area and play a more essential part.

All this will become plainer when we trace this great transformation through in detail. Yet however radically we may be forced to alter the picture of a sudden change, there is no doubt that the development of new methods of growing and garnering and utilizing food altered man's relation to the entire habitat, and placed at his disposal vast resources of food and vital energy he had never before been able to tap. With this, getting a living ceased to be an adventure: it fell into a settled routine. The hunter had either to change his habits of life, or retreat to the jungle, the steppe, or the arctic tundra, thwarted as he was by the steady encroachments of cultivated fields and human settlements, and the inevitable shrinkage of both hunting grounds and game.

If one reads the record correctly, different hunting groups succeeded at all three forms of survival. But the hunter fared best when he entered into symbiotic relations with the new peasants and builders, and helped to create a new economy and a new technics—a technics based on weapons, in which, with his imagination and his audacity, he could as an aristocratic minority gain control over a large population.

3 : FROM COLLECTING TO PLANTING

As evidence heaps up it becomes plain that mesolithic domestication, with the year-round habitation of a single site, marks a necessary transition—in widely scattered areas and at different times—between the paleolithic and neolithic periods. In the later development of culture, the northern lands of Europe were always two or three thousand years behind the Near Eastern territory where the final innovations in domesticating cattle and grain took place: so it is not unreasonable to find in the well-established mesolithic culture of Denmark suggestive evidence that points to many much earlier developments elsewhere.

For the long series of experiments needed for the cultivation and improvement of plants, one must assume a margin of safety from starvation: and only large runs of fish, like salmon, which can be caught in weirs, as in the Pacific Northwest, or a steady supply of shellfish, would meet the earliest requirements for continuous local occupation. In such tropical or sub-tropical territory an additional source of food would be available from trees like the coconut and date palms, the banana and the breadfruit. The length of time required for cultivation of fruit and nut-bearing trees, which sometimes take thirty years or more to mature, was far greater than it is necessary to allow for the hybridization of annuals. This would argue, as Oakes Ames pointed out, that the protection and tending of trees began at a much earlier date. In other words, horticulture—with its prizing of single fine specimens—preceded agriculture, with its emphasis on larger yields, and in no small measure made it possible. The main tropical food staples, taro, manioc, coconut, breadfruit, to say nothing of the banana, the mango, and the durian, have the widest provenance in the Pacific and the South

Seas; while yams, the most widely distributed of all the root vegetables, even reached South America.

Though these pieces of evidence are thinly scattered and ill-assorted, they still add up to a reasonably firm conclusion: with mesolithic culture we find the beginnings of a stable occupancy of the land, *through all seasons*—the very condition essential for the exhaustive observation of the habits of plants that exhibit sexual reproduction and must be cultivated from broadcast or planted seed. The resultant increase in security must sometimes have appealed to the hungry hunter. But the partnership between hunter and cultivator worked both ways, for when crops failed, hunting and fishing might tide the community over a bad season. During the depression of the 1930's in the United States many families in destitute rural mining communities managed to survive partly by fishing and hunting.

With this new security derived from a regular food supply came a new regularity in life; and with this regularity, a new tameness, too. The little mesolithic communities became as rooted as the tubers and the mollusc beds themselves. This indeed was a favorable condition for further experiments in domestication.

The knowledge needed to stimulate such experiments probably moved along the same routes that spread the favored types of stone. Such stone travelled far, and testifies to keen prospecting and comparative testings of quality. Evans, in 'Man's Role in Changing the Face of the Earth,' notes that "certain bluestone axes found through the British Isles came from a small exposure of porcellanite on Tievebulliagh . . . so small indeed as to have escaped the notice of the Geological Survey of Ireland."

With the first steps toward plant domestication Hahn postulated a correlative advance in animal domestication: the domestication of the dog and the pig. About the fact that the dog was the earliest creature to be domesticated, biologists and ethnologists are pretty well agreed; and it seems also plain that it was not due to any original usefulness as a hunting companion. The dog's ancestors, the jackal and the wolf, were rather attracted to human settlements by the same appetite that causes my country neighbors' dogs, however well fed, to haunt my garbage compost: a taste for bones and offal.

In time the dog identified himself with the human community, becoming a watchman who, like another early domesticate, the goose, gave warning of intruders. Only later did he become a protector of the young and an ally in hunting and herding. But his principal first use was probably as a scavenger; and in that capacity he and his companion, the pig, served all through the development of close communities, even down to the nineteenth century, and in cities as big as New York and Manchester. Significantly the pig and the fish, right into historic times in Mesopotamia, remained sacred animals: both form part of the original mesolithic constellation.

The most important aspects of this long process of domestication can be described without reference to any new tools except the axe; but the axe, which had long been in use, was certainly improved in shape, and attached, however infirmly, to a handle, and a new technique in making other cutting implements was introduced, by utilizing small sharp stones, microliths, set in clay or wood, as saw-toothed cutting edges.

One of the reasons indeed why the technique of preparing the earth for cultivation and for keeping the soil friable went on even more slowly than the improvement of plants, was the restrictive lack of adequate tools. Though neolithic cultivation is often referred to as hoe culture, the hoe is a relatively late contribution. There could be no cheap, efficient hoes till the Iron Age. The chief means of working the soil, right into Egyptian and Sumerian times, was the digging-stick, sometimes with a stone attached near the bottom to weight it. Even after the plow had been invented, it was actually a digging-stick in traction, not the furrow-turning plow that came in only at a nearer point in the Iron Age. The rather late Sumerian 'Dialogue Between the Plow and the Pickaxe' as to their respective merits suggests that the plow did not become dominant at once.

The enormous increase in crop yields that the records establish for the Near East was based upon the utilization of humus-rich soils, once swamps, on manuring, on irrigation, and above all on seed-selection: it owed little or nothing to improvements in tools. As for the ox-drawn plow, its great advantage was that it made extensive cultivation possible with less manual labor. This mode of labor-saving expanded acreage, but did not of itself produce higher crop yields per acre.

The botanical knowledge that was achieved through long-pursued plant cultivation was not based on any precise system of symbolic abstractions: so the modern observer would hesitate to call it scientific. But could it have succeeded so wel! if it had not in fact been the outcome of causal insight and relevant correlations, passed on by speech? If some of the magical prescriptions that survived for thousands of years in proverbs and folk-say have proved to be erroneous, there is still a greater bulk of observations that show a remarkable facility to put two and two together. The keeping in mind of the necessary sequences by proverbial saws was not the least feat of these archaic cultures. Fortunately for this interpretation, in Hesiod's 'Works and Days' some of these traditional observations found their way eventually into writing.

Those who are still contemptuous of the errors of pre-scientific lore overlook the large accretions of positive knowledge that justified it; and this knowledge was often more important than the physical tools employed. Long before Bronze Age technics had fully utilized the earlier improvements in horticulture and agriculture, archaic man had done the preliminary work of exploration so well that except for a few plants like the cultivated strawberry and the boysenberry, all our present domestic plants and animals are neolithic end-products. Civilized man has refined the early breeds, quantitatively increased their productivity of food, improved their form and taste and texture, exchanged plants between distant cultures, and encouraged endless variations. Granted: but he has not brought any important new species into cultivation.

Except for the length of time necessary to make these first steps, the magnitude of the achievement itself is fully comparable to that of the scientific advances that have finally led to the fission of the atom and the enlargement of astronomical space.

Long before the metal-using civilizations had taken form, early man had identified the most useful varieties of plants, animals, and insects out of the thousands of species—themselves singled out of hundreds of thousands of species—that he must have sampled. All man's food resources and most of the materials for clothing, shelter, and transportation were identified and utilized before the introduction of metallurgy. Though bitter tastes are repulsive, early man experimentally learned ways of depriving potentially useful foods of their poisonous alkaloids or acids; and though starchy, hard-husk grains are not digestible in their raw state, our neolithic predecessors learned to pulverize them and make a paste for baking a digestible bread on a flat stone.

The use of the horse as a draft or mounted animal came late, well after the onager had been harnessed. And we know that the Egyptians, in historic times, tried to domesticate some of the more ferocious felines for household pets and for assault in war, only to fail in both efforts, for the still wild creatures in the panic of battle too often turned against their masters. The primitives' use, in the Amazonian forests, of the sap of the rubber tree, that essential contribution to modern motor transport, for making balls and raincoats may have come at a relatively late date, like the use of infusions of the coffee bean as a stimulant. Who can say? But the important thing to remember is that, early or late, all these innovations were direct derivatives of 'neolithic' horticulture. And without the endless searching and sampling that characterized man's primeval food-collecting economy the last stage—selection and cultivation—could never have been reached.

The exhaustiveness of this original series of discoveries is almost as astonishing as the variety that was achieved through sexual selection and hybridization. Edgar Anderson pointed out in 'Plants, Man and Life' that "there are five natural sources of caffeine, tea, coffee, the cola plant, cacao, yerba mate and its relatives. Early man located all of these five and knew that they reduced fatigue. Biochemical research has not added a single new source."

Similarly it was not the busy chemist of today's pharmaceutical laboratories, but the primitive Amerindians, who first discovered that snake-root (reserpine) was a useful herb for tranquillizing people in manic states. That was a far more improbable discovery than penicillin: only a spirit of experiment and a gift for close observation could have made this correlation: even then it remains astonishing, indeed mysterious, like the folk belief, justified in the case of the cinchona bark, that the natural remedy lies in the area of a particular disease.

The kind of knowledge demanded by domestication was not a simple identification, then, of nourishing plants: but rather an insight into soils, seasonal successions, climatic changes, plant nutriments, water supply: an exceedingly complex group of variables, different for different plants even though they occupy the same habitat. Much of this observation antedated neolithic practices; for grain-gathering Australians, still living under paleolithic conditions, were watchful enough in their wanderings to observe that the grain grew better if well-watered, and would turn the course of a brook in order to irrigate the patch of wild grain that they utilized.

In the first instance, then, I must emphasize, it is not the changes that actually took place in the manufacture of tools and utensils that identifies the neolithic phase; for the decisive technical advances—through boring and grinding, and the eventual transformation of reciprocating motion ᵼto rotary motion, in the bow-drill and the fire-drill—were essentially late paleolithic mutations. Even modelling in clay produced paleolithic animal sculpture and figurines long before it formed Mesopotamian houses and pots (art again coming before utility). To understand the technics of domestication we must rather take account of a religious change that centered increasingly on life, growth, sexuality, in all their manifestations.

That new cultural pattern spread throughout the planet from about 6000 or 7000 B.C. onward. The individual inventions that accompanied this social change were transferred erratically, leaving many products to be invented or domesticated on the basis of purely local resources and opportunities: but the pattern as a whole forms the underlayer on which all higher civilizations until now have been based.

4: THE DAILY GRIND

On this interpretation, the improvement of tools alone, except for the axe and the later pickaxe or mattock had little to do with the neolithic advances in domestication. But there is one aspect of neolithic tool-making that throws a significant light upon every other aspect of the culture. And this is the fact that, apart from the early development of the original microlithic 'saw-tooth,' the main method of making neolithic tools was by grinding, boring, polishing.

This practice of grinding began in paleolithic times, as Sollas did well to bring out half a century ago; but the shaping of tools by grinding is a general neolithic improvement. By itself it expresses a definite characteristic of the whole culture. The patient application to a single task, reduced to a single monotonous set of motions, advancing slowly, almost imperceptibly, toward completion, was far from characteristic of food-collectors or hunters. This new trait became visible first among the skilled flint-knappers who made the finely chipped Solutrean and Magdalenian spearheads and gravers. But the grinding of even soft stones is a tedious and laborious process; granite or diorite, both extremely hard, demand a willingness to endure drudgery that no human group had ever imposed on itself before. Our very word to express ennui, 'boring,' derives from—boring. Here was ritual repetition pushed almost beyond endurance.

Only groups that were prepared to remain long in the same spot, to apply themselves to the same task, to repeat the same motions day after day, were capable of gaining the rewards of neolithic culture. The restless ones, the impatient and adventurous ones, must have found the daily routine of the neolithic hamlet intolerable, as compared with the excitement of the chase, or of fishing with net and line. Such people reverted to the hunt or became nomadic herdsmen.

One does not push the evidence too far if one says, briefly, that the neolithic tool-maker first invented 'daily work' in the sense that all later cultures were to practice it. By 'work' one means industrious devotion to a single task whose end product is socially useful but whose immediate reward might be small to the worker himself, or might even, if too prolonged, turn into a penalty. Such work could be justified only if its ultimate use to the community proved greater than could be achieved by a more fitful, capricious, 'amateurish' attitude toward the job.

One of our common expressions for work, 'the daily grind,' would not have been a figure of speech in the early neolithic community. But it was not only grain that needed daily grinding. With the first paleolithic stone utensils, the mortar and the stone lamp, went a decisive contribution to all later technology—circular motion. And with the translation of this motion from the hand to the wheel came the next important machine, after the bow-and-arrow, the potter's wheel.

In grinding, steady application counts for more than the fine sensory-motor coordination needed for flint-knapping. Those who were ready to submit to this discipline would, it seems likely, also have the patience to watch the same plants, through all the processes of growth, season by season, and of repeating the same processes, year after year, to achieve the same anticipated result. These repetitive habits proved to be immensely productive. But there is hardly any doubt that in some degree they dulled the imagination, and tended to select and advance the more submissive types, while by providing a better food supply they in turn ensured their multiplication and survival.

The process of grinding had the advantage of freeing the fabricator from a few types of stone, like flint, which are especially amenable to chipping: tools could be made of other hard stones, like granite, and utensils, like pots or vases, could be ground out of soft sandstones and limestones, before baked clay pots were invented. But the great incentive to grinding came from the domestication of the cereals, for to use these, before pots for boiling became available, it was necessary to grind them, so that they might be made into a paste and baked on a stone. The mechanical process and the functional need, plus the botanical skill in plant selection and cultivation developed together.

With the cultivation of grains, a new order of settlement became possible in parts of the planet not favored by tropical luxuriance and equable climate: for the grain-bearing grasses have as wide a distribution as grass itself; and though the systematic cultivation of grains seems to have begun in the few great sub-tropical river valleys, barley, wheat, and rye eventually made it possible to get large yields of storable food in much colder areas. With this, the march of agriculture polewards began in both northern and southern hemispheres.

The domestication of grain was accompanied by an equally radical innovation in the preparation of food: the invention of bread. In an endless variety of forms, from the unleavened wheat or barley of the Near East to the corn tortillas of the Mexicans and the yeast-risen bread of later cultures, bread has been up to now the center of every diet. No other form of food is so acceptable, so transportable, or so universal. "Give us this day our daily bread" became a universal prayer, and so venerated was this food, as the very flesh of a God, that to cut it with a knife is still, in some cultures, a sacrilege.

Daily bread brought a security in the food supply that had never before been possible. Despite seasonal fluctuations in yield due to floods or droughts, the cultivation of grains made man assured of his daily nourishment, provided he worked steadily and consecutively, as he had never been certain of the supply of game or his luck in killing it. With bread and oil, bread and butter, or bread and bacon, neolithic cultures had the backbone of a balanced diet, rich in energy, needing only fresh garden produce to be entirely adequate.

With this security, it was possible to look ahead and plan ahead with confidence. Except in the tropical areas, where soil regeneration was not mastered, groups could now remain rooted in one spot, surrounded by fields under permanent cultivation, slowly making improvements in the landscape, digging ditches and irrigation canals, making terraces, planting trees, which later generations would be grateful for. Capital accumulation begins at this point: the end of hand-to-mouth living. With the domestication of grains, the future became predictable as never before; and the cultivator not merely sought to retain the ancestral past, but to expand all his present possibilities: once the daily bread was assured, those wider migrations and transplantations of men, which made the country town and the city possible, speedily followed.

5: THE RITUALIZATION OF WORK

With grain cultivation, the daily grind took over a function that only ritual had performed before: indeed, it would be nearer the truth, perhaps, to say that ritual regularity and repetition, through which early man had learned in some degree to control the mischievous and often dangerous outpourings of his unconscious, was now at last transferred to the sphere of work, and brought more directly into the service of life, in application to the daily tasks of the garden and the field.

This brings me to a point that has been too little recognized by machine-minded technologists, concentrated mainly on the dynamic components of technology. The radical neolithic inventions were in the realm of containers: and it was here that the effect of the tedious processes of grinding was partly overcome by the utilization of the first great plastic—clay. This earth is not merely easier to shape than stone, but it is lighter and more convenient to transport. And if baked clay is more fragile than stone, it is likewise far easier to replace. The creation of moisture-proof, leak-proof, vermin-proof clay vessels to store grain, oil, wine, and beer was essential to the whole 'neolithic' economy, as Edwin Loeb emphasized.

Many scholars who have no difficulty in recognizing that tools are mechanical counterfeits of the muscles and limbs of the male body—that the hammer is a fist, the spear a lengthened arm, the pincers the human fingers—seem prudishly inhibited against the notion that woman's body is also capable of extrapolation. They recoil from the notion that the womb is a protective container and the breast a pitcher of milk: for that reason they fail to give full significance to the appearance of a large variety of containers precisely at the moment when we know from other evidence that woman was beginning to play a more distinctive role as food-provider and effective ruler than she had in the earlier foraging and hunting economies. The tool and the utensil, like the sexes themselves, perform complementary functions. One moves, manipulates, assaults; the other remains in place, to hold and protect and preserve.

In general, the mobile, dynamic processes are of male origin: they overcome the resistance of matter, push, pull, tear, penetrate, chip, macerate, move, transport, destroy; while the static processes are female and reflect the predominant anabolism of woman's physiology: for they work from within, as in any chemical transformation, and they remain largely in place, undergoing qualitative changes, from raw meat to boiled meat, from fermenting grain to beer, from planted seed to seeding plant. It is a modern solecism to regard stable states as inferior to dynamic ones: yet savants who smile at the fact that the ancients regarded the circle as a more perfect form than the ellipse, make the equally naive discrimination in favor of the dynamic over the static, though both are equally aspects of nature.

Cooking, milking, dyeing, tanning, brewing, gardening are, historically, female occupations: all derive from handling the vital processes of fertilization, growth, and decay, or the life-arresting processes of sterilization and preservation. All these functions necessarily enlarge the role of containers: indeed are inconceivable without baskets, pots, bins, vats, barns; while true domesticity, with its intimate combination of sexuality and responsible parenthood, comes in only with the permanent dwelling house, the cattle-fold, and the settled village. As with the other components of neolithic culture, this changeover was no sudden revolution: it had long been in preparation. The village itself, I must remind the reader, was a paleolithic mutant at least twenty thousand years ago, possibly earlier, though it only became a dominant after the glaciers receded.

As home-maker, house-keeper, fire-tender, pot-molder, garden-cultivator, woman was responsible for the large collection of utensils and utilities that mark neolithic technics: inventions quite as essential for the development of a higher culture as any later machines. And she left her personal mark on every part of the environment: if the Greeks held that the first 'patera' was molded on the breast of Helen, the Zuñi women, to corroborate the fable, used to make their pitchers in the actual form of the female breast. Even if, instead, the round gourd could be plausibly regarded as the original model, that fruit, too, came within woman's province.

Protection, storage, enclosure, accumulation, continuity—these contributions of neolithic culture largely stem from woman and woman's vocations. In our current preoccupations with speed and motion and spatial extension, we tend to devaluate all these stabilizing processes: even our containers, from the drinking cup to the recorder tape, are meant to be as transitory as the materials they contain or the functions they serve. But without this original emphasis on the organs of continuity, first provided by stone itself, then by neolithic domesticity, the higher functions of culture could not have developed. As work begins to disappear in our society under automation, and the daily grind becomes personally meaningless, we shall perhaps for the first time realize the part that neolithic culture played in the humanization of man.

CHAPTER SEVEN

Garden, Home, and Mother

1: DOMESTICATION ENTHRONED

The earliest animal to come under domestication was man; and the very word we use to describe the process reveals its point of origin. For 'domus' means home; and the first step in domestication, which made all the later ones possible, was the establishment of a fixed hearth with a durable shelter: possibly in the midst of a forest clearing, where the first cultivated plants could be watched by the womenfolk, while the men continued to range abroad in search of game or fish.

Daryll Forde points out that among surviving peoples living under much the same conditions "patches on which wild yams grew abundantly were protected, partially cleared, and transmitted from mother to daughter among some Australian aborigines." When the hunter returned empty-handed, perhaps cold and wet, too, he would find a fire still burning for him, and a store of edible roots or nuts would stay his hunger.

Garden culture, different from later field culture, is pre-eminently, almost exclusively, woman's work. Clearly the first steps in domestication were taken by her. If this culture was not politically a matriarchal one, its emphasis was nevertheless maternal: the care and nurture of life. Woman's old role as a discriminating collector of berries, roots, leaves, herbs, 'simples,' has continued among peasants down to our own time, in the old woman (witch doctor) who knows where to find the herbs for medicine and how to apply their 'virtues' in curing an ache, lowering a fever, or healing a wound. Neolithic domestication enlarged this role.

With a more abundant and regular food supply, other results followed that increased the importance of hearth and home: the richer and more varied diet not only increased sexual appetite, but likewise, we now know, increased the likelihood of conception; while a fixed shelter and plenty of food contributed to the survival and better care of infants, partly because, in stable village groups, more women of various ages were at hand to keep an eye on the growing children.

So if the grinding of stone was tedious, and if remaining in a single spot added to the monotony, there were compensations. Under these new conditions of security, there was longer expectation of life: this gave more time for knowledge to accumulate and be passed on; and just as more infants survived, old men and women, too, became more numerous, and served as repositories of oral tradition. Now not youth and daring, but age and experience counted as never before. The democratic Council of Elders was an essentially neolithic institution.

But there is a marked difference between the early and late phases of neolithic culture, which corresponds roughly to the difference between horticulture and agriculture, between growing flowers, fruit, and vegetables, and growing grain. Except for the house itself and the hamlet, this is a period of small containers: hearth and altar and shrine, baskets and pots and vegetable pits; while the latter is the period of large containers, ditch and canal, field and cattlefold, temple and town. Small or big, the emphasis falls, except for one important new tool, the axe, upon the container.

In the second phase, leadership shifts, by reason of heavy demands on human muscle, back to the masculine occupations and the masculine roles; though even after the hunter has re-asserted his dominance by command of the citadel and by rule over the city, in religion and the practices of daily life woman for long played a coordinate role with man, as the written records in Egypt and Babylonia testify. But if one wishes to focus on the most critical neolithic advances, it is within the circle of woman's interests that one will find them: above all, in the new mutant, the garden.

2: THE INFLUENCE OF THE GARDEN

Central to the whole process of domestication was the garden: it was the bridge that united the perennial care and selective cultivation of tubers and trees with the clearing of wilds and the planting of early seeding annuals, emmer, einkorn, and barley. The wholesale cultivation of cereals was only the culminating point in this long experimental process; and once that step was taken fixation and stabilization began to set in.

The first successful domestication of cereals could not have taken place in the grasslands or swamps of the Near East. With the existing implements, it was easier to make a clearing in the forest, which the neolithic axe was capable of effecting, than to break the heavy grass roots of the open plain: anyone who has tried to tackle such roots only with the aid of a steel hoe or spade will know why. Oakes Ames cites the use in New Guinea of a digging-stick ten to fourteen feet long, with eight men working it to break up the tough sod; but though such cooperative effort could offset the lack of good tools, it was obviously too costly of energy to come into general use.

Since the earliest gardens must have developed from the mere guarding of wild patches that produced edible leaves or fruit, some of that wildness surely remained, as Edgar Anderson suggests, in the early gardens: he has found their variety and informality in contemporary Mexican gardens in small villages. Such neolithic gardens contained a mixture of different botanic species, some on their way to cultivation, others intruding as weeds, with those most resembling the more cultivated seedlings often mistaken for the desired species—something that still happens to every gardener—so that this mixed garden was specially favorable to crossings, often with the aid of 'volunteers.'

In this early stage of cultivation, no cattle were needed to maintain fertility: if burning and composting were not sufficient, a shift in the clearings sufficed. Within the neolithic garden, if one follows Anderson's shrewd reconstruction, foods, condiments, aromatics, medicines, useful fibers, flowering plants admired for their color, perfume, or beauty of form or their place in religious ritual, all grew side by side, sometimes, like the nasturtium, equally valued for salad greens as for decoration. Observe the variety and the lack of specialization, along with the concern for quality rather than quantity: and it is not perhaps an accident that some of the most useful plants to be cultivated were those that may have been prized at first for their brightly colored flowers, like the mustard, the squash, the pepper tree, the broad bean, the wild pea, or even for their perfume, like most of the aromatics.

Anderson, in separating garden from field culture goes so far as to identify the second with a conspicuous lack of interest in flowers and ornamental plants. If we confined our idea of plant domestication to the cereals, we should completely forget this vital esthetic contribution, not merely of floral colors and forms, but of a whole variety of delicate tastes and odors—so different from the rankness of animal foods that many vegetarian people, like the Japanese, find even the body odors of the meat-eating Westerners repulsive. Good taste, at least in clothes and food, is a distinctly neolithic contribution.

From Indonesia, where tropical horticulture probably first arose, a whole series of neolithic inventions based on bamboo might have spread over a large part of the world, even if clay and stone and metal had not become central to a progressive technology. Alfred Russel Wallace, in his first explorations of the Malay Archipelago, pointed out the many uses of bamboo alone: "split and shaved thin, [it] is the strongest material for baskets; hen-coops, bird-cages, and conical fishing traps are very quickly made from a single joint. . . . Water is brought to the houses by little aqueducts formed of a large bamboo split in half and supported on crossed sticks of various heights so as to give it a regular fall. Thin long joined bamboos form the Dyaks' only water vessels. . . . They are also excellent cooking utensils; vegetables and rice can be boiled in them to perfection." To which one can add many other uses discovered by the Japanese and the Chinese, including use as pipes to convey natural gas in China.

In little garden patches, then, long before large-scale systematic field cultivation took place, the first food plants were deliberately planted, harvested, and their surplus seeds planted again. The wide distribution of the pulses and the squashes indicates the antiquity of this mode. By the time the record becomes clear—perhaps four or five thousand years after the last Ice Age, three or four thousand before the cities of Mesopotamia—the chief food and fiber plants of man had already been domesticated, and some food plants like *Camelina*, 'gold of pleasure,' which have now dropped out of cultivation, were then used for oil. Flax, too, may have been grown for vegetable oil (linseed) before its fibers were macerated to produce linen thread: the Russian peasant's practice of pouring linseed oil on potatoes may have had a neolithic counterpart.

The plenitude of nourishment that came in with the domestication of cereals and animals would not have been possible to people who lived on the produce of a garden alone. But these early gardens may well have made up in variety and quality, in the vitamins available from freshly cut leaves and from berries, what was lacking in quantity; and for the first time in history neolithic peoples achieved a continuously balanced diet, adequate through the year because some of it could be dried and stored.

If I have emphasized the regulative effect of the grinding operations in disciplining the neolithic cultivators to monotonous practices of every kind, one must temper this characterization by remembering that organic processes, not least the growing of plants, are full of subtle changes and present unexpected problems. So if a constant watchfulness is demanded, a certain alertness to small variations is necessary, too; and this would have held particularly during the first stages of domestication and acclimatization.

In the close garden patch overcrowding would lessen the yield, but overspacing might permit too many weeds; while selectivity was the condition for achieving and maintaining variety. The protection of favored plants was an essential part of the whole effort to protect and foster and appreciate the forces of life. If hunting is by definition a predatory occupation, gardening is a symbiotic one; and in the loose ecological pattern of the early garden, the interdependence of living organisms became visible, and the direct involvement of man was the very condition for productivity and creativity.

Behind all these varied changes in domestication was an inner change whose significance has been slowly, indeed reluctantly, dawning upon the students of early man: the change that went on in his mind and was translated, long before he made further practical use of them, into the forms of religion, magic, and ritual: the consciousness of sexuality as a central manifestation of life itself, and of woman's special role in both effecting and symbolizing sexual delight and organic fecundity.

This sexual transformation, this erotization of life, is still visible in the early Egyptian and Sumerian legends: Enkidu has to be deliberately won away from his barbarous bachelor preoccupation with hunting by being enticed and seduced by a town prostitute. But by the time sexuality finds expression in ritual or legend, many unrecorded aspects of it had probably disappeared. What is left in the record are the rituals of Osiris or the sacred ritual union of the King and the Goddess, in the form of a priestess, at the Babylonian New Year Festival. A later rite, the Hellenic dance-orgies of the wild women, after Bacchus, probably points to a much older manifestation.

3: THE CLIMAX OF DOMESTICATION

The overwhelmingly material preoccupations of our own age, its impatient efforts to turn pinched subsistence economies into affluent industrial economies, tempt us to regard the whole process of domestication as a more or less deliberate effort to increase the supply of food. Only belatedly has it dawned on a few scholars that primitive man did not look at the world in this way; and that what is a primary motive for us played only a secondary role, if any at all, in his life.

In reconstructing the process of domestication, we would do well to treat the increased consciousness of sexuality, an essentially religious consciousness, as the dominating motive power in this whole change: from later data we may plausibly reconstruct a religious cult, exalting the body and the sexual functions of woman as the ultimate source of all creativity. The first evidence of this deepened sexual consciousness, as I have already pointed out, may have borne fruit in the paleolithic ivory carvings of female figures with all their sexual traits greatly enlarged. But until we come to historic times, it is notable that both the woman's male mate and her child are missing: these enter the picture for the first time in Jericho, where, as Isaac points out, "cultic figurines in groups of three occur, each containing man, woman, and child."

With the cultivation of plants, woman's special sexual characteristics become symbolically significant: the menstrual onset of puberty, the breaking of the hymen, the penetration of the vulva, the milking of her breasts, make her own life a model for the rest of creation. All these activities, in becoming concentrated and magnified, also become sacred. The interest in woman's central role intensified the consciousness of sex in many other aspects.

Birds, which are almost absent from cave paintings, were everywhere in evidence in tropical regions, and multiplied in the temperate clearings, with the spread of berry bushes and vines. Birds became the very model of human sexuality by their pre-marital preening and courting, by their tidy nesting in a fixed habitat, by their calls and songs, by their unflagging care of the eggs and fledglings. Feathers, which remained, with flowers, the dominant form of body decoration in Polynesia, may well be remnants of this early identification and appreciation of the role of beauty in sexual activity. Birdsong itself may have awakened the latent musical gifts of man.

One of the signs of domestication still visible in art is the part that birds and insects both begin to play in the human imagination: an interest that may have been stirred by the important role one plays in scattering seeds, the other in fertilizing annuals. The beetle's change from chrysalis to winged creature became a symbol of the passage and release of the human soul; and Egyptian paintings of birds vie with Audubon's both in quality of observation and in beauty. Not merely do hawk- and ibis-headed gods

play a part in Egyptian religion, but in far Siberia engravings have been
found on a stone block, dated 3000 B.C. by a Russian archeologist, with
birds' heads; while the appearance of wings on early human or godlike
figures indicates a later association of bird flight with command and with
swift communication. With some reason Aristophanes chose birds to symbolize both aspects in his Cloudcuckooland utopia.

That birds and insects are indispensable to the cultivation of gardens
and fields, and that the birds are needed to keep down the overpopulation
of insects, such as the locust invasions that still recur in Mesopotamia,
must have been discovered by the ancient cultivators who likewise found
out how to improve the yield of date palms by hand fertilization.

The heightened consciousness of her sexual role not merely gave a new
dignity to woman, no longer merely a drudging camp-follower of the
hunter, committed to the dirty work of separating and chewing the guts for
thread, and scraping and tanning the skins: it filtered through the imagination into her other activities, shaping vases, coloring textiles, ornamenting
the body, perfuming with blossoms the air.

Certainly the lunar rhythm that governs woman's own menstruation
was transmitted to planting, for to this day the phases of the moon are
still piously respected by primitive cultivators all over the world. If Alexander Marshack prove right in interpreting the inscriptions on ancient
reindeer bones, thirty-five thousand years ago, which correlate with similar signs in Azilian rock wall paintings, as a lunar calendar, this would only
re-enforce the view that the initial steps leading to domestication go back
to the food-collecting stage.

The world of plants was woman's world. With much better reason than
one can speak of the agricultural or the urban revolution, one might call
this essential change, which was the prelude to all the other great changes
that came in with domestication, the sexual revolution. All the daily acts of
life became sexualized, eroticized. So emphatically concentrated was this
image that, in a whole series of figurines and paintings, woman herself, as
represented in paleolithic art, disappears: only the sexual organs are left.

With this change, one must identify, on shadowy but widespread evidence, the myth of the Great Mother. But there is a dark side to woman's
dominance, plainly revealed in the late Babylonian epic of Marduk's bloody
struggle with Tiamat, the savage *Urmutter:* for in taking the lead in this
cultural transformation, woman's latent masculine animus must often have
come to the fore. In more than one later religious myth, she is bodied
forth as a powerful figure, attended by lions, as the vengeful fury, the
destroyer goddess, like Kali, the devourer, in Hindu religion; while the
male principle, in the Great Mother myth, is represented as a minor lover,

an accessory, but hardly an equal mate. To forget this other side of woman's triumphs in domestication would be to prettify and falsify the whole story.

4: THE MYSTERY OF SACRIFICE

As the anthropologist repeatedly finds, the intermingling of practical knowledge and causal insight with magical prescriptions often based on fantastic associations, characterizes so-called primitives of our own time; and must have been equally true in earlier cultures. No myth, however life-oriented, is wholly rational in its promptings; and the steady accumulation of empirical knowledge that accompanied the earliest garden culture was not sufficient to ward off spurious, often perverse, suggestions of the unconscious, fostered originally by some accidental success.

Perhaps the most mysterious of all human institutions, one that has been often described but never adequately explained, is that of human sacrifice: a magical effort either to expiate guilt or promote a more abundant yield of crops. In agriculture, ritual sacrifice may have resulted from a general identification of human blood with all the other manifestations of life: possibly derived from the association of menstruation and blood with fecundity. Such a belief might have been given a factitious support from the gardener's empiric knowledge of the fact that in order to produce a few lusty plants a hundred seedlings may have to be rooted out. In gardening, such sacrifice is a means of ensuring growth; and the effect of thinning and pruning would not have escaped the sharp eyes that had discovered the function of seeds and had selected and cultivated many plants.

But at the point where causal insight might have been sufficient to establish the wholly rational practices of mulching, watering, thinning, and weeding, the unconscious may have misread the process and offered an infantile improvement of its own, as a surer and quicker way of arriving at the same results: killing not a few plants, but a human being whose blood would ensure more abundant fruits. Was not blood the essence of life? Even this might have been based on observation of a rich vegetational growth above a shallow human grave, and to that extent the sacrificial offering may have sometimes been as effective as the dead fish that the American Indians used to plant under a hill of corn.

These are unverifiable conjectures, but they are not entirely baseless.

There is better evidence for human sacrifice in the neolithic community than there is for anything that could be called war. Along with all the vast gains achieved through domestication, rooted in the cult of the mother, one must take into account the possibility that the perversion of huma sacrifice then came into existence.

At this point one must call in the scholar of religions. "According to the myths of the early horticulturalists of the tropical regions," Mircea Eliade observes in 'City Invincible,' "the edible plant is not given in nature; it is the product of a primordial sacrifice. In mythical times, a semi-divine being is sacrificed in order that tubers and fruit trees may grow out of his or her body." There is similar evidence from the Near East in the early myths of Osiris and Tammuz, as well as in the later one of Dionysus.

The solemn doing to death of one or more victims, often a young girl, at the beginning of the vegetative season, in many widely scattered regions of the earth, is an historically attested fact. And though this practice was gradually shifted to animals, fruits or plants with the oncoming of civilization, human sacrifice was never wholly dropped. In such advanced cultures as those of the Maya and the Aztecs, sacrifice remained right down to the Spanish conquest. Among the cultivated Maya, slaves were even sacrificed at an upper-class feast, merely to give it a properly genteel elegance. Significantly, the sacrificial offerings of fruits by Cain, the farmer, were less acceptable to Jehovah than those of Abel, the pastoralist, who sacrificed an animal.

Human sacrifice, then, is the dark shadow, vague but ominous, that accompanied the myth of maternity and the superb technical and cultural feats of domestication. And as so often happens, this particular mutation, quantitatively restricted in the culture where it originated, dominated and debased the urban civilization that grew out of it, by taking another collective form: the collective sacrifice of war, the negative counterpart of the life-promoting rituals of domestication.

But if the sacrificial altar was one derivative from the domestic hearth, the oven, the kiln, and the furnace were others: from this source came the firing of bricks and clay vessels, and eventually the transmutation of sand into glass and stones into metals. And here again art preceded utility: glass was first used for decorative beads and iron for finger rings, while in early Jericho the clay figure of a cow preceded pottery—the paleolithic clay bisons had by many thousand years preceded the neolithic milch cow.

5: THE VENERATION OF ANIMALS

As the archeological finds now show, the domestication of the herd animals came in at the same level as seed agriculture, and one would hardly have been possible without the other, though in time herding spread to the grasslands as a specialized nomadic culture. Carl Sauer has brought cogent arguments to show that mixed farming antedates pastoralism, and in the absence of contrary evidence, his argument seems decisive.

But here again it is doubtful if the first steps in animal domestication were due to any desire to increase the food supply. As with the earlier domestication of the dog and the pig, even their use as scavengers was probably secondary to a playful sense of companionship, such as still exists among the Australian aborigines with their opossums and their wallabies. And with what eventually proved to be highly useful cattle, the ox, the sheep, and the goat, sexuality itself may well have first set them apart as expressive symbols, employed in religious and magical rituals.

Erich Isaac has pointed out that "in view of the size and fierceness of the animal, the original domesticators must have had a strong motive for overcoming the difficulties of the task. That this motive was economic is unlikely, since it would not have been possible to foresee the uses to which the animal might be put, and the only obvious use, that of the animal as meat, would not have warranted the effort of capturing the animal, keeping him alive in captivity, and feeding him. . . . The most sensible explanation remains that of Eduard Hahn, who argued that the *urus* was domesticated for religious, not for economic reasons. Although the reason for the religious significance of *urus* is not certain, it probably lay in the animals' horns, which were considered to correspond to the horns of the moon, which in turn was identified with the Mother Goddess." Hathor, the Egyptian moon goddess, was a cow. Long before she appeared in Egypt a human figure holding a crescent horn appeared on the walls of a paleolithic cave.

If sexual power was exalted by the myth of the Great Mother, it is obvious that the bull was both sexuality and power incarnate, with his mighty chest, his flagrant testicles, and his sudden-spearing penis. Not merely did the bull appear in later ages, on the Narmer palette, for instance, as the symbol for the king, but the bull was frequently sacrificed in historic times instead of the Divine King. If the slaying or castration of the male was possibly the ultimate expression of woman's sexual dominance in neolithic culture, the domestication of *urus* might even be explained as a defensive measure by the male to transfer the sacrifice to an animal. One cannot overlook the fact that the chief fertility myths of later periods, like those of Osiris or Dionysus, involve the murder and the brutal dismemberment of a male deity, whose death and resurrection result in the emergence of plant life.

Animal domestication may well have begun, then, with the capture of rams and bulls for purposes of ritual, and eventual sacrifice. Conceivably this went hand in hand with the utilization, also for religious purposes, of the surplus milk of the ewes and cows necessary to propagate the captured stock. The fondling and petting of the young of the species, treated as 'members of the family,' forming part of the house-and-byre, possibly re-enforced the general process of gentling: like what happened to Remus and Romulus, but in reverse. The preservation of the urine and excrement of sacred cattle, as is still done in India, very likely had the same religious origin. Hocart, following Hahn, is not overstating the case when he says that "manuring is . . . difficult to explain on what is called 'rational grounds.' . . . The first use of excreta on fields may have been as a purifient life-giver."

Here again, as with milking, practices begun as religious ritual had results that probably did not escape the sharp eyes of the neolithic cultivators, long before their value for agriculture was so well established that, in an Akkadian poem, the farmer welcomes the pasturing of the herdsman's animals on his untilled land. Even the eating of domestic animals may have first had a religious significance, which set it apart from the eating of the products of the chase or of fishing: one was eating the body and blood of a god, or at least the sacrificial substitute for a god.

Once the domestication of animals reached the stage of utilizing their milk, their blood, or their meat, this new art brought into further use a custom derived directly from ritual sacrifice: the deliberate slaughter in cold blood of man's playmate, companion, and friend. Only the dog and the horse, the earliest and the latest of the domesticates, usually escaped this fate—but in Mexico even the dog did not escape.

Civilized man, who has long been a beneficiary of domestication, habitually effaces this ugly practice from his consciousness. When the hunter goes after big game, he often risks his life to get the food: but the cultivator and his descendants risk nothing but their humanity. This killing in cold blood, this suppressing of pity toward creatures man had hitherto fed and protected, even cherished and loved, remains the ugly face of domestication, along with human sacrifice. And it set a bad precedent for the next stage of human development; for, as Lorenz's study of the rabbit and the pigeon helps explain, the savagery and sadism of domesticated man has, time and time again, surpassed that of any carnivore. Hitler's satanic accomplice in mass torture and extermination was known as 'a good family man.'

The originally sexual and religious motivations of animal domestication were supported by the mechanical inventions which, in many parts of the world, proved helpful, indeed essential, to seed cultivation. It is significant that the first known harnessing of cattle to sleighs or wagons was in religious processions; just as the earliest vehicles that have survived were not farm wagons or even military chariots, but hearses, which are found buried with draught animals and human attendants in the royal tombs at Kish, Susa, and Ur. So, too, the plow, as Hocart suggests, may have first been a purely religious instrument, drawn by a sacred ox, held by a priest, penetrating Mother Earth with its masculine tool, getting ready the soil for fertilization: so that gardens and fields, opened actually by digging-stick or pickaxe, as yet never worked by any plow, might benefit by the ritual. "The plow is from its earliest development," Isaac again observes, "associated with cattle in ritual usage."

As with every other aspect of culture, domestication was a cumulative process; and in tracing all the changes brought about by this new habit of life, one must give due attention to holdovers as well as novelties, and also take note of cultures where parts of the new institutional complex for long were missing. In Sumer, despite the enormous crops available from seed agriculture, animal husbandry did not completely suffice to provide meat. As S. N. Kramer notes, "there are texts recording the deliveries of deer, wild boars, and gazelles"—something we need hardly wonder at, for in due season supplies of venison and grouse and hare still come into fashionable metropolitan markets of London or Paris.

But although the pre-Columbian peoples of the New World domesticated dogs, guinea pigs, llamas and vicuñas, they never achieved the kind of mixed farming economy we note in the Old World; and as a result, Gertrude Levy has pointed out, that failure "deprived these peoples of those undertones of the pasture and sheepfold . . . that had united the protective aspect of the Mother Goddess with the deeply remembered service of the hunted beasts."

6: THE 'NEOLITHIC' SYNTHESIS

What 'neolithic' domestication did, once cattle were incorporated into farming, was to bring together at a higher level the two oldest economies, those of the food-gatherers and the hunters. And though mixed farming did not spread to every part of the world, many of its subsidiary inventions did so: not least, the institutional complex of the archaic village.

The initial stages of domestication, though slow if one judges them by the pace of the last three centuries of mechanization, were full of adventurous adaptations and useful surprises. Each new addition to the diet, each increase in the size and quality of fruit, each new fiber that proved useful for spinning into thread and weaving, each new medicinal plant that dulled pain, healed wounds, or overcame fatigue, must have given far more genuine reason to rejoice and marvel than the latest motorcar or rocket model.

Not merely the growing of food but its preparation became the subject of thought and art: with the clay vessels that began to appear around the Eighth Millennium B.C., boiling as well as roasting and baking became feasible—though possibly in the tropics green bamboo vessels preceded them. With a variety of foods and condiments available, cookery, the proper putting together of this variety, became an art, at least in seasonal feasts.

During this phase of domestication, the free imaginative forms of paleolithic art disappeared. The first decorated pottery confined itself to geometric figures, doubtless symbolic, but skeletonized; and the weaving of textiles, a slow, deliberate, repetitious art, was probably long void of decoration. Yet in textiles, many neolithic plants, which had provided dyes once used in body decoration, finally contributed to the color of the cloth, while in later geometric symbols the order and regularity of neolithic culture was externalized.

Though there was no richness of invention in tools and utensils till the end of the neolithic phase in the Near East when the loom, glazed pots, the plow, and the potter's wheel came in, this supposed lack is due to the current habit of restricting the term 'invention' to mechanical appliances. Against that misleading habit this whole book is a reasoned protest. On a more realistic interpretation, there had never been a period before the nineteenth century more rich in inventions: for every new plant that was selected or hybridized or made to yield more heavily was a new invention. Now that plant hybrids can be patented in the United States, quite as much as the new antibiotics, perhaps this fact may be more generally recognized. Infinitely more was accomplished in this line of inventions in the five thousand years before the Bronze Age than has been achieved by civilization in an equivalent period since.

The agricultural utilization of herd animals and grain, which took place between 5000 and 2000 B.C., in the area Breasted called the Fertile Crescent, swinging from the Nile around to the delta of the Euphrates River, completed the process of domestication and magnified every new possibility. But in so far as it brought about a radical improvement, it was the pattern and the process, not any single set of tools or any single species of plant or animal that made this change so effective: for the skilled cultivators of Luzon, in the Philippines, the Igorots, who practice irrigation and terracing, still do not use the plow, and the oasis of Jericho supported a town before pottery was introduced.

The result everywhere was a great burgeoning of life accompanied, one must suppose, by a sense of well-being and security. With a plentiful supply of grain for bread and beer, with the possibilities of storage in bins and barns and granaries, protected against rodents by cats and snakes, as well as by baked clay walls, large populations could be secured against famine, except when visited by some dire natural calamity. Where only a handful of fishermen, hunters, and trappers could once live, now many times that number of cultivators could flourish. Villages might grow into country towns, or even cities, as Jericho and Çatal Hüyük now indicate.

But this last stage of domestication had one unforeseen result: it overcame the dominance of woman. For the first effect of animal domestication was to restore the balance of the sexes, even before patriarchal pastoral specialization set in. Carl Sauer sums it up neatly: "Cattle, cart, plow, broadcasting, and drilling, all began as ceremonies of a rising fertility cult of the Near East, in which the officiants were males, and henceforth the care of the cattle, the hand at the plow, the sower, were male. The husbandman thereafter takes over the agricultural operations, the women return to the house and to garden work." Not herding and plowing alone, but castrating, slaughtering, butchering, became masculine occupations: all essential to the new economy.

While henceforward goddesses, queens, and priestesses appear side by side with their masculine counterparts, the repressed male element recovered lost ground in every part of the economy. But woman, freed from her masculine obligations to work and govern, no longer crippled physically by excessive muscular effort, became more enchanting not just for her sexuality but for her beauty. With all the knowledge of breeding that domestication brought in, it would be strange if it did not have some effect in human sexual selection. The subtle contours and undulating lines of a woman's body were a perpetual delight to the Egyptian sculptor: so that even today the delicate carving of a lovely nude on a sarcophagus lid tempts the masculine hand to stroke it, as the polished *mons veneris* of more than one figure in the Louvre testifies.

7: ARCHAIC VILLAGE CULTURE

By now it should be plain that neolithic domestication produced a mixed economy, which combined different forms of plant cultivation and animal breeding in different regional patterns: but beneath all these superficial changes was the enrichment of sexuality, and the expression of life in harmony with the processes of seasonal growth and fruition. In this mixed economy, the seed cultivator, at the end, is the dominant figure: the quarryman, the fisherman, and the trapper are persistents; while the food-collector and the hunter are remnants. Yet almost unnoticed till the transition to the Bronze Age are two figures in the background: the woodman and the miner, though it is doubtful if either of these was at the beginning a specialized occupation.

The woodman, as feller of trees, opened the forest for seed cultivation: as the maker of dams and irrigation ditches, the provider of fuel for pottery kilns and metal furnaces, the builder of rafts and boats, sledges and wagons, he plays an obscure part in the earliest phases, since his special tools and products, unlike stone, survive only by the happiest accident. But the woodman is in fact the primitive engineer; and his work was essential to all the metallurgical and engineering activities that grew out of the neolithic economy. The first great power machines of modern industrialism, the watermill and the windmill, were made of wood; and even the boilers of the first steam-engines and locomotives were made of wood.

Neolithic village culture drew from every part of the valley-section for its resources and its technics. Though the village was rooted in the soil, even the earliest villages reached out for stone or wood or minerals, even as they reached out for mates in marriage, beyond the usual ambit of daily life; and while technical changes were introduced slowly, a steady infiltration of inventions nevertheless took place: the farmer's almanac still records the peasant's earliest debt to the astronomical advances of the Bronze Age, while the iron hoe, the spade, and the plowshare bear witness to the farmer's later debt to the Iron Age. But if the basic neolithic culture absorbed many of the later advances of civilization, it did so at its own good time. Unless overcome by brute force, it did not willingly trade certified goods for dubious gains.

The customs of this archaic neolithic culture were handed down in a more or less continuous tradition that reaches back into the mesolithic phase; and that tradition spread at an early date to every part of the earth. The archaic village was a rooted community; and because its roots went deep it tapped still deeper sources in man's past, preserving, like cultivated flowers whose wild forms are no longer known, some of the earliest —though of course unidentifiable—human experiences, in folklore, proverbs, riddles, songs, dances, even children's games, whose original meanings can now only be caught in hints, beyond any possibility of accurate translation.

With protection and continuity ensured by the village itself, there was more time for the watchful guardianship and instruction of the young; and in all probability a greater number of them survived the diseases of infancy, sustained, too, by a more sufficient diet. If more survived, there would be a larger number of siblings in the same nest; and this would accelerate the pace of learning almost as much as grandparental care and example— for more of the old, too, would be likely to survive. Though the first dolls were probably paleolithic, the appearance of children's toys indicates not merely a greater margin for playful activity but a growing interest in children's needs. Little children are of no use in the hunt; on the contrary, they handicap free movement. But they could now be enjoyed for their own sake, like puppies and kittens; and what is more, they could be used, too, for picking and hulling vegetables as well as minding, at a later stage, the herds and flocks.

Intermingled with this economy from the beginning was a whole array of ancient magical rituals and religious notions, tightly bound up with many practical achievements. They formed what André Varagnac has characterized as archaic culture, whose beliefs, superstitions, observances, ceremonies are worldwide, and still pop out under sophisticated later behavior. Evans has identified many of these neolithic practices that still continue among the peasants of Ireland: "Summer is welcomed on May Eve by ceremonies of decking with flowers the house, byres, middens and springs which are thus linked in a golden chain of fertility, for the flowers chosen— marigolds, primrose, and gorse—are golden yellow, the color of fresh butter. Clearly there is in this an element of sympathetic magic. Thus the dandelion, which is blessed with both a golden head and a milky stem, is 'the plant of Bride,' associated with the favorite Irish saint, who was a milkmaid, protector of cows, and successor of a pagan goddess."

Wherever the seasons are marked by holiday festivals and ceremonies: where the stages of life are punctuated by family and communal rituals: where eating and drinking and sexual play constitute the central core of life: where work, even hard work, is rarely divorced from rhythm, song, human companionship, and esthetic delight: where vital activity is counted as great a reward of labor as the product: where neither power nor profit takes precedence of life: where the family and the neighbor and the friend are all part of a visible, tangible, face-to-face community: where everyone can perform as a man or woman any task that anyone else is qualified to do —there the neolithic culture, in its essentials, is still in existence, even though iron tools are used or a stuttering motor truck takes the goods to market.

The institutional accompaniments of neolithic culture made as important a contribution to civilization as any of its technical inventions. The reverence for ancestral ways and ancestral wisdom preserved many customs and rituals that could not be committed to writing, including the basic principles of morality: the nurture of life, the sharing of communal goods, the practice of forethought for the future, the maintenance of social order, the establishment of self-discipline and self-control, ungrudging cooperation in all the tasks needed to maintain the integrity or the prosperity of the local group.

This pattern seems to have been solidly established before written records were made: so solidly that it has preserved its continuity while civilizations have risen and fallen and written records have been made and destroyed and made again. Whatever rational criticisms remain to be made, this culture had two outstanding characteristics: it was universal, and it survived under every sort of catastrophe. In an age whose inordinate scientific triumphs have brought on grave doubts of its own capacity for survival, these traits are perhaps worthy a more searching analysis, and a more ready appreciation. Are we sure that these surviving archaic traditions are mankind's worst curse—or the greatest obstacle to man's continued development?

Until the present period of urbanization, the greater part of the world's population, as the French geographer, Max Sorre, has pointed out—some four-fifths at the time of his writing—still lived in villages and practiced a routine of life from birth to death that closely resembled the ancestral neolithic mode, in all but the use of stone tools. Even beneath the new universal religions, like Christianity, the old gods and demons of the household and the shrine remained, in Italy and France, quite as much as in Mexico, Java, or China.

The extraordinary durability of neolithic village culture, as compared with the more daring transformations of later urban civilizations, bears witness to its having done justice to natural conditions and human capabilities better than more dynamic but less balanced cultures.

Once this culture had reached a plateau, its further achievements were small: the new peaks will be found in the metal-using civilizations that took their rise afterward. But the total amount of culture needed to ensure its continuity could be mastered within the span of youth and transmitted to a community holding as few as fifty families; and the multiplication of such communities throughout the planet made it possible for these fundamental human achievements to survive every natural disaster or human crisis. Great cities might be levelled to the ground, their temples ransacked, their libraries and records burned: but the village at least would spring up again, like fireweed, in the ruins.

The secret of this social and technological success was twofold. Every member of the community had access to the entire cultural heritage, and could ordinarily master every part of it; and there was no order of authority, no hierarchy of precedence, except the natural one of age, since in such a community, he who lived longest knew most. The easy interchange of skills and occupations, with a minimum amount of specialization, gave village culture a flexibility and range that counterbalanced its eventual conservatism, once the first great experiments in domestication had been made. Even the specialists who became a necessary part of such communities, the potter or the blacksmith, the miller or the baker or the weaver, could on call take part in communal work at harvest time.

In short, every member of the village community, of every age from childhood onward, had an active part in its whole economic and social life, each contributing his effort and skill to the extent of his ability. In his admirable study of the Trobriand Islanders who lived in so many ways at the same level as the earliest neolithic farmers, Malinowski brings out this happy relationship. "Tiny children," he noted, "actually did make their own gardens; the heavier labor is of course done for them by their elders, but they have to work seriously for many hours at clearing, planting, and weeding, and it is by no means a mild amusement to them, but rather a stern duty and a matter for keen ambition." This daily participation in a meaningful activity is exactly what is missing in the modern machine economy, and probably accounts in large measure for juvenile boredom and juvenile delinquency.

In neolithic agriculture, man had for the first time a round of work equally varied, equally demanding, equally enjoyable, in which the whole community could participate, on a far higher level of well-being than had been possible in a mainly food-gathering economy. This daily work not merely unified the 'reality-principle' and the 'pleasure-principle,' making one a condition of the other; but it brought the outer and inner life into harmony, making the most of man's powers, but neither taxing them too heavily, nor over-emphasizing one set of functions at the expense of another. For both security and enjoyment, the cultivators performed more work than was strictly necessary for obtaining crops.

"The gardens of the community," Malinowski emphasizes, "are not merely a means to food; they are a source of pride and the main object of collective ambition. Care is lavished upon the effects of beauty . . . upon the finish of the work, the perfection of various contrivances, and the show of food." If the art and decoration of neolithic peoples is less imaginative than that of the paleolithic hunters, it is perhaps because so many of their esthetic needs were served directly through their daily work, through sexual play, and through their enjoyment of the forms and perfumes of the flowers. Some of this delight disappeared, perhaps, with the large-scale cultivation of cereals and still more with dingy urban overcrowding. But the joy of working together as families, producing and sharing abundance, made regular work a ceremony and a sacrament, a source of health and sanity, not a penalty and a curse.

Through the daily round, every member of the archaic village was in conscious touch with all the operations of field, garden, pasture, and swamp: a witness and a willing participator in the plantings, breedings, and matings of field and fold; and, finally, in the begetting and nurturing of his own offspring: at one with all the generative forces of life, all the more because the most intense and heady joys open to the organism—those of sex—permeated his daily rituals, as promise or fulfillment. Work and play, religion and education were united. That aspect of the archaic culture is still visible in such villages as remain close to the old ways: an American physician in East Africa writes me that his native women patients, despite all the hardships of their lives, still wear on their faces the "lineaments of satisfied desire."

The unashamed sexuality of the village community, which remained exuberantly evident in historic times in Greece in the phallic post, the Hermes, which stood outside the door of the dwelling house, often with the carved figure showing an erect penis, was the antithesis of the sexual exhaustion that flagellates itself with pornography in the dissolute metropolis of today. Eating and copulating, singing and dancing, conversing and telling stories, were integral parts of the working life: so however repetitious the routine, the inhabitants, like the peasants described by Tolstoi in Anna Karenina, had the joy of being at one with themselves and their world: not like the growing mass of unfortunates today, alienated by the sterile environment, the sordid routine, and the faked excitements and amusements of a modern city.

"All had been drowned," as Tolstoi describes it, "in the sea of their joyful common toil. God had given them the day and the strength, and both the day and the strength had been devoted to labor which brought its own reward." They did not feel themselves "strangers and afraid" in a world they never made. Their ancestors had helped to make the world for them; and they in turn would conserve that world and pass it on, renewed and sometimes improved, to their children.

Most of the equipment that makes for domestic comfort, the hearth, the chest, the closet, the storeroom, beds, chairs, cooking utensils, drinking vessels, blankets, woven clothes and hangings—in short, the whole furniture of domestic life—are neolithic or chalcolithic inventions: mostly before 2000 B.C. If some wicked fairy were to wipe out this neolithic inheritance, leaving us only vacuum cleaners, electric washing machines and dishwashers, electric toasters, and an automatic heating system, we should no longer be able to keep house: indeed we should not even have a house to keep—only unidentifiable and uninviting space-units, now, alas! massively achieved in current bureaucratic housing projects from Paris and New York to Singapore and Hong Kong.

All this may be said on behalf of the archaic neolithic synthesis; but once seed cultivation was achieved, its greatest days were over and all the adventurous experiments of domestication had reached a terminus. By the Fifth Millennium in the Near East the neolithic community had achieved a basis for stability and security: life had now become predictable and manageable. This economy, as long as it remained on a subsistence level, with a sufficient holdover of food to ensure against the unexpected, was self-regulating and self-maintaining. Its motto was: Enough is plenty. When its customary wants were fulfilled, its members had no impulse to work harder in order to achieve other goals. The household gods demanded no inordinate gifts or sacrifices. If threatened with a surplus, such communities disposed of the excess easily, by free gifts or periodic feasts.

Despite all its basic human advantages, the archaic village nevertheless had too narrow a province: there was nothing of the heroic in its ways, nothing of the saintly or the self-transcending in order to achieve some higher good. As in the terminal stage of the nineteenth-century utopian community at Amana, Iowa, its very prosperity and its generosity in communal distribution may often have brought about a slackening in effort and a falling off in productivity. Through having favored equally the zealous worker and the drones, even the industrious might in time do a little less than their best. The very stability and fruitfulness of such a community might cause it prematurely to cease experiment and settle down. Isolation, in-group loyalty, self-sufficiency—these archaic village traits do not make for further growth. The smugness of 'Main Street' began long ago.

In short, the neolithic village community had to pay the penalty for its success: its own virtues arrested it. The horizon was too confined, the routine too limited, the religion too closely bound to petty ancestral gods, the village itself too complacent in its isolation, too narcissistic and self-absorbed, too suspicious of the stranger, too hostile to invading customs—its little local good a stubborn enemy to any foreign best. Even the language of such villages tended to become so inbred that a local dialect might be unintelligible a day's walk away. In surviving tribal communities all these vices have become ingrained by five thousand years of repetition and protective isolation and perverse elaboration: the creative moment has long passed.

All these traits made for persistence and endurance: but at a low level. Once formed, the neolithic culture lacked the very qualities that had made it so attractive in the beginning—its exploratory curiosity and its adventurous experiments. In many parts of the world, an elaboration of neolithic technics took place; but further human development, though it always fell back on the neolithic stabilities when threatened with extinction, took a different route, exploiting not sex but power: the route of civilization.

Yet it is not, perhaps, a mere coincidence that the occupational therapy now used to restore neurotic patients to normal activities and mental balance utilizes the chief neolithic arts—weaving, modelling, carpentry, pottery-making. The repetitive nature of these formative tasks helps control the erratic unchannelled impulses of the personality and provides in the end a gratifying reward for submitting to a constructive routine. Perhaps this was not the least contribution of neolithic culture: it taught man the importance, not only of sex and parenthood, but of regular work. That lesson we forget at our peril.

CHAPTER EIGHT

Kings as Prime Movers

1: THE ROLE OF SOCIAL ORGANIZATION

During the Third Millennium B.C. a profound change took place in human culture. History, in the sense of a transmissible written record of events in time, came into existence; and a new set of institutions, which we associate with 'civilization'—a term I shall later re-define and qualify—sprang up in a few great river valleys. Archeologists have attempted to portray this transformation mainly as the result of technological changes: the invention of writing, the potter's wheel, the loom, the plow, the making of metal tools and weapons, and the large-scale cultivation of cereals in open fields. V. Gordon Childe even introduced the dubious notion of the "Urban Revolution" as the culminating stage of the previous "Agricultural Revolution."

All these technical improvements were important; but behind them was a more central motive force that has been neglected: the discovery of the power of a new kind of social organization, capable of raising the human potential and bringing about changes in every dimension of existence—changes that small, down-to-earth communities, on the early neolithic scale, could hardly contemplate even in the imagination.

In attempting a hypothetical reconstruction of prehistory, I have sought to show that every technical advance was intermeshed with necessary psycho-social transformations both before and after: the emotional communion and rigorous discipline of ritual, the beginnings of ideated communication in language, the moralized ordering of all activities, under the discipline of taboo and rigorous custom, to ensure group cooperation.

On these three foundation-stones—communion, communication, and cooperation—the basic village culture was erected. But outside the restricted

territory of the tribe or village, these essential modes of socialization operated only sporadically and ineffectively. The communal pattern itself was
universal: but each group was a social island cut off from other groups.
Wherever this village culture was left to itself, it eventually became fos-
silized; and if later it continued to develop, it was either by being coerced
into association with a larger society, or by assimilating institutions filtering
down from the higher civilizations.

Out of the early neolithic complex a different kind of social organization arose: no longer dispersed in small units, but unified in a large one:
no longer 'democratic,' that is, based on neighborly intimacy, customary
usage, and consent, but authoritarian, centrally directed, under the control
of a dominant minority: no longer confined to a limited territory, but
deliberately 'going out of bounds' to seize raw materials and enslave helpless men, to exercise control, to exact tribute. This new culture was dedicated, not just to the enhancement of life, but to the expansion of collective power. By perfecting new instruments of coercion, the rulers of this
society had, by the Third Millennium B.C., organized industrial and military
power on a scale that was never to be surpassed until our own time.

At this point, human effort moves from the limited horizontal plane
of the village and the family to the vertical plane of a whole society. The
new community formed a hierarchic structure, a social pyramid, which
from base to pinnacle included many families, many villages, many occupations, often many regional habitats, and not least, many gods. This
political structure was the basic invention of the new age: without it,
neither its monuments nor its cities could have been built, nor, one must
add, would their premature destruction have so persistently taken place.

I have already outlined some of the beneficent cultural results of this
change in 'The City in History,' so here I shall concentrate on its technological outcome. The new social organization seems to have been brought about
by the meeting and fusion of the two cultural complexes whose prehistoric
course we have been trying to plot: and it is not surprising that the juncture
took place in the hot valleys of the Jordan, the Euphrates, the Tigris, the
Nile, and the Indus rivers. From the Palestinian, the Iranian, and the
Abyssinian highlands came the hunter and the woodcutter, likewise, it now
seems, the first domesticators of grain. In the lowland valleys, in whose
drying swamps and lakes green islands of cultivation began to appear,
there was still enough game left to excite the hunter and hamper the
farmer: so they were temporarily in a happy symbiotic association.

But from the south and east came the mesolithic garden and orchard
culture, whose special products were the necessary complement, with their
sugars and oils and starches and spices, to the grains that would help

support much larger populations. Peoples who thrive on the date, the coconut, or the breadfruit, know a freedom from importunate labor that may well have made their original habitats seem like a Garden of Eden, as they still seemed to Herman Melville only a century or so ago. We have corroborative evidence of this mingling in the discovery of Mesopotamian artifacts in Harappa and Mohenjo-Daro on the Indus River, while under the flood silt of Ur, Woolley found two amazonite beads, a stone whose nearest known place of origin is the Nilgiri Hills of Central India; and it is possible that the interchange of their cultivated plants took place at an even earlier stage.

Both the technical and the social components of 'civilization' made their appearance at almost the same time in the classic river valleys from the Nile to the Hwang Ho; and if the mixture of a diversity of needs and inventions was responsible for the immense explosion of power that actually took place, no better geographic conditions for such a mixture could be found. For until wheeled vehicles were invented, and horses and camels domesticated—indeed right down to the end of the nineteenth century—the river was the backbone of both transportation and communication; even the wide ocean was a smaller obstacle to human intercourse than mountain and desert.

The great rivers were drainage basins, not only of water, but of culture, not only of plants, but of occupations and technical inventions; and the existence of a river guaranteed the water supply necessary for large crops from their heavily silted soils. In Mesopotamia two, sometimes three, crops of barley or wheat were possible every year. Under proper management, which was forthcoming, the mainly subsistence economy of the village would be turned into an economy of abundance.

The new flood of energy from food, which rivalled that from coal and petroleum in the nineteenth century, provided both the groundwork and the incentive for a new kind of political society. But no tool or machine, in the ordinary sense, was responsible for the form that this organization assumed, since the new institutional and ideological complex took hold, certainly in Egypt and probably in Mesopotamia and elsewhere, before wheeled vehicles and plows were invented, or even a written language. What ordinary mechanical inventions did was to speed and facilitate the new form of organization.

2 : THE CHANGE OF SCALE

Viewed from our present technical perspective, the passage to 'civilization' is hard to interpret. While no single technical factor marked the transition from the neolithic economy to the typical forms of a power-centered economy, abundant power was available, power sufficient to build if not to move mountains, before metals had been smelted and hard-edged metal tools had been put to work. Yet 'civilization' from the beginning was focussed on the machine; and it will help us to understand what was new in the post-neolithic technics, if we place the new inventions side by side, along with the institutional controls that they demanded. We shall then see how the might of an invisible machine anticipated the machine itself.

When we turn to the early records of Sumer and Egypt, the chief source of power still derives from agriculture: the large-scale planting of grain in bounded and measured fields, whose boundaries must be restored by public authority if effaced by flood. Grain cultivation takes place under public control, for the land and its products belong to the local god, and the surplus is duly stored in centralized granaries within the fortified citadel of the newly built cities. As population increased in the river valleys and the available land got taken up, irrigation and canalization, once done sporadically on small scale in the village, gave way to a wider system of public organization; and it was in the orderly exercise of this over-all control by the temple and the palace that writing was first invented, to keep account of quantities of produce received or disbursed. The political agents that collected and distributed the grain could control the entire population.

In all these operations two changes become increasingly evident, a change of pattern and a change in scale. The common factor that underlies these activities is an increase in mechanical order, mathematical exactitude, specialized skill and knowledge, and, above all, centralized intelligence. These new qualities derived directly from the systematic observation of the heavens, and the careful plotting of the movements of the planets and the procession of the seasons.

Though our knowledge of Babylonian astronomy and mathematics comes from very late documents, the formulation of the Egyptian calendar, at the beginning of the Third Millennium B.C., indicates the culmination of a long and widespread process of exact observation and some kind of mathematical notation. Concern with the heavenly bodies and the discovery of a dynamic pattern of order in their seemingly random distribution may have been one of civilized man's earliest triumphs.

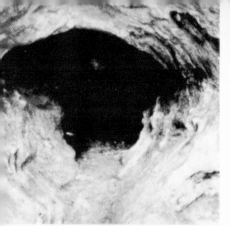

1: In the Beginning

From Peking Man onward, some five hundred thousand years ago, caves have served as the womb and the tomb of human culture. All over the world, caves and grottoes became sacred places, reserved for ceremonials and for memorials of the dead. The paleolithic cave (top left) at La Magdeleine has two female figures at either side of the entrance and a horse in the foreground at the right, not visible here. The Temple of Siva at Elephanta, one of the many examples in India of temples and statues hollowed out of a stone mountain, repeats that ancient arrangement of symbols. But early man also appropriated caves for shelter, security, and storage: witness (top right) this later Amerindian habitation within a cliff. (Top left) La Magdeleine (Tarn). *From Sigfried Giedion, 'The Eternal Present.' Photograph by Achille Weider.* (Top right) Gila National Monument. New Mexico. *Courtesy of United States Department of the Interior, National Park Service Photo.* (Bottom) Siva Temple, Elephanta, c. Eighth century. *Courtesy of Museum of Fine Arts, Boston.*

2-3: Representations of Life

Paleolithic art extends over a period of almost twenty thousand years: four times the total span of historic civilizations. To interpret its varied manifestations as examples of a single paleolithic culture would be absurd, all the more because dates and even time sequences are often obscure. But two recurrent themes pervade this art, whether mobile or stable, whether incised on bone tools or carved in little portable figures or spread monumentally over rough walls and natural vaults. One is the power and grace of the animals these varied peoples trapped or hunted: the bison, the hairy mammoth, the deer, the horse, the ibex; and the other is the wonder and mystery of sex, centered in woman's organs of generation. The hunted animals were a source of food, clothing, tools, and ornaments, a center of organized work, and a conspicuous feature of the habitat, at once a means of material

sustenance and an agent of psychic power through the sacramental transference of blood: not least daily companions, whose breeding and feeding habits the hunter closely studied. Animals dominated the minds of paleolithic hunters, much as mechanical power, in guns, motorcars, planes, rockets, automatic machines, dominates modern man. We can only guess, without any assurance of coming near the mark, what these images meant to succeeding generations of paleolithic peoples. But the discovery of the mature paintings of the Altamira caves opened up a new era in the interpretation of Aurignacian and Magdalenian man—proving he not merely had time for art but ambitions, devotions, aspirations commensurate with our own. *Photograph of mural from 'The Lascaux Cave Paintings,' by Fernand Windels (see Bibliography).*

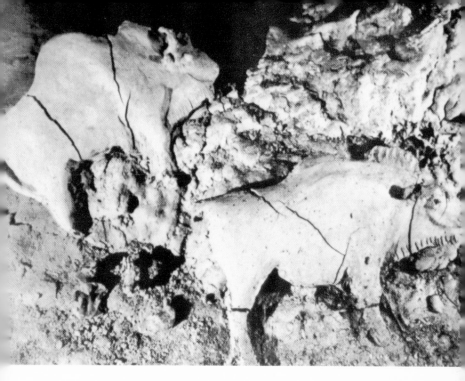

4: The Enchantments of Sex

Some of the earliest paleolithic art is far more accomplished, both technically and esthetically, than comparable images done tens of thousands of years later in Azilian, Halafian, or Cycladic cultures. The emphasis on the sexual organs and secondary sexual characteristics like the buttocks, as in the 'Venuses' of Willendorf or Lespugue, has been sometimes too glibly taken as showing their equivalence to comparable figures in later 'Mother Goddess' religions. More prudently, one may interpret these figures, along with detached vulvas and phalluses often found nearby, as demonstrations of a positive conscious interest in sex—and possibly in sexual rites. But the woman of Laussel, carved in the rock, holds a bison's horn, itself a sexual symbol long in use in Italy as a taunt of cuckoldry. The combination of crescent horn and female form recurs in the Egyptian Moon Goddess, Hathor; and may indicate an early astronomical interest. This in turn would make still more plausible Marshack's reading of a moon calendar in the markings on a multitude of paleolithic bones and pierced cylinders (see Bibliography). Such inscriptions would imply a capacity for patient, orderly scientific observation that would match the esthetic advances in paleolithic art.

5: Anticipations in Clay

Since there is no evidence of clay being used for pot-making till many thousands of years later, these modelled clay bisons stand out not merely as noteworthy examples of art, but as a first experiment with this natural plastic, so important later in neolithic pots and buildings. As so often with technological advances, the playful or religious use anticipated the utilitarian. (Top) Bisons from Tuc d'Audoubert. (Left) The 'Venus' of Laussel. In the Bordeaux Musée des Arts. *Both photographs from A. Leroi-Gourhan's 'Préhistoire de l'Art Occidental.' Courtesy of Editions d'Art Lucien Mazenod.*

6: The World of Dream

7: Ancestors, Gods, Masks

6: THE WORLD OF DREAM

The ambivalent nature of the dream, its bestiality and its divinity, has long been recognized. No better example could be found than Matthias Grünewald's Isenheim Altar: one panel showing a prostrate St. Anthony, beset by diabolic beast-forms standing for lusts, appetites, fears, hatreds, springing partly from man's pre-human past; the other, in the sun-haloed Christ, representing the emergence of the human and divine, released symbolically from the stone tomb and from his brutal but lethargic captors. Panels from the Isenheim Altar, Colmar, by Matthias Grünewald.

7: ANCESTORS, GODS, MASKS

Among the demons, spirits, powers, gods that haunted early man, one constant presence seems manifested almost everywhere: the Ancestors, whether as ghosts or carved idols (top). To heed the ancestors' admonitions, to conform to their practices, to walk obediently in their ways, enabled the group to remain safely human. Rigid conformity was an absolute necessity, since even small departures might break the chain of cultural continuity.

But the social elaborations of 'primitives' were often as regressive and stultifying as technological ceremonialism and scientific exhibitionism have become today. The mask (lower right) may have originated as a labor-saving device for quickly altering personality or exchanging roles (lower left). The powerful abstract image of a supernatural being foreshadows some of C. S. Lewis's or Olaf Stapledon's fantasies: an exquisite rendering—realistic even as to the vestigial legs—of the Big Brain! These forms, often termed 'primitive,' are actually terminal stages of an elaborate, involuted tribal culture. (Top) Painted wooden figures of The Ancestors shown in sacred Men's House (New Guinea). (Lower left) Group of tribesmen under an umbrellalike mask, representing a sacred being, worn collectively by initiates, Angoram, New Guinea. *Both illustrations from Douglas Fraser, 'Primitive Art.'* (Lower right) Secret Society initiation mask of the Bena Biombo tribe of the Congo (Kinshasha). *Courtesy of The Smithsonian Institution (No. 204,314).*

9: SIGNS AND SYMBOLS

The hand preceded the face as a worldwide symbol in cave painting, sometimes sacrificially mutilated by the removal of one or more finger joints. Cave art likewise presents more abstract signs, doubtfully named 'tectiforms,' and diversely interpreted as huts, traps, animal corrals. But the serried dots and the 'plan' (bottom), seemingly mathematical notations, have even been interpreted as sexual symbols. However elusive the explanation, these signs surely indicate not only abstract thought but the earliest beginnings of the permanent record. (Upper left) Left hand on low ceiling, El Castillo. *From Sigfried Giedion, 'The Eternal Present.' Photograph by Hugo P. Herdig.* (Upper right) Strong Wind, Ojibway, by George Catlin. *U.S. National Museum, courtesy of The Smithsonian Institution.* (Lower) Abstract signs, sometimes called tectiforms, El Castillo (Santander). *From Sigfried Giedion, 'The Eternal Present.' Photograph by Achille Weider.*

9: Signs and Symbols

10: Dance Before Drudgery

After the glaciers retreated, wave after wave of cave art washed over Spain and Africa, and continued in existence, through the Bushmen, almost into recent times. The hunters shown here are highly stylized, but animated, indeed intensely dynamic, moving over the wall as in a dance. Were such fluent compositions perhaps a last expression of the hunters' freedom and esthetic release before the monotonous neolithic tasks took over? Painting from Rock Shelter of Cuevas del Civil. Abocacer, Castellan, Spain. *Photograph courtesy of the American Museum of Natural History.*

11: The Human Emphasis

The changeover from hunting and gathering to food production and food preservation was probably responsible for the disappearance, in time, of the old animal art. But James Mellaart's recent excavations at Çatal Hüyük in southwestern Turkey reveal how much of the paleolithic life lingered, at least in symbol and probably in sport and play, even in the emerging towns. These murals have the advantage of the new technology: the rough irregular cave wall gives way to the smooth plaster surface, bounded by a rectangle. But the painting itself shows continuity with the older cave paintings, and contrasts with the stodgily abstract cult figurines or the geometric decorations of neolithic pottery. Dr. Kathleen Kenyon has associated these esthetic and social changes with the amelioration of the climate and environment that followed the last retreat of the glaciers. Wall painting found in dwelling at Çatal Hüyük. Inset: reconstruction of whole painting by Anne Louise Stockdale. *Courtesy of James Mellaart.*

12: Neolithic Economy

The formation of village settlements, once considered a direct result of the 'agricultural revolution,' had its beginnings at a much earlier stage. Paleolithic houses and hearths have been uncovered in southern Russia from perhaps more than ten thousand years before Jarmo or Jericho. Diggings by a joint Turkish-American expedition at Çayönü in Turkey—preceded by Pumpelly's pathbreaking finds in Turkestan in 1904—seemed to indicate that the beginnings of food production may now be pushed back to around 11,000 years ago. Similarly the finding here of the earliest metal artifacts around 9,000 B.C., including pins, apparently made from cold hammered copper, still further blurs the old distinction between the Stone and the Copper-Bronze Ages. These new datings make the early appearance of Jericho and Çatal Hüyük more credible. The unearthing of the bones of wild sheep, goats, and pigs in this area would indicate an early start for mixed farming. In tropical areas, 'neolithic' agriculture—based on rice, taro, yams, breadfruit, coconut—tells still another story. With its security of food and continuity of site, neolithic culture had perhaps six thousand years to spread via small independent relatively self-contained communities before any attempt was made to unify their activities, increase the tempo of production, or extract the surplus products by coercion or conquest, for the benefit of a ruling class. Excavation at Çayönü by Turkish, Arab, and American archeologists, with Robert J. Braidwood, of the Oriental Institute, Chicago, as co-director. *Courtesy of Robert J. Braidwood.*

13: Neolithic Continuities

Though pottery no longer serves to identify 'neolithic' culture, the need for storing liquids, produced by brewing, wine-making, and oil-pressing, hastened the development of baked clay containers. Clay pots were lighter than stone, and easier to make, especially after the invention of the potter's wheel. The Cretan pots shown here survived fire and earthquake; and over three thousand years later a similar type of pot is still used in Vietnam for making fish sauce. (Upper) Wine jars found in ruins of the Palace at Knossos in Crete. *Photograph from Ewing Galloway, New York.* (Lower) Jars for making fish sauce. *From 'Vietnam!' by Felix Greene. Courtesy of D. R. V. Information Department.*

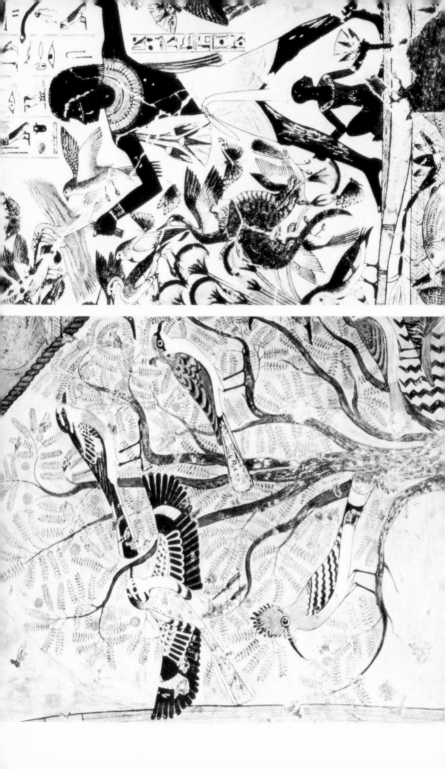

14: Birds, Songs, and Flowers

The flowering of neolithic agriculture is properly symbolized by the richness of bird life, not merely on river or canal bank, but in the swamps and tilled fields, where they kept down the insects that so often made mass assaults on the crops. The new delight in birds was both esthetic and practical: as in Italy today, even small birds were welcome additions to the peasant's diet. The Egyptian Audubon who painted the lower picture was no careless observer: the ornithologist identifies one hoopoe, three shrikes, and a redstart. (Upper) Fowling in the Marshes. 1420-1375 B.C. (Lower) Birds in an Acacia Tree. Tomb of Khnemhotpe. C. 1920-1900 B.C. *Both pictures in Nina Davis, 'Ancient Egyptian Paintings.' Courtesy of The Oriental Institute of the University of Chicago.*

15: Vital and Mechanical Polarities

The upper bird picture in 14 gains much by juxtaposition to the stone carving of Ramses I, receiving offerings in death. In the above, the king is a colossus, indeed a god, patently towering above the little people, yet the composure and order of this carving show an esthetic sensitiveness and human feeling that have not been completely suppressed by the strict external regimentation introduced in the Pyramid Age. The athletic activity and winged excitement of the bird catchers painting (left), like the superb modelling of Egyptian nudes, which surpass even the Greeks in subtlety, modify the established clichés of Egypt as static, immobile, rigid, death-centered; and soften the impression made by the early megamachine. Bas relief from the Temple of Ramses I at Abydos. *Courtesy of the Metropolitan Museum of Art, gift of J. Pierpont Morgan, 1911.*

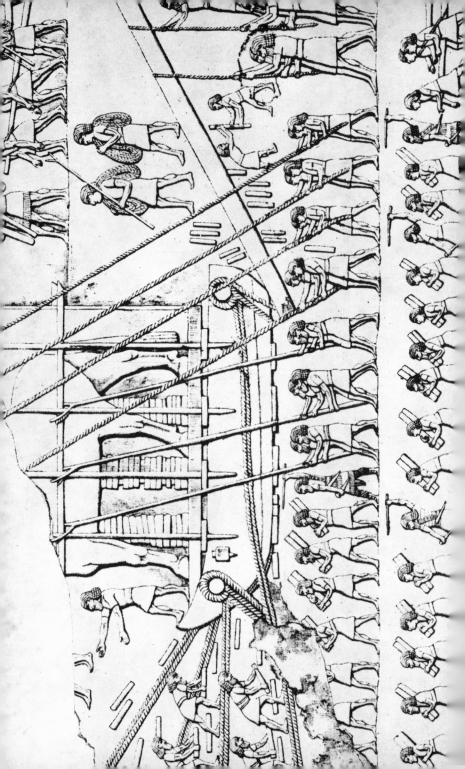

16-17: Megamachines in Operation

(Left) Hauling a monumental sculpture of a bull on a sledge. Note gang bosses with uplifted staves. (Right) Assyrians building a mound for the erection of a palace. Note the frieze of soldiers. Both taskmasters and soldiers were essential parts of the megamachine. *Illustrations from Austen Henry Layard's 'The Monuments of Nineveh,' 2 vols. (London: 1853).*

18: Divine Kingship—Above and Below

(Upper) A King conferring with his God on equal terms: the fact that the ideograph for God, a star, hovers equidistant over both is significant. (Lower) The Stela of the Vultures. The Sumerian phalanx: a pedestrian tank, grinding over the bodies of its victims. *Collections du Musée du Louvre: both photographs by courtesy of the Director of the Museums of France.*

19: 'Civilized' Destruction

The Assyrians, like the later Mongols and Aztecs. were notorious for their in-
sensate destructiveness and ingenious sadism. But the demolition and sacking of
Hamanu only prefigured on a modest scale the psychotic ABC plans and massively
inhumane practices of reputedly 'civilized' governments today. Marble slab: The
Sacking of Hamanu, Kouyunjik. *Photograph courtesy of The Mansell Collection,
London.*

20: Achievements of the Megamachine

The first exhibition of megatechnics comes from the Step Pyramid at Sakkara, constructed under the architect, engineer, scientist, and physician, Imhotep, who well earned his later deification. This Pyramid of King Zoser, the dominant feature of a whole city dedicated to the dead, surpassed all contemporary works, and has defied time almost as well as the colossal later pyramids of Chephren and Cheops. The megamachine was re-assembled repeatedly for military purposes under the institution of divine kingship. But it displayed its best form and highest efficiency in large-scale engineering activities. The Great Wall of China, built in the third century B.C., to keep Mongol invaders at bay, pushing across a difficult terrain for 1,500 miles, gives evidence of the megamachine's massive constructive powers. *Photographs from Ewing Galloway, New York.*

21: From Ancient to Modern

The Corinth Canal, well begun by the Romans in A.D. 67 and completed in 1893, would have been inconceivable without the megamachine: as a mechanical feat, it still takes the breath away when one passes by steamer between its towering walls of solid stone, broken by an occasional Roman figure or inscription. *Photograph from Ewing Galloway, New York.*

22-23: Handicraft and the Human Scale

Despite the prestige of megatechnics the lesser arts and crafts spread far and wide, accumulating skills, perfecting traditional methods, utilizing local resources, making regional adaptations, establishing standards of workmanship, and making endless innovations in design. The majority of crafts were carried on in the home or the farmstead, dividing their tasks according to age, skill, and strength, calling for both active participation and personal responsibility.

These Japanese paintings of typical 'medieval' crafts, reputedly by Yoshinobu Kano, c. 1620–1630, are roughly contemporary with similar European wood engravings by Jost Ammann and others. The whole series was conceived with

rare sociological insight, showing not merely the workers, the materials, the work
processes, but also the domestic setting. Only lack of space prevents me from
showing the range of individual habitations, from the humble cottage of the
matmaker, with its old thatched roof, to the elegant and commodious apartments
of the favored trades, gold lacquer, embroidery, armor, swords, that served the
military aristocracy. (Left) Weaving and spinning. Note different ages from the
elderly spinster to the child winding thread. (Upper) Metal-working. Characteristically the charcoal fire also supplies water for domestic tea. (Lower) Leather-
workers and glovemakers. *From paintings in the Kitain, a Buddhist temple in
Kawagoye. Courtesy of Professor Tsutomu Ikuta.*

24: Animal and Water Power

The belief that power machines date from the eighteenth century is historically as unrealistic as the notion that the age before 1920 was the 'horse-and-buggy' age, despite the fact that the many-sided transportation network then in existence, based on railroads, electric trolleys, steamboats, ferries, private motorcars, and pedestrians, was definitely more efficient, flexible, reliable, and far more rapid in densely populated areas than the crippled, ecologically disruptive system of monotransportation via public plane and private auto favored today. But the emancipation of human labor from exhausting drudgery began long before with the use of animal power, as in the cattle-driven pump still used in Egypt, and with the use of wind power for sailboats.

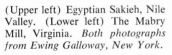

(Upper left) Egyptian Sakieh, Nile Valley. (Lower left) The Mabry Mill, Virginia. *Both photographs from Ewing Galloway, New York.*

(Above) Battery of windmills: man-powered galley with auxiliary sails, foreground; sea-going three-masted sailing ship, rear. *From 'The Nürnberg Chronicle.'* (Right) Windmill on the Vecht River, The Netherlands. *Photograph from Ewing Galloway, New York.*

25: Wind Power

The invention of the Greek watermill (third century B.C.?) and the Persian invention of the windmill (eighth century B.C.?) not merely served agriculture and land reclamation, but improved mining technology. The widespread adoption of this cheap form of energy probably contributed as much as moral improvement to wipe out the more ancient forms of slavery. By the fifteenth century batteries of windmills and fleets of sailing ships were witnesses to the new power age. Cotton mills and paper mills in New England and elsewhere were still using water power at the end of the nineteenth century, and even later.

26: Industrial and Military Arts

Quite early medieval book illuminations and wood and stone carvings in churches pictured the homely crafts and vocations. This growing interest in the work process competed with the iconography of the Trinity, the Virgin Mary, and the Saints, though the two interests mingled in the guild processions and the mystery pageants and plays. Such symbolic representation of the arts and crafts indicates not only vocational pride but a recognition of the role of work in forming and expressing a balanced human personality.

The most spectacular advances in medieval technics, apart from power mills, were made in the mining and working of metals: lead roofs for public buildings, copper and iron pots, brass and iron cannon, steel swords from Toledo. Significant innovations appeared both in long-distance destruction (cannon) and long-distance oceanic transportation, by means of the new three-masted sailing ship, guided by the magnetic compass, and later, the sextant, the chronometer, and the increasingly exact navigation charts and maps. What is usually treated as the technological backwardness of the six centuries before the so-called Industrial Revolution represents in fact a curious backwardness in historical scholarship. Significantly, the great technical advances of the eighteenth century took place in the earliest neolithic or chalcolithic industries, textiles, pottery, metallurgy, canal-building. Large capital investments in the metallurgical (military) industries spurred the pre-'Industrial Revolution' economies of Europe as much as nuclear weapons and rockets do currently in the United States. Capitalism, mining, militarism, and mechanization—along with megalomania—went hand in hand. (Upper) Metal-workers' shop, showing mine (left) and forge (right). (Lower) Heavy siege cannon with cannon-balls, first stone, later metal. Also three small-bore guns mounted for rapid fire: a first step toward the machine gun.

27: Domestic Improvements

Technical innovations in the domestic arts, from the piping of water and the invention of the water-closet and the iron cook-stove and oven, to the wider use of washable linens and cotton underclothes, have spread slowly. But from the sixteenth century on, there was a notable increase of inventions directly concerned with child care: the potty stool, the high chair, the toddler, the rocking cradle, all shown here, indicate increased concern with the welfare of children: a prelude to Europe's population explosion. Many important inventions were first conceived as children's toys: automatic figures and miniature automatic workshops, magic lantern projectors, gyroscopes, and talking dolls—forerunners of the adult toys of our own age, the tape-recorder, the motion picture camera, and the computer.

28: Playful Automation

Fantasy and play have constantly had a part in mechanization and automation. Early clocks like the ancient Cathedral clock in the Marienkirche in Lübeck not merely recorded the hours and days, but ostentatiously paraded animated automata when the hour struck. This side of automation reached its climax in the lifelike contraptions of Vaucanson, whose famous duck ate and even defecated, and in the Jaquet-Droz draughtsman shown here, who executed four drawings in pen and ink. Behind this playful automation was a deeper motive, only now visible: the desire to create life by purely mechanical means—or at least to place every living function under mechanical direction and control. Le Dessinateur, automaton by Henri Jaquet-Droz (fabricated 1772-1774). *Courtesy of the Museum of Art and History, Neuchâtel, Switzerland.*

29: From Manuscript to Typeface

From 3000 B.C. onward, writing was the chief means of converting the kinetic energy of mind into potential energy: therewith greatly enlarging the stored resources of culture and vastly widening the scope of human association in space and in time. No other single invention deserves to rank with writing except the collective achievement, over the ages, of intelligible speech. The 'European' invention of the printing press with movable type, centuries after the Chinese application of carved wood blocks to book printing, was the next advance of comparable importance. The printing press, with its use of uniform, replaceable parts (movable type) was the model for other forms of standardized mass production, centuries before Samuel Bentham and Eli Whitney. But note: before the printing press was invented handicraft was esthetically 'mechanized,' for the first uniform typefaces were already in existence in medieval manuscripts and Roman monuments. Comparing the lettering in this fifteenth-century Book of

Hours (inset, upper left) with the Gutenberg Bible, 1456, the notable fact is that nothing has altered except the mode of production: even the elaborate initial letter and the floral border for long remained. But the printed page had already been standardized. *Inset is from a Book of Hours in the Edinburgh University Library. The page of the Bible is courtesy of The Art Institute of Chicago.*

30-31: Subterranean Mechanization

The series of inventions which made deep mining feasible paved the way for many of the major technical innovations of the last five centuries. The railway, the subway, the mechanical lift, the steam engine, air-conditioning, daytime artificial lighting, all derive from mining practice. The illustration at right demonstrates

a closed system of automation, without workers, in which a watermill is attached to a pumping system to carry off underground water. (Left) Transport and washing of ores. (Right) Undershot waterwheel and pump for automatic mine drainage. *From Agricola's 'De Re Metallica' (see Bibliography).*

32: The Second Coming of the Megamachine

Two panels from the mural by José Clemente Orozco, in the Baker Library at Hanover, N.H. *By permission of the Trustees of Dartmouth College.*

The cultivation of this new language gave its possessors, the early priesthood, an exceptional power of astronomical and then meteorological prediction. This was a source of their supernatural authority, as interpreters of cosmic influences and their human consequences. That skill was not so prone as magic to be confounded by incorrigible events. The orderly cosmos thus disclosed satisfied one of man's deepest needs—though that need was perhaps ultimately a product of this order. The Voltairean notion that the offices of priesthood were created merely to perpetrate fraud and blackmail upon gullible believers without performing any tangible services, overlooks the fact that the temple by its command of this higher knowledge made an essential contribution to large-scale farming, by synchronizing agricultural operations.

The earliest stages of this religious transformation antedates writing and can only be inferred from the later documents. But there is general evidence of a shift in interest and authority from the gods of vegetation and animal fertility—subject to human weaknesses, to suffering, misfortune and death—to the gods of the sky: the moon, the sun and the planets, the lightning and the storm wind—powerful and implacable, awful and irresistible, not to be swayed from their course. Atum and Enlil, like Marduk and Zeus later, were incarnations of cosmic power. In a Hittite ritual for the building of a new royal palace we still find: "To me, the King, have the gods, the Sun God and the Weather Gods, entrusted the land and my house."

These earth gods and sky gods remained side by side in most cultures; but if the vegetation gods continued to be the more sympathetic, lovable, and popular, there is no doubt which were the more powerful.

The regularity and order that had first come in with neolithic grinding and polishing, and became visible in geometric patterns and decoration, now spread over the whole landscape: rectangles, triangles, pyramids, straight lines, bounded fields, testify to both astronomic order and strict human control. Standardization was the mark of the new royal economy in every department. Confucius was characterizing a much earlier achievement of this culture when he observed: "Now all over the empire carriages have wheels of the same size, all writing is with the same characters, and for conduct there are the same rules."

But, above all, there was a change in scale. Quantification and magnification are the marks of the new technology. Instead of the little neolithic shrine, there stands a towering temple, the 'Mountain House,' and nearby a huge granary: instead of the cluster of frail, mud-walled village houses, for a score of families, a wall-engirdled city, with a thousand or more families, no longer merely a human home, but the home of a god:

indeed a replica of Heaven. And the same change of scale shows itself in
every department, not least in the tempo of life. Changes that once would
have needed scores of years to consummate now took place almost overnight, not because better tools and equipment were at the disposal of the
builders and fabricators, but because a highly efficient type of social organization, hitherto unknown, had taken command.

Since our documents come mainly from the brief Bronze Age and the
following Iron Age, scholars have been tempted to throw the emphasis on
the many technical improvements that the use of copper and bronze first
made possible. But the radical changes I am calling attention to antedated
the age of metals by many centuries, possibly by millennia.

V. Gordon Childe's attempt to explain this vast explosion of power and
confident human command mainly through inventions like the plow and
the military chariot, neglected the most important fact—namely, that the
technological exhibitionism that marks the beginning of the Pyramid Age
was effected with only small, modest, mechanically primitive instruments:
chisels, saws, mallets, ropes. The huge stones that were transported for miles
to the pyramids at Giza were borne on wooden sledges, and raised into
position without the aid of a wheel, a pulley, a windlass, or a derrick, or
even any animal power except that of mechanized men.

3: THE CULT OF KINGSHIP

The increase in the food supply and the population that marked the dawn
of civilization may well be characterized as an explosion if not a revolution; and together they set off a train of minor explosions in many directions, which have continued at intervals over the entire course of history.
But this outburst of energy was subjected to a set of institutional controls
and physical compulsions that had never existed before, and these controls
rested upon an ideology and a myth which perhaps had their faint beginnings in the magical ceremonies in paleolithic caves. At the center of this
whole development lay the new institution of kingship. The myth of the
machine and the cult of divine kingship rose together.

Until the nineteenth century conventional history had remained largely
a chronicle of the deeds and misdeeds of kings, nobles, armies. In revolting
against this studious inattention to the daily affairs of ordinary people,
democratically oriented historians swung to the opposite extreme and be-

littled the part that kings and institutions derived from kingship have actually played. Today, both historians and anthropologists have been looking at kingship with a more open eye, if only because the massing of centralized economic and political power in every modern state, totalitarian or quasi-totalitarian, has cast a fresh light on the earliest similar assemblages.

The institution of kingship, as Henri Frankfort, its brilliant modern interpreter, pointed out, is one of the early innovations to which we can assign an approximate date, place, and agent, fairly closely in Egypt, a little more loosely in Mesopotamia. The historic effort, as recorded on two famous Egyptian palettes, begins at the point where the paleolithic hunting chief, the first among equals, passes over into the powerful king, who takes to his own person all the powers and prerogatives of the community.

As to the origin of the king's unconditional supremacy and his special technical facilities, there is no room for doubt: it was hunting that cultivated the initiative, the self-confidence, the ruthlessness that kings must exercise to achieve and retain command; and it was the hunter's weapons that backed up his commands, whether rational or irrational, with the ultimate authority of armed force: above all, the readiness to kill.

This original connection between kingship and hunting has remained visible all through recorded history: from the stelae upon which both Egyptian and Assyrian kings boast their prowess as lion-hunters, to the preservation of vast hunting forests as the inviolable domains of kings in our own epoch. Benno Landsberger notes that with kings in the Assyrian empire hunting and fighting were virtually interchangeable occupations. The unscrupulous use of the weapons of the hunt to control the political and economic activities of whole communities was one of the effective inventions of kingship. Out of that a whole series of subsidiary mechanical inventions eventually came.

In the mixture of paleolithic and neolithic cultures, there was doubtless an interchange of psychological and social aptitudes as well, and up to a point this may have been of common advantage. From the paleolithic hunter the neolithic cultivator may have gained those qualities of the imagination that the dull, thrifty, sober round of farming did not awaken. So far no weapons of the hunt, still less weapons of warfare, have been unearthed in the earliest neolithic villages, though they became common enough in the Iron Age; and this lack of weapons may account for the primitive peasant's docility and for his easy surrender and virtual enslavement: for he had neither the tested courage nor the necessary weapons, nor yet the means of mobilizing in large numbers for fighting back.

But at the same time the punctual, prudent, methodical life of the agricultural community gave the incipient rulers some share in the neolithic habits of persistence and orderly drill, which the hunter's mode of life, with its fitful spurts of energy and uncertain rewards, did not encourage. Both sets of aptitudes were needed for the advance of civilization. Without the assurance of a surplus yield from agriculture, kings could not have built cities, maintained a priesthood, an army, and a bureaucracy—or waged war. That margin was never too large, for in ancient times war was frequently suspended by common consent till the harvests were gathered.

But sheer force could not by itself have produced the prodigious concentration of human energy, the constructive transformation of the habitat, the massive expressions in art and ceremonial, that actually took place. That demanded the cooperation, or at least the awed submission and passive consent, of the entire community.

The agency that effected this change, the institution of divine kingship, was the product of a coalition between the tribute-exacting hunting chieftain and the keepers of an important religious shrine. Without that combination, without that sanction, without that luminous elevation, the claims that the new rulers made to unconditional obedience to their king's superior will, could not have been established: it took extra, supernatural authority, derived from a god or a group of gods, to make kingship prevail throughout a large society. Arms and armed men, specialists in homicide, were essential; but force alone was not enough.

Even before the written record becomes available, the ruins from the earliest pre-dynastic al'Ubaid period of Ur indicate that the transformation had already been effected: here, as elsewhere, Leonard Woolley found a temple, within a sacred enclosure, where the royal granary, at once a food storehouse and a bank, would likewise be placed. The authority, priestly or royal, that collected, stored, and allocated the grain, held the means of controlling a large, dependent population—provided the granary was constantly guarded by walls and warriors.

Under the protective symbol of his god, housed in a massive temple, the king, who likewise served as high priest, exercised powers that no hunting chief would have dared to claim merely as the leader of his band. By assimilation, the town, once a mere enlargement of the village, became a sacred place, a divine 'transformer,' so to say, where the deadly high-tension currents of godhead were stepped down for human use.

This fusion of sacred and temporal power released an immense explosion of latent energy, as in a nuclear reaction. At the same time it created a new institutional form, for which there is no evidence in the simple neolithic village or the paleolithic cave: an enclave of power, dominated by an élite who were supported in grandiose style by tribute and taxes forcibly drawn from the whole community.

The efficacy of kingship, all through history, rests precisely on this alliance between the hunter's predatory prowess and gift of command, on one hand, and priestly access to astronomical lore and divine guidance. In simpler societies these offices were long represented separately in a war chief and a peace chief. In both cases the magical attributes of kingship were grounded on a special measure of functional efficiency—a readiness to take responsibility and make decisions in government. This was fortified by the priesthood's observation of natural phenomena, along with the ability to interpret signs, collect information, and ensure the execution of commands. The power of life and death over the whole community was arrogated by the king, or imputed to him. This mode of ensuring cooperation over a wider area than was ever ordered before contrasts with the petty ways of the farming village, whose ordinary routine is carried on by mutual understanding and consent, guided by custom, not command.

In Egypt almost from the beginning, in Mesopotamia at intervals, the king was conceived as a god in his own right. Egyptian history, as a tale transmitted, begins at this point. By this union of cosmic and earthly powers the ruler became at once a living person and an immortal: he was born and died like other men, yet he would be reborn, like his other self, Osiris, even as his power was renewed each day, like that of the returning sun, Atum-Re, after effecting a safe passage through the night and emerging again in the East.

As with Ptah, the primal Egyptian deity, the words coming forth from the king's mouth brought a world into existence; and when he uttered a command, he must be obeyed. Not merely did he hold the powers of life and death over the community, but he was the living incarnation of that community: they were one, as Ptah himself was one with all he had created. The Pharaoh's life was the community's life, his prosperity its prosperity, his health its health. The community lived and flourished vicariously, through the king: so in piously saluting every mention of his name with the words, "Life, Prosperity, Health," they were also ensuring these benefits for themselves.

The earlier chieftains and their followers, well-armed, contemptuous of bodily injury or hardship, dissociated from the laborious routines of cultivators and herdsmen, and disinclined to systematic work, were probably already using these proto-military traits to exercise power and draw tribute, in food or women, from their intimidated, compliant, unarmed village neighbors. The weapon that established this new rule of force was not (pace Childe!) the Bronze Age war chariot, still many years away, but a far more primitive weapon, the mace. Such a club, with a heavy head of stone, which was useful for killing a wounded animal with a single blow on the skull, must have proved equally efficient in doing the same job at close quarters with frightened, weaponless peasants, or with the captured chief and warriors from a rival band, who appear as cowering captives on surviving palettes and stelae. Witness the final act of Marduk's battle with the primeval goddess Tiamat: "With his unsparing mace he crushed her skull."

Should we be surprised, then, to find that the period of political unification of the Upper and Lower Nile Valley, which marks the beginnings of kingship in Egypt, coincides with mass graves in which are found an unusual quantity of cracked skulls? The significance of this weapon, particularly the time and place of its appearance, has been curiously overlooked. In the Sixth Millennium at Hacilar, James Mellaart notes, the economy showed a great decline in hunting and an absence of hunting weapons; but significantly the mace and sling survived. Thus it is no accident that the mace, in only a slightly sublimated form, the scepter, has remained the symbol of royal authority and unchallengeable power throughout the ages. When the British Parliament is in session, a gigantic specimen lies on the Speaker's table.

So much for the shadowy events that must have led up to the establishment of kingship. The next step, which set it on foundations that sustained it, with occasional lapses, for more than five thousand years, comes properly within the bounds of history, more especially sacred history, for it was based on the application of supernatural powers to the control of human behavior.

The clue to this second stage lies in the first reputed act of Menes, the unifier of Upper and Lower Egypt, the earliest Pharaoh of record, repeated again and again by kings throughout history: the establishment of a temple to sustain his own claim as the divine Sun God, Atum-Re, at the site of what was doubtless an already venerable shrine in Memphis. The document that tells of this event establishes the pre-existence of an all-embracing god, Ptah, whose energies pervade creation.

By the time kingship took form as a unifying agent that transcended local limitations, a multitude of deities, male and female, great and small—'personified,' or rather 'animalized,' in the form of falcons, beetles, cows, hippopotamuses, lions—were already in existence, each with a different character and different social functions, often related to a different aspect of the environment. Out of this fecund, multiheaded family of gods, with its swarm of distant relatives in every little village, the Sun God had in Egypt become pre-eminent; and the new authority of kingship was sustained, not by brute strength alone, but by its representation of the eternal power and order of the cosmos.

Here a new kind of science, different from the close observation and intimate association that fostered domestication, came into existence: now based on an abstract impersonal order: counting, measurement, exact notation—attributes without whose early development no such consummate monuments as the pyramids could have been built. The counting of the days, the observations of the lunar months and the solar year, the determination of the rise and the flood stage of the Nile—all this was the task of the priestly caste. This new power and order were effectively symbolized, as I have already noted, by the establishment of the first Egyptian solar calendar.

Though heavily overlaid by dramatic legend, sensuous metaphor, and infantile magic, astronomical order spread into every department. The emerging institutions of civilization were power-minded, cosmos-centered, mechanically regulated and regimented. Space and time, power and order, became the main categories of a divinely regulated existence: the recurrent movements of the moon and the sun, or great expressions of natural power, such as flood and storm and earthquake, left a profound impression on the mind and awakened, it would appear, at least in the dominant minority, an interest in exerting their own physical power in imitation of the gods themselves.

In the ancient Chinese 'Book of Changes' (I Ching) one reads: "We may be ahead of heaven, but heaven will not violate its course; we must follow, and adapt ourselves to its time and its seasons." Everywhere, early and late, this laid the basis of the new regimen, and became a source of an even more rigorous regimentation. "In ancient China," Joseph Needham observes, "the promulgation of the calendar by the Emperor was a right corresponding to the issuing of minted coins, with image and superscription, in [later] western countries."

Both modes were symbols of rational order and coercive physical power, and both significantly remained a royal or priestly monopoly through the ages: for the exclusive right to coin money and establish uniform weights and measures is an emblem of all state sovereignty; while the calendar most of the world now follows was first put forward by the Emperor Julius Caesar, and rectified by the Roman Pope, Gregory XIII. Without this widely shared reverence for undeviating cosmic order, the great technical achievements of early civilization would have lacked the mathematical precision and physical mastery they actually showed.

By identifying the person of the king with the impersonal, above all implacable, order of the heavens, royal power received an immense supercharge of energy: the king's political authority, based on weapons and military exertion, was vastly augmented by the inordinate supernatural powers he wielded. This was fortified in Egypt, where royal power became most absolute and most assuredly identified with the divine, by close association with older organic vitalities—with an even more ancient manifestation of the son, Horus, and his father, Osiris, the god of vegetation and the teacher of agriculture and the handicrafts to men. In symbolism, the king was originally a bull, the very incarnation of physical strength and sexual fertility, doubtless tied in the unconscious to the Sacred Cow, Hathor, who was both cattle-goddess and moon-goddess.

Moon-gods, sun-gods, storm-gods were based, Eliade notes, on "hierophanies of the sky (high, bright, shining, sky, rain): The Sumerian sky god Anu became the chief god of the Babylonians, and his temple at Uruk was called a Sky House." This fixation on cosmic forces was a sound interpretation of the condition of man: his dependence upon physical events beyond his control. While the paleolithic identification with the animal world and the neolithic immersion in sexuality were not abandoned, the sky religions struck a lofty note. The contemplation of the distant heavens, the consciousness of long passages of time, could have had only a faint start in earlier cultures: except for circular amulets and incised bones there is no legible sign of them in cave art.

This new interest in the lofty, the distant, the regular, the predictable, the calculable coincided with the birth of kingship: but older interpretations of the world were not easily sloughed off. On the contrary, the most delusive practices of sympathetic and verbal magic still remained and attached themselves to the gods of the sky: the word still seemed so important that in one legend the goddess Isis tries to secure power for herself by using magic to learn Atum's secret name. The Turin papyrus from the Nineteenth Dynasty (1350–1200 B.C.) from which we learn this story was itself a charm whose words were repeated as a cure for snake-bite.

But in the end the sky gods prevailed. The phenomena of the sky, measured with increasing care and exactitude, gave the assurance of an orderly world, part of which at least was lifted above primal chaos or human caprice. As the chief representative of these heavenly powers, at least in his own territory, the king could maintain order everywhere. Once confined mainly to tribal ritual and articulate speech, order now became universal.

Eventually the Babylonians introduced the same concept of predetermined order into the seemingly irregular events of daily life: they plotted the course and position of the planets and associated this with the hour of a person's birth in order to predict the entire course of his life. The biographic data needed for such plotting was based on systematic observation. Thus scientific determinism not less than mechanical regimentation had their inception in the institution of divine kingship. Long before the Ionian scientists of the sixth century B.C. the fundamental mathematical and scientific foundations had been laid in astronomy. This, then, was the constellation of rational insights and irrational presumptions that produced the new technology of power.

Before we follow through the consequences of this change, let us witness the same phenomena operating under quite different geographic and social conditions. If the myth of divine kingship was to capture the forces of civilization and become further enlarged through the derivative 'Myth of the Machine,' it was essential that it should be capable of over-riding purely local circumstances and taking advantage of different cultural settings. As a matter of fact, during later ages kingship not merely seeped back into more primitive tribal societies, but its technical and institutional complex spread, by one means or another, over the whole planet, from China and Cambodia to Peru and Mexico.

At this point I shall use the same evidence Henri Frankfort handled so competently—only to come to a different conclusion.

4 : MESOPOTAMIAN CORROBORATION

In Mesopotamia, kingship arose roughly at about the same time as in Egypt; though in neither case can the earliest manifestation be dated. The ancient Sumerian King List expresses no doubt about the origins of kingship: it was "lowered down from heaven." This means that kingship was from the beginning a religious phenomenon, not just an assertion of physical prowess and organized manpower, nor yet a mere enlargement of venerable ancestral authority.

From the beginning, kingship in Sumer or Akkad combined both authority and power, understanding and command: the same qualities that Breasted noted as attributes of the Sun God after 3000 B.C. With this new office came a comparable enlargement of the sense of time. Egyptian, Mesopotamian, Hindu, and later Mayan religions all embrace cycles of thousands of years; and a single king, Lugalbanda, was assigned a twelve-hundred-year reign, which would have been long for a whole dynasty. Even if these years were months, as early commentators on Manetho's chronology supposed, that was still a long span. To kings, therefore, was attributed not merely an enlarged cosmic power but a more intense and prolonged vitality. Every dimension of existence was magnified, whether on earth or in heaven.

In Mesopotamia, it is true, it was only at intervals that the king, beginning with Naram-Sin, identified himself as a god. Henri Frankfort, leaning over backward to emphasize the genuine differences in the physiognomy of Sumerian and Egyptian cultures, nevertheless counts almost a score of such instances in a span of some eight hundred years. But that the claim could have been made at all shows that the underlying presuppositions were not in fact so different. Certainly kingship everywhere partook of divinity; all kings exercised their extraordinary authority by 'divine right,' for the king was a necessary executioner of the gods' decrees, as well as the chief agent for establishing great collective enterprises such as the building of cities and canal systems.

Significantly, it was during the Third Dynasty of Ur—a period of vigorous constructive activity—that all the kings except the founder claimed divinity. This evidence decisively couples divine kingship with the characteristic public works program of the megamachine. Little tasks might still be left to little men, but big tasks belonged to the king by reason of the special powers he commanded: above all, the unique power to create a colossal labor machine.

As in many primitive tribal communities examined during the last few centuries, and among later historic nations, the king shifted back and forth between a sacred and a secular role: now a religious, now a military head. This duality is still in evidence among various primitive tribes, and it holds all through civilized history. Even today, the wearer of the British crown is the titular head of the Established Church, whose archepiscopal sanction in turn is a necessary condition, as Edward VIII discovered, for holding royal office. That was the archetypal relationship from the beginning. So, too, the upstart Napoleon, to confirm his legitimacy as Emperor, sought the offices of the Roman pontiff to sanctify his coronation—though in seizing the crown from the Pope's hands to place it on his own head, his overweening ego committed a sacrilege that any ancient Babylonian would have told him was sure to bring a heavenly curse down on his pretensions.

Both to establish and maintain kingship, an infusion of divine power was essential. But the constant intercourse with Heaven, necessary for the guidance of the king, demanded professional aid from priests, magicians, soothsayers, interpreters of dreams, and readers of cosmic signs, who in turn were dependent upon the king's secular power and wealth for their own status and office.

This essential coalition between royal military power and often dubious supernatural authority anticipated a similar alliance between scientists and mathematical games-theorists with the higher agents of government today; and was subject to similar corruptions, miscalculations, and hallucinations. Again and again dependence upon unverifiable data from Heaven vitiated the ability to make rational decisions in battle, for instance, on the basis of the locally visible circumstances. Too often the soothsayer's addled counsels counted more heavily than the soldier's professional knowledge, as the Mari letters testify.

On the theological basis of kingship, then, the testimony of Mesopotamia is as clear as that of Egypt for all their historic and geographic differences of culture. And the words uttered by the earliest kings of both lands continue to ring through history both in the claims of 'legitimate' kings like Louis XIV, and in the no less extravagant assertions of a Hitler, a Stalin, or a Mao, whose abject and adoring followers have imputed omniscience to them. The words that the young Babylonian god Marduk uttered, before becoming the chief defender of his fellow gods against the ancient goddess of the primal waters, Tiamat, are the words kings learned to utter long before Marduk took his place in the Babylonian pantheon. The gods are in fact the kings of the unconscious, enlarged, as the kings in turn became incarnate dream gods, exercising visible supremacy over waking life, transmitting their claims of inviolable sovereignty to the whole apparatus of the State.

As a condition of taking office, Marduk insists that when he gives a command, he must be obeyed by his fellow gods without question. "Let my word instead of you, determine the fates: unalterable shall be what I may bring into being: neither recalled nor changed shall be the command of my lips." These words are worth noting. They set forth the terms on which the new collective mechanism was brought into existence.

This new emphasis on unqualified power of command was in some measure, it would seem, a necessary reaction to the disorders and difficulties that multiplied with the growth of population. Regularity and security now became a political desideratum; for while small groups of people may migrate elsewhere when threatened, a whole city or a closely settled countryside cannot be evacuated for a season when overwhelmed by flood or starved by drought. The problems of regulating river flow and repairing flood damage, of apportioning water for irrigation, of storing sufficient food annually to avert famine before the next crop could be gathered —all these proved increasingly beyond the competence of any local community. There was genuine need for some unifying authority in these great valleys; and kingship, for lack of more rational cooperative authority, met that need.

Though neolithic agriculture had produced a hitherto unheard of abundance of food, this very affluence bred new anxieties. "Throughout Mesopotamian history," Frankfort notes, "the kingship of the gods was believed to have originated, not as a natural concomitant of an orderly

society, but as the product of confusion and anxiety." But as I have sometimes observed, in the case of once poverty-stricken friends who have
become affluent, wealth and security themselves may bring on a state of
anxiety not experienced when their possessors did not know where the next
day's food was coming from.

While the forces of nature often worked catastrophically in Mesopotamia, beginning with the historic Biblical flood, similar anxieties arose
even in smiling Egypt, quite apart from the seven lean years reported in the
story of Joseph in the Bible. Egypt provides still other documentary refer-
ences to a failure of crops and to famine, not only through plagues of
locusts, but through an insufficient summer flooding of the Nile. During such
crises there was need for an unchallengeable authority that could marshal
the manpower of many communities and ration their resources equitably;
and to the extent that such efforts brought relief, the ruler who took responsibility would earn gratitude, and gain support on other occasions.

The association of kingship with anxiety, fear, and crisis has been, unfortunately, a long-continued one. Thorkild Jacobsen has shown that the
oldest known political institution that can be identified, through a Mesopotamian text, is the urban assembly of all free men. This assembly in turn
left the power to deal with current matters in the hands of a group of
elders, but in times of emergency they chose a king to "take charge for a
limited period." Millennia later, Herodotus pictures a similar delegation of
power, among the Medes and Persians: so, again, when the community
was threatened, the Romans appointed a temporary dictator; and the same
temporary concentration of power in a 'national emergency' still characterizes the constitutional prerogatives of the President of the United States,
though only today has a President dared to fake an emergency in order to
wield such power and politically sanctify his cumulative errors of judgement and his inhuman acts in Vietnam.

The power that was thus concentrated in kingship in turn produced a reliable source of fresh crises in the institution of war. Everywhere war took
precedence over even hunting as the main prerogative and the dominant
activity of kingship. For in the act of establishing law and order within the
sacred territory of their gods, kings came into conflict with rival kings and
foreign gods, equally arrogant in their imputed divinity, who also claimed
the same kind of blind loyalty and awestruck obedience. Too often they
were tempted to assert their superior power by encroaching on neighboring
states and despoiling their inhabitants.

Even when nature smiled on a community, the man-made catastrophe
of war was ever at hand, to create disorder and to bolster up the absolute
power of kings. Conflicts between cities over water rights and land bound-

aries recur in the chronicles of Sumer and Akkad: but beyond these claims, whose settlement might have been a subject for reasonable compromise, were the irrational efforts of ambitious and tyrannous 'gods,' to secure abject submission.

Here again Jacobsen re-enforces Frankfort's interpretation. "An orderly world [for the Mesopotamian] is unthinkable without a superior authority to impose his will." One can match Jacobsen's emphasis on the belief in superior authority by quoting the Egyptian 'Satire on the Trades': "There is no profession free of a boss." The Mesopotamian feels convinced that the authorities are always right, or at least that there is no use arguing with them. "The command of the palace, like the command of Anu, cannot be altered. The King's word is right; his utterances, like that of a god, cannot be changed." These words resound with sickening familiarity in our present totalitarian states, whether 'democratic' or 'communist.'

This dictum—the first clear expression of what in present-day political systems is known as the 'party line' or the 'consensus'—lessens the importance, in this context, of the many differences that distinguish the ancient civilization of the Nile from that of the Euphrates and the Tigris. In both situations the king exercised godlike powers; and pragmatically, it hardly matters that while in Egypt he was also a divinity in his own right, in Mesopotamia he served usually as a surrogate, yet with full 'powers-of-attorney,' as long as his luck held, to act in the god's name. His mission was to take part in the constant struggle between order and chaos, which, as Ephraim Speiser reminded us, "was the fatal drama renewed at the turn of each year."

In the face of such fearful anxieties, and such violent means of relief from anxiety, one need hardly wonder that the village surrendered both its autonomy and its relative self-sufficiency to these higher powers: the divine king and the officers of the state, the governors and tax collectors, who rigorously carried out the king's commands. The scattered villages took their orders from the control centers.

On earth, only one who was appointed by the gods a king could command such unconditional obedience: only one who could back such presumptuous claims with armed force, in the face of skepticism or active dissent, could have broken down the habits of self-government that small communities had evolved on the basis of ancestral customs and their own limited capabilities for taking counsel and exercising the common sense through prudent actions.

Common sense was exactly what kingship, almost by definition, lacked: when the king's orders were executed no one dared to tell him honestly how they had turned out. With the absolute powers bestowed by kingship came an arrogance, a ruthlessness, an inflexibility, a habit of compulsion, an unwillingness to listen to reason, that no small community would have endured from any of its members—though the aggressive and humanly disagreeable qualities that make for such ambitious leadership might be found anywhere—as Margaret Mead discovered among the Mundugumor, whose leaders were known to the community as "really bad men," aggressive, gluttonous for power and prestige.

But once kingship, and the institutions that supported kingship, were firmly established they remained the chief political model of civilized society right down to the end of the nineteenth century; and in the course of five thousand years they spread to far more primitive tribal communities, like those of the Shilluks in Africa, where all the magical prescriptions and ideological premises have been preserved intact in much the same form we discover at the very beginning, along with the same breed of crumpled-horn cattle that the ancient Egyptians delighted in.

In so far as kingship was in time somewhat humanized, moralized, reduced to more modest dimensions, this came about largely through the stubborn resistance of village communities, many of whose ways and modes of life had spread, along with the migrant villagers themselves, into the new cities. Under the surface, we shall find that this struggle between a democratic and an authoritarian technics has been going on all through history.

5 : THE TECHNIQUE OF DIVINE

RULERSHIP

Unfortunately, most of our data on kingship come from documents put into writing centuries, even millennia, after the original events. Taking the lowest layers of Jericho as evidence, one finds, before there is any visible trace of kingship, an economy with a surplus sufficient to build a large town and keep its inhabitants permanently employed. Under those circumstances one must surmise that the 'Primitive Democratic Community,' as Frankfort calls it, had evolved to a fairly high level of technical cooperation and skill without kingship. Possibly this happened under some milder, more persuasive form of government, such as might have arisen, upon Kathleen Kenyon's hypothesis, during the benign and more favorable climatic conditions that prevailed in this area immediately after the melting of the glaciers. The recent unearthing of Çatal Hüyük in Turkey gives further color to this hypothesis.

Before its own specialized institutions based on coercion and punishment had fully taken form, kingship had sprung up as a mutation in agricultural communities that were still without any permanent division of labor or any strict separation of castes, with only a minimum of economic differentiation through vocational specialization, private property, or enslavement: a state more or less corresponding to Hesiod's Golden Age. If so, this might explain a persistent feature of early kingship, visible at a much later date in New World culture like that of the Incas: namely, its authoritarian communism: state-controlled, but benignly reproducing for the larger community the common sharing of labor and the products of labor as in the village. Basically, the same genial animus and the same type of coercive organization underlies contemporary communism.

When kingship replaced the powers of the agricultural communes, their local functions were taken over on a grand scale by the central authority, in either the temple or the palace. Public property remained public property, but now it belonged to the god, in the person of the king. And when the ruler distributed either this property or fresh booty among his followers, it became 'private' property, about which there has hung, throughout history, an aura of regal, if not divine, sacredness. The god's share in the produce of the land, apportioned by the temple, was the first claim upon it: an ancient custom continued in the tithe into medieval Christendom. Yet every member of the community had a customary share. As long as one served the gods and obeyed the king, one enjoyed security and a portion of the divine bounty. The Welfare State today retains—or should one say 'has regained'?—many of these features.

This state-administered communism seems to mark the earliest stage of kingship: the land itself, the common functions and the common rights, come under the control of the king, and when necessary, his edicts and laws replace the immemorial customs of the local community. For it is through the king that the community enjoys the favors of the gods; and so long as the people pay their taxes in grain and in labor, their security is assured. That basic communism is attested in Egypt and Mesopotamia, and again in Peru; and if kingship rested on such traditional modes, merely extending them and ensuring them, it perhaps explains why the harsher aspects of the system were accepted, though gross inequalities, as between slaves, freemen, nobles, which attended the growth of private property, all too soon crept in.

What is still called the sovereignty of the state maintains without abate-

ment the original royal claims of power and privilege, of ultimate owner-
ship and unconditional obedience, along with such punishments or sacrifices as the sovereign may see fit to exact in the name of national welfare.

This solidarity between a divinely empowered ruler and his community
achieved classic expression in Egypt: as Frankfort puts it, "the trust which
people put in their 'shepherds' was . . . that Pharaoh should wield to the
full the absolute power to which his divinity entitled him, and which enabled him—as nothing else could—to ensure the well-being of the whole
community." But the same relation holds equally, if not so fulsomely,
elsewhere: the words of Wilson about Egypt, "it is the king who built temples and cities, who won battles, who made laws, or who provided
bounty for the tombs of his nobles," are echoed by Kramer describing the
activities of the King of Lagash. Even the forms of speech traditionally used
by kings declare this: as in Shakespeare's 'Henry the Fifth' where sovereigns
address each other as 'England' and 'France.'

The relationship between king and community transcended the loyalties
of clan, family, neighborhood; and it explains why kings, or even upstart
counterfeits, tyrants, so often won popular support, as against such minor
contenders for power and authority as magnates and nobles. Under this
mystique of absolute power, functions that would later be taken over by
the machine were at first conceived and executed solely through the unique
offices of kingship.

At the beginning, such power was associated with the idea of steward-
ship and responsibility to the gods. By 2000 B.C. no Pharaoh could hope for
immortality unless he had served the cause of righteousness and justice
(Ma'at). In a text from the Middle Kingdom, Atum declares: "I made the
great inundation that the poor man might have his rights therein like the
great man. I made every man like his fellow." In this declaration one
sees a recognition of a persistent pressure not only to legalize but to moralize
power: to keep it within bounds and make it respect the human condition.

Thus the divine head of this hierarchy of power returned, at least in
principle, to the egalitarian social and moral ideals of the village. There was
always, it is true, an ambivalence in this relationship: the kindliness of the
ruler that is emphasized in Egyptian texts goes along with his emphatic
capacity to rouse terror and inflict death. Yet the memory of the earlier
communal attributes of kingship partly offset, perhaps, the daily reminders
of personal caprice and collective harshness. But too often, the documents
tell us, the officials who executed the king's orders identified themselves with
the source of authority, and characteristically overdid royal insolence without a compensating display of royal grace.

More primitive types of community operate effectively in unison by clinging tenaciously to well-established habits and 'immemorial customs': conformity is the price of mutual toleration, and ostracism is the severest type of punishment that is usually invoked. But to ensure the performance of royal commands through a long human transmission belt, operating often at a great distance from the center of power, kingship required a more exacting kind of compliance. If the state organization was to work smoothly as a unit, conformity must be automatic and complete.

This mechanical obedience was achieved by various symbolic and practical devices; and first, by establishing an insuperable psychological distance between the king and all who came near him. His person was not merely inviolate, untouchable; but those who came within his presence were commanded to prostrate themselves, as if dead, fully conscious that if they offended the king nothing would stand between them and death. In the presence of the king the highest functionary grovelled like a slave: witness one of the Tell el Amarna letters (c. 1370 B.C.): "At the two feet of the King, my lord, the Sun God from heaven, seven times seven times I fall, both prone and supine." The suppliant was "the first on which thou dost tread."

Such submission, such abject self-humiliation, never had a counterpart among the humble members of any village community until 'civilized' institutions filtered down from above. But this drill had the effect of turning human beings into 'things,' who could be galvanized into a regimented kind of cooperation by royal command, to perform the special tasks he assigned them, however stultifying to their family life and incompatible with normal village routines.

Not the least evil of this system was the sense of human degradation produced by its officers' obligation to take orders, slavishly, from a superior authority. M. I. Finley points out that free citizens, in later Greece and Rome, were extremely reluctant to accept high administrative offices for this very reason; and as a result even posts of ministerial or military command were often assigned to slaves, who were too thoroughly conditioned to obedience to recognize their humiliation. Individual initiative and responsibility had no place in the megamachine; for such freedom might mean countermanding faulty orders or disobeying immoral ones. The dedicated members of the megamachine, early and late, remain Eichmanns: doubly degraded because they have no consciousness of their own degradation.

With the ideas of submission and absolute obedience which were essential to the composition of a human machine, came, however, the possibilities of disobedience, treachery, and rebellion. To ensure that the heavenly sanctions of kingship were sufficiently respected, kingship in the end must be ready to fall back on force: not merely naked force, but force in ferocious, sadistic forms, repeatedly magnified into nightmarish extravagances of cruelty, as dehumanized as those we have witnessed in the last generation in the ingenious horrors perpetrated by 'civilized' governments in Warsaw, Auschwitz, Tokyo, and Vietnam.

Here, again, the gods served as a model for kings: for Marduk, in fighting Tiamat, his ancient rival, used "the Evil Wind," the Whirlwind, the Hurricane: he yokes his storm chariot to a team: "the Killer, the Relentless, the Trampler, the Swift." "Sharp were their teeth, bearing poison." Again this is not a trait that belongs only to the notably quarrelsome and violent Babylonians, Assyrians, or Hittites. In the metaphors used in the most ancient Pyramid text, describing a deified Pharaoh, one encounters a kind of unrestrained cannibal lust in dwelling on the scope and power of the divine king. As pictured there, kingship was actually a man-eating device. To match these frightful symbols one would have to turn to a playwright of our own day who presents a woman eating the genitals of her lover.

Lest my characterization seem extravagant, let me support it with Erman's translation:

"He it is that eateth men; that liveth on Gods, that possesseth the carriers and despatcheth messages. . . . The Runner-with-all-Knives . . . he that strangleth them for him; he draweth out for him their entrails, he the messenger whom he sends death to. . . . He it is that eateth their magic and swalloweth their lordliness. Their great ones are for his morning meal, their middle-sized ones for his evening meal, and their little ones for his night meal. . . . He hath broken up the backbones and the spinal marrow, he hath taken away the hearts of the Gods, he hath eaten the Red Crown, he hath swallowed the Green One. He feedeth on the lungs of the Wise Ones: he is satisfied with living on hearts and their magic."

To take such a characterization as a graceful rhetorical device is to close one's senses to the social context of the words. Brutal compulsion was the necessary accompaniment of the large-scale organization and the extensive order introduced by kingship. Herodotus' history is full of revolting descriptions of the rabid violence of kings, such as the tale he told of Cambyses. One of the king's dearest friends at court suggests that he is in the habit of drinking too much wine for his own good. To prove that the wine could have no effect, Cambyses drinks more recklessly than ever, then draws a bow and aims it at his friend's son, placed at the other end of the hall, shooting him through the heart, and finally tearing out the lad's heart to prove to the father that his arrow had pierced it.

Primitive society recognizes by and large only two serious crimes: the breaking of the incest taboo and murder. But with the new system of administration and codes of law introduced by kingship, the number of possible crimes increased and the punishments became more terrifying. Disobedience to the orders of a superior was the worst of sins; and even 'answering back' was a serious offense. If we may judge from the practice of the Cheyenne Indians, this may have been a paleolithic hunting trait; for one of the three punishable crimes among them was disobedience to the leader's orders on the buffalo hunt.

Woolley cites a Hittite law: "And if ever a servant vexes his master, either they kill him, or they injure his nose, his eyes, or his ears." Such mutilations were the favorite forms of punishment. When one compares these practices with the decent, humane customs of surviving primitive peoples, one understands how the repressive cult of power introduced a degree of ferocity and debasement unknown among earlier groups, who inflicted bodily mutilations only on *themselves,* and then chiefly for magical reasons.

Even under the relatively benign code of Hammurabi, the systematic infliction of punishment by torture and the permanent maiming of the body was sanctioned, though such practices were quite foreign to the archaic small community before the Iron Age. These sadistic methods applied equally to education and left a mark that is only now being lifted. Functionaries with a whip to keep order were, Kramer tells us, a standard feature of the Sumerian school; and in Egyptian the verb 'to teach' is also the verb 'to punish.' The latter usage indeed survived into our own age, for parents in punishing a child—before 'permissiveness,' with a kind of inverted sadism, visiting punishment on everyone *but* the child, swung wildly to the opposite extreme—would often say: "I'll teach you to behave!"

Apart from murder and rape, the most horrendous crimes punished by civilized authority stem back to the 'unpardonable sin' of kingship: disobedience to the sovereign. Murderous coercion was the royal formula for establishing authority, securing obedience, and collecting booty, tribute, and taxes. At bottom, every royal reign was a reign of terror. With the extension of kingship, this underlying terror formed an integral part of the new technology and the new economy of abundance. In short, the hidden face of that beautiful dream was a nightmare, which civilization has so far not been able to throw off.

6: CIVILIZATION AND 'CIVILIZATION'

With kingship, power as an abstraction, power as an end in itself, became the chief identifying mark of 'civilization,' as opposed to all the earlier norms and forms of culture.

Civilization, still often used as a word of eulogy and admiration, in comparison with what used to be called savagery and barbarism, is taken as a general term to cover law, order, justice, urbanity, civility, rationality; and it currently implies a cumulative effort to further the arts and sciences, and to improve the human condition by continued advances in both technology and responsible government. All these terms of admiration and praise, which seemed in the eighteenth century self-evident and self-justifying, except to an occasional dissident like Rousseau, have now become ironic: at best they represent a hope and a dream that have still to be fulfilled.

Here and hereafter I use the term 'civilization' in quotation marks in a much narrower sense: to denote the group of institutions that first took form under kingship. Its chief features, constant in varying proportions throughout history, are the centralization of political power, the separation of classes, the lifetime division of labor, the mechanization of production, the magnification of military power, the economic exploitation of the weak, and the universal introduction of slavery and forced labor for both industrial and military purposes. These institutions would have completely discredited both the primal myth of divine kingship and the derivative myth of the machine had they not been accompanied by another set of collective traits that deservedly claim admiration: the invention and keeping of the written record, the growth of visual and musical arts, the effort to widen the circle of communication and economic intercourse far beyond the range of any local community: ultimately the purpose to make available to all men the discoveries and inventions and creations, the works of art and thought, the values and purposes that any single group has discovered.

The negative institutions of 'civilization,' which have besmirched and bloodied every page of history, would never have endured so long but for the fact that its positive goods, even though they were arrogated to the use of a dominant minority, were ultimately of service to the whole human community, and tended to produce a universal society of far higher potentialities, by reason of its size and diversity. Even immediately their symbols perhaps attracted those who were only spectators of these achievements. This universal component was present from the beginning, because of the cosmic foundations of royal power: but the efforts to create a universal society were delayed, until our own day, by the lack of adequate technical instruments for rapid transportation and instantaneous communication.

Yet the claim of universality itself, from Naram-Sin to Cyrus, from Alexander to Napoleon, was repeatedly made: one of the last of the 'all-powerful' monarchs, Genghis Khan, proclaimed himself the sole ruler of the entire world. That boast was at once an aftermath of the myth of divine kingship and a prelude to the new myth of the machine.

CHAPTER NINE

The Design of the Megamachine

1 : THE INVISIBLE MACHINE

In doing justice to the immense power and scope of Divine Kingship both as myth and active institution I have so far left one important aspect for closer examination, its greatest and most durable contribution—the invention of the archetypal machine. This extraordinary invention proved in fact to be the earliest working model for all later complex machines, though the emphasis slowly shifted from the human operatives to the more reliable mechanical parts. The unique act of kingship was to assemble the manpower and to discipline the organization that made possible the performance of work on a scale never attempted before. As a result of this invention, huge engineering tasks were accomplished five thousand years ago that match the best present performances in mass production, standardization, and meticulous design.

This machine escaped notice and so naturally remained unnamed until our own day, when a far more powerful and up-to-date type, utilizing a congeries of subordinate machines, came into existence. For convenience, I shall designate the archetypal form by more than one name, in reference to a specific situation.

Because the components of the machine, even when it functioned as a completely integrated whole, were necessarily separate in space, I shall for certain purposes call it the 'invisible machine': when utilized to perform work on highly organized collective enterprises, I shall call it the 'labor machine': when applied to acts of collective coercion and destruction, it deserves the title, used even today, the 'military machine.' But when all the components, political and economic, military, bureaucratic and royal, must be included, I shall usually refer to the 'megamachine': in plain words, the Big Machine. And the technical equipment derived from such a megamachine thence becomes 'megatechnics' as distinguished from the more modest and diversified modes of technology, which until our own century continued to perform the larger part of the daily work in the workshop and on the farm, sometimes with the help of power machinery.

Men of ordinary capacity, relying on muscle power and traditional skills alone, were capable of performing a wide variety of tasks, including pottery manufacture and weaving, without any external direction or scientific guidance, beyond that available in the tradition of the local community. Not so with the megamachine. Only kings, aided by the discipline of astronomical science and supported by the sanctions of religion, had the capability of assembling and directing the megamachine. This was an invisible structure composed of living, but rigid, human parts, each assigned to his special office, role, and task, to make possible the immense work-output and grand designs of this great collective organization.

At its inception no inferior chief could organize the megamachine and set it in motion. And though the absolute assertion of royal power rested on supernatural sanction, kingship itself would not have prevailed so widely had these claims not in turn been ratified by the colossal achievements of the megamachine. That invention was the supreme feat of early civilization: a technological exploit which served as a model for all later forms of mechanical organization. This model was transmitted, sometimes with all its parts in good working condition, sometimes in a makeshift form, through purely human agents, for some five thousand years, before it was done over in a material structure that corresponded more closely to its own specifications, and was embodied in a comprehensive institutional pattern that covered every aspect of life.

To understand the point of the machine's origin and its line of descent is to have a fresh insight into both the origins of our present over-mechanized culture and the fate and destiny of modern man. We shall find that the original myth of the machine projected the extravagant hopes and wishes that have come to abundant fulfillment in our own age. Yet at the same time it imposed restrictions, abstentions, and compulsions and servilities that, both directly and as a result of the counter-reactions they produced, today threaten even more mischievous consequences than they did in the Pyramid Age. We shall see, finally, that from the outset all the blessings of mechanized production have been undermined by the process of mass destruction which the megamachine made possible.

Though the megamachine was first assembled during the period when copper for tools and weapons came into use, it was an independent innovation: the mechanization of men had long preceded the mechanization of their working instruments, in the far more ancient order of ritual. But once conceived, this new mechanism spread rapidly, not just by being imitated in self-defense, but by being forcefully imposed by kings acting as only gods or the anointed representatives of the gods could act. Wherever it was successfully put together the megamachine multiplied the output of energy and performed labor on a scale that was never conceivable before. With this ability to concentrate immense mechanical forces, a new dynamism came into play, which overcame by the sheer impetus of its achievements the sluggish routines and the petty inhibitions of small-scale village culture.

With the energies available through the royal machine, the dimensions of space and time were vastly enlarged: operations that once could hardly have been finished in centuries were now accomplished in less than a generation. On the level plains, man-made mountains of stone or baked clay, pyramids and ziggurats, arose in response to royal command: in fact the whole landscape was transformed, and bore in its strict boundaries and geometric shapes the impress of both a cosmic order and an inflexible human will. No complex power machines at all comparable to this mechanism were utilized on any scale until clocks and watermills and windmills swept over Western Europe from the fourteenth century of our era on.

Why did this new mechanism remain invisible to the archeologist and the historian? For a simple reason already implied in our first definition: because it was composed solely of human parts; and it possessed a definite functional structure only as long as the religious exaltation, the magical abracadabra and the royal commands that put it together were accepted as beyond human challenge by all the members of the society. Once the polarizing force of kingship was weakened, whether by death or defeat in battle, by skepticism or by a vengeful uprising, the whole machine would collapse. Then its parts would either regroup in smaller units (feudal or urban) or completely disappear, much in the way that a routed army does when the chain of command is broken.

In fact these first collective machines were as subject to breakdown, were ultimately as frail and vulnerable as the theologico-magical conceptions that were essential to their performance. Hence those who commanded them were in a constant state of anxious tension—often with good reason, fearing heresy or treason from their near-equals and rebellions and reprisals on the part of the submerged masses. Without submissive faith and unqualified obedience to the royal will, transmitted by governors, generals, bureaucrats, taskmasters, the machine would never have been workable. When these attitudes could not be sustained, the megamachine collapsed.

From the beginning the human machine presented two aspects: one negative, coercive, and too often destructive; the other positive, life-promoting, constructive. Yet the second factors could not readily function unless the first were in some degree present. Though a primitive form of the military machine almost certainly came before the labor machine, it was the latter that achieved an incomparable perfection of performance, not alone in quantity of work done, but in the quality and complexity of its organized structures.

Now to call these collective entities machines is no idle play on words. If a machine be defined, more or less in accord with the classic definition of Franz Reuleaux, as a combination of resistant parts, each specialized in function, operating under human control, to utilize energy and to perform work, then the great labor machine was in every aspect a genuine machine: all the more because its components, though made of human bone, nerve, and muscle, were reduced to their bare mechanical elements and rigidly standardized for the performance of their limited tasks. The taskmaster's lash ensured conformity. Such machines had already been assembled if not invented by kings in the early part of the Pyramid Age, from the end of the Fourth Millennium on.

Just because of their detachment from any fixed external structures, these labor machines had much fuller capacities for change and adaptation than the more rigid metallic counterparts of a modern assembly line. In the building of the pyramids we find not only the first indubitable evidence of the machine's existence, but the proof of its astonishing efficiency. Wherever kingship spread, the 'invisible machine,' in its destructive if not its constructive form, went with it. This holds as true for Mesopotamia, India, China, Yucatan, Peru, as for Egypt.

By the time the megamachine had taken form, all the preliminary stages had been obliterated: so we can only guess at the way in which its members were assorted, assigned their places, and trained in their duties. At some point in the process an inventive mind, or more probably a series of inventive minds, following through the opening gambit, must have been able to grasp the essential problem—that of mobilizing a large body of men and rigorously coordinating their activities in both time and space for a predetermined, clearly envisaged and calculated purpose.

The difficulty was to turn a random collection of human beings, detached from their family and community and their familiar occupations, each with a will or at least a memory of his own, into a mechanized group that could be manipulated at command. The secret of mechanical control was to have a single mind with a well-defined aim at the head of the organization, and a method of passing messages through a series of intermediate functionaries until they reached the smallest unit. Exact reproduction of the message and absolute compliance were both essential.

This grand problem may well have been first worked out in quasi-military organizations in which a relatively small body of hunters, roughly disciplined to obey their leader, were addressed to the task of controlling a much larger body of unorganized peasants. At all events, the type of mechanism created never operated without a reserve of coercive force behind the word of command; and both the method and the structure have been passed on, almost without change, to all military organizations, as we now know them. *Through the army, in fact, the standard model of the megamachine was transmitted from culture to culture.*

If one single invention was necessary to make this larger mechanism operative for constructive tasks as well as for coercion, it was probably the invention of writing. This method of translating speech into graphic record not merely made it possible to transmit impulses and messages throughout the system, but to fix accountability when written orders were not carried out. Accountability and the written word both went along historically with the control of large numbers; and it is no accident that the earliest uses of writing were not to convey ideas, religious or otherwise, but to keep temple records of grain, cattle, pottery, fabricated goods, stored and disbursed. This happened early, for a pre-dynastic Narmer mace in the Ashmolean Museum at Oxford records the taking of 120,000 prisoners, 400,000 oxen, and 1,422,000 goats. The arithmetical reckoning was an even greater feat than the capture.

Action at a distance, through scribes and swift messengers, was one of the identifying marks of the new megamachine; and if the scribes formed the favored profession it was because this machine could not be effectively used without their constant service, to encode or decode the royal messages. "The scribe, he directeth every work that is in this land," an Egyptian New Kingdom composition tells us. In effect, they probably played a part not too dissimilar to that of the political commissars introduced into the Soviet Russian army. They made possible the constant 'report to political headquarters' essential for a centralized organization.

Whether the military or the labor machine came first, they had the same organization. Were the Egyptian and Mesopotamian raiding-parties and mining-gangs military or civil organizations? At first these functions were undistinguishable, or rather, interchangeable. In both cases the fundamental unit was the squad, under the supervision of a gang boss. Even in the domains of the rich landed proprietors of the Old Empire, this pattern prevailed. According to Erman, the squads formed into companies, marching or parading under their own banner. At the head of each company of workmen stood the chief workman, who bore the title, Chief of the Company. Nothing like this, one may venture to say, had ever been seen in an early neolithic village. "The Egyptian magistrate," Erman observes, "cannot think of these people otherwise than collectively; the individual workman exists for him no more than the individual soldier exists for our higher army officers." Precisely: this was the original pattern of the archetypal megamachine and has never been radically altered.

With the development of the megamachine, the broad division of labor between functions and offices, long familiar to us in the army, was likewise applied at an early date to highly specialized parts of the work process. Flinders Petrie notes that in the realm of mining—a sphere wherein, both in Egypt and Mesopotamia, the work army, I repeat, can hardly be distinguished from the military army—a minute division of labor had been established. "We know from the mummy records," Petrie observed, "how minutely work was subdivided. Every detail was allotted to the responsibility of an individual; one man prospected, another tested the rock, a third took charge of the products. There are over fifty different qualities and grades of officials and laborers named in the mining expeditions."

These divisions, inevitably, became part of the broader social organization that operated beyond the closed domain of the megamachine. And by the time Herodotus visited Egypt in the fifth century B.C. the over-all division of labor and the minute subdivision into specialisms—no longer confined to the megamachine—had reached a point comparable to that which it has come to in our own time; for he records that "some physicians are for the eyes, others for the head, others for the teeth, others for the belly, and others for internal disorders."

But note the difference between the ancient human machine and its more efficiently de-hominized modern rivals, in both its method and its underlying purpose. Whatever the actual results of their employment, all modern machines are conceived as labor-saving devices: they attempt to perform the maximum amount of work with the least immediate expenditure of human effort. Labor-saving had no part in the institution of the earliest machines: just the contrary, they were *labor-using devices,* and their inventors had reason to exult over the increased number of workers that they could, by efficient design and organization, bring to bear on any given task, provided the job itself were big enough.

The total effect of both types of machine was the same: they were designed to perform with efficiency and undeviating exactitude, and with copious power, tasks that could never be performed by individual tool-users more loosely organized. Both types of machine achieved a hitherto unattainable level of efficiency. But instead of freeing labor, the royal megamachine boasted of imprisoning and enslaving it.

If purely humane modes of work, which men would undertake voluntarily to fulfill immediate needs, had prevailed, the colossal achievements of early civilizations would probably have remained inconceivable: this must be granted. And it is even possible that the modern non-human machine, powered by extraneous energies, meant to economize labor, might never have been invented, for the mechanical agents had first to be 'socialized' before the machine itself could be fully mechanized. But at the same time, if the collective machine had not been able to utilize forced labor—either by periodic conscription or slavery—the colossal miscarriages, perversions, and wastages that so constantly accompanied the megamachine might not have taken place.

2: MECHANICAL STANDARDS OF PERFORMANCE

Let us now examine the human machine in its archetypal form. As so often happens, there was a certain clarity in this first demonstration that was lost when the machine was diffused and worked into the more complex patterns of later societies, mingling with familiar, humbler survivals. And if the megamachine never achieved a higher peak of performance than in the Pyramid Age, this is perhaps not only because of the singular engineering talents that designed and operated these early machines, but also because the myth that held the human parts of the machine together could never again exert such a massive attractive power, unstained as it was until the Sixth Dynasty by any serious letdowns and failures. Until then its triumphs were indisputable, its chronic perversities still unexposed.

Among all the feats of construction in which the megamachine excelled, the pyramid stands forth as an archetypal model. In its elemental geometric form, in the exquisite accuracy of its measurements, in the organization of the entire working force, in the sheer mass of construction involved, the final pyramids demonstrate to perfection the unique properties of this new technical complex. To exhibit the properties of this system,

I shall concentrate upon the pyramid alone: in particular, the Great Pyramid at Giza.

The Egyptian pyramid was conceived as a tomb to hold the embalmed body of the Pharaoh and secure his safe passage into the after-life. The king alone, at first, had the prospect of such a godlike extension of his existence. In mummy and pyramid, time symbolically stood still forever. This heavenly destination of the king altered every earthly prospect: but as with the conquest of space today the common man played no part in it—except to pay the bill in taxes and forced labor.

Between the first stone pyramid—built in the step form we find later in Meso-America—and the mighty pyramid of Cheops' Fourth Dynasty, the first and the most enduring of the Seven Wonders of the Ancient World, lies less than a century and a half: a change comparable in speed to the development of steel-frame building construction in our own age. On the ancient time-scale for inventions the most primitive form and the final one, never again to be equalled, were practically contemporary.

The swiftness of this development indicates a concentration of physical power and technical imagination. That transformation is all the more striking because the Pharaohs' tombs did not stand alone: they were part of a whole city of the dead, a complex structure with buildings that housed the priests who conducted the elaborate rituals deemed necessary to ensure a happy future existence for the departed divinity.

The Great Pyramid is one of the most colossal and perfect examples of the engineer's art at any period or in any culture. Even without allowing for the primitive character of the tools available in the Third Millennium B.C., no construction of our own day surpasses this in either technical virtuosity or audacity. And yet this great enterprise was undertaken by a culture that was just emerging from the Stone Age, and was long to continue using stone tools, though copper was available for the chisels and saws that shaped the massive building-stones for the new monuments. All the operations were performed by hand.

National conscription, if not serfdom or slavery, was an essential part of the system: essential as a source of sufficient energy. Even the priesthood, Erman tells us, was not absolved from forced labor. The actual operations were performed by specialized handicraft workers, aided by an army of unskilled or semi-skilled laborers, drafted at quarterly intervals from agriculture. The whole job was done with no other mechanical aids than two 'simple machines' of classic mechanics: the inclined plane and the lever, for as yet neither wheel nor pulley nor screw had been invented. We know from graphic representations that large stones were hauled on sledges, by battalions of men, across the desert sands. But note: the single stone slab that covers the inner chamber of the Great Pyramid where the Pharaoh lay weighed fifty tons. An architect today would think twice before calling for such a mechanical exploit.

Now the Great Pyramid is more than a formidable mountain of stone, 755 feet square at the base, rising to a height of 481.5 feet. It is a structure with a complex interior, consisting of a series of passages at different levels that lead into the final burial chamber. Yet every part of it was built with a kind of precision that, as Breasted properly emphasized, belongs to the watchmaker's art rather than that of the modern bridge builder or skyscraper constructor. Blocks of stone were set together with seams of considerable length, showing joints of one ten-thousandth of an inch; while the dimensions of the sides at the base differ by only 7.9 inches, in a structure that covers acres. In short, fine measurement, undeviating mechanical precision, and flawless perfection are no monopolies of the present age. The Pharaonic social organization had leaped ahead five thousand years to create the first large-scale power machine: a machine whose total output was from 25,000 to 100,000 manpower, at the very minimum the equivalent of 2,500 horsepower.

Obviously no ordinary human hands, no ordinary human effort, no ordinary kind of human collaboration such as was available in the building of village huts and the planting of fields, could have mustered this superhuman force, or have achieved this almost supernatural result. Only a divine king could demand such a massive act of the collective human will and effect such a large-scale material transformation. Was it possible to accomplish such massive engineering feats without the aid of a machine? Emphatically not. Only a complex power machine could fabricate these immense constructions. The ultimate product itself showed that it was not only the work of a machine, but of a highly refined type of machine. Though the material equipment of dynastic Egypt was still crude, patient workmanship and disciplined method made good these shortcomings. This megamachine was composed of a multitude of uniform, specialized, interchangeable but functionally differentiated parts, rigorously marshalled together and coordinated in a process centrally organized and centrally directed: each part behaving as a mechanical component of the mechanized whole.

In some three centuries, possibly in half that time in Egypt, the human machine had been perfected. The kind of mind that designed the pyramids and the massive temples and walled cities represented a new human type, capable of effecting the abstract organization of complex functions in a structural design whose final form determined every stage in the work. Not merely mathematical calculations but meticulous astronomical observations were necessary for the siting of these great structures, so that each side was oriented exactly in line with the true points of the compass. These fine measurements called for a rigorous standard of workmanship—unsurpassed until our own era. Since at inundation the pyramid site is only one quarter of a mile from the river, a rock foundation—which demanded the removal of sand—was needed. In the Great Pyramid the perimeter of its bed deviates from true level by little more than half an inch.

The minds that solved these problems and carried out these designs were obviously minds of the highest order, with a unique combination of theoretic analysis, practical grasp, and imaginative foresight: Imhotep, who built the first stone pyramid at Sakkara, was a minister of state, an architect, an astronomer, and a physician. No narrowly trained specialists or 'experts' these, but men who moved freely over the entire area of existence, like the great artists of the Italian Renascence. Their prowess and their self-confidence were equal to any occasion: indeed sometimes defied prudence and outstripped the powers of their mighty machines, as later in the embedded Assouan obelisk, weighing 1,168 tons, never finally detached from the solid rock.

Now the workers who carried out these designs also had minds of a new order: mechanically conditioned, executing each task in strict obedience to instructions, infinitely patient, limiting their response to the word of command. Machine work can be done only by machines. These workers during their period of service were, as it were, stripped down to their reflexes, in order to ensure a mechanically perfect performance. Their leaders, nevertheless, could read written orders; and the men employed may have recognized a few signs, for they left their names in red ochre, Edwards tells us, on the blocks of the Meidum Pyramid. "Boat Gang," "Vigorous Gang," "Craftsmen Gang," and so forth. In their inurement to mechanical order they would have felt at home today on any assembly line. Only the naked pin-up girl was lacking.

Alike in organization, in mode of work, in rapid tempo of production, and in product, there is no doubt that the machines which built the pyramids and the great temples, and which performed all the great constructive works of 'civilization' in other areas and cultures, were true machines. In their basic operations, they collectively performed the equivalent of a whole corps of power shovels, bulldozers, tractors, mechanical saws, and pneumatic drills, with an exactitude of measurement, a refinement of skill, and an even output of work that would still be a theme for boasting today. These characteristics were no monopoly of Egypt: "The German excavators at the site of Uruk reckoned that the construction of only one of the Proto-Literate temple complexes there must have taken 1,500 men, each working a ten-hour day, five years to build."

This extension of magnitude in every direction, this raising of the ceiling of human effort, this subordination of individual aptitudes and interests to the mechanical job in hand, and this unification of a multitude of sub-ordinates to a single end derived from one source alone: the divine power exercised by the king. The king, or rather kingship, was the prime-mover. In turn, the staring success of the enterprise confirmed that power.

Such strict, all-embracing order began, one must remember, at the top: consciousness of the predictable movements of the sun and the planets or, if Zelia Nuttall's old surmise was right, the even more steadfast and predictable position of the Pole-star. In giant collective works, as in the temple ceremonials, it was the king who gave forth the original commands: the king who demanded absolute conformity and who punished even trivial disobedience. It was the king who alone had the godlike power of turning men into mechanical objects and assembling these objects in a machine. The order that was transmitted from Heaven through the king was passed on to every part of the machine, and in time created an under-lying mechanical unity in other institutions and activities: they began to show the same regularity that characterized the movements of the heavenly bodies.

No older vegetation myth, no fertility god, could establish this abstract order or detach so much power from immediate service to life. But note: only the minority who were closely attached to the megamachine could fully share this power; while those who resisted it courted death—as well might they resist the stars in their courses. Despite repeated setbacks and failures, these cosmic fantasies have remained intact to this day: indeed, they have come back again in the guise of 'absolute weapons' and 'absolute sovereignty'—the far from innocent hallucinations of the 'Nuclear Age.'

3: THE MONOPOLY OF POWER

To understand the structure or the performance of the human machine, one must do more than center attention upon the points where it material-ized. Even our present technology, with its vast reticulation of visible machines, cannot be understood on those terms alone.

Two devices were essential to make the machine work: a reliable organization of knowledge, natural and supernatural; and an elaborate structure for giving orders, carrying them out, and following them through. The first was incorporated in the priesthood, without whose active aid the institution of divine kingship could not have come into existence: the second, in a bureaucracy. Both were hierarchical organizations at whose apex stood the high priest and the king. Without their combined efforts the power complex could not operate effectively. This condition remains true today, though the existence of automated factories and computer-regulated units conceals both the human components and the religious ideology essential even to current automation.

What would now be called science was an integral part of the new machine system from the beginning. This orderly knowledge, which was based on cosmic regularities, flourished, as we have seen, with the cult of the sun: star-watching and calendar-making coincided with and supported the institution of kingship, even though no small part of the efforts of the priests and soothsayers was, in addition, devoted to interpreting the meaning of singular events such as the appearance of comets, eclipses of the sun or moon, or erratic natural phenomena such as the flight of birds or the state of a sacrificed animal's entrails.

No king could move safely or effectively without the support of such organized 'higher knowledge,' any more than the Pentagon can move today without consulting its specialized scientists, technical experts, games theorists and computers—a new hierarchy supposedly less fallible than the entrail-diviners, but, to judge by their gross miscalculations, not notably so.

To be effective, this kind of knowledge must remain a secret priestly monopoly. If everyone had equal access to the sources of knowledge and to the system of interpretation, no one would believe in their infallibility, since their errors could then not be concealed. Hence the shocked protest of Ipu-wer against the revolutionaries who overthrew the Old Kingdom in Egypt was based on the fact that the "secrets of the temple lay unbared"; that is, they had made 'classified information' public. Secret knowledge is the key to any system of total control. Until printing was invented, the written word remained largely a class monopoly. Today the language of higher mathematics plus computerism has restored both the secrecy and the monopoly, with a consequent resumption of totalitarian control.

Not the least affiliation of kingship with the worship of the sun was the fact that the king, like the sun, exerted force at a distance. For the first time in history, power became effective outside the immediate range of hearing or the arm's reach. No military weapon by itself sufficed to convey such power. What was needed was a special form of transmission gear: an army of scribes, messengers, stewards, superintendents, gang bosses, major and minor executives, whose very existence depended upon their faithfully carrying out the king's orders, or, more immediately, those of his powerful ministers and generals. In other words, a well-organized bureaucracy is an integral part of the megamachine: a group of men, capable of transmitting and executing a command, with the ritualistic punctilio of a priest, the mindless obedience of a soldier.

To fancy that bureaucracy is a relatively recent institution is to ignore the annals of ancient history. The first documents that attest the existence of bureaucracy belong to the Pyramid Age. In a cenotaph description at Abydos, a career official under Pepi I, in the Sixth Dynasty, c. 2375 B.C., reported "His majesty sent me at the head of this army, while the counts, while the seal-bearers of the King of Lower Egypt, while the sole companions of the Palace, while the nomarchs [governors] and *mayors* of Upper and Lower Egypt, the companions and chief dragomans, the chief prophets of Upper and Lower Egypt, and the Chief Bureaucrats were [each] at the head of a troop of Upper or Lower Egypt, or of the villages and towns which they might rule."

Not merely does this text establish a bureaucracy: but like Petrie's evidence, quoted earlier, it shows that the division of labor and specialization of functions necessary for efficient mechanical operation had already taken place.

This development had begun at least three dynasties earlier: not by accident with the building of the great stone pyramid of Djoser (Zoser) at Sakkara. John Wilson observed, in 'City Invincible' that "we credit Djoser, not only with the beginnings of monumental architecture in stone in Egypt, but also with the setting up of a new monster, the bureaucracy." This was no mere coincidence. And W. F. Albright, commenting upon this, pointed out that "the greater number of titles found in sealings of the First Dynasty . . . certainly pre-supposes an elaborate officialdom of some kind."

Once the hierarchic structure of the human machine was established, there was no theoretic limit to the number of hands it might control or the power it might exert. The removal of human dimensions and organic limits constitutes indeed the chief boast of such an authoritarian machine. Part of its productivity is due to its use of unstinted physical coercion to overcome human laziness or bodily fatigue. Occupational specialization was a necessary step in the assemblage of the human machine: only by intense concentration of skill at every part of the process could the super-human accuracy and perfection of the product have been achieved.

The large-scale division and sub-division of labor throughout modern industrial society begins at this point.

The Roman maxim that the law does not concern itself with trivial matters applies likewise to the megamachine. The forces that were set in motion by the king demanded collective enterprises of a commensurate order: great earth-moving operations that turned rivers, dug canals, built walls. As with modern technology, the machine tended increasingly to dictate the purpose to be served, and to exclude other more intimate human needs. These human machines were by nature large and impersonal, if not deliberately dehumanized; they had to operate on a big scale or they could not work at all, for no bureaucracy, however efficient, could hope to govern directly a thousand little workshops and farms, each with its own traditions, its own craft skills, its own willful personal pride and sense of responsibility. So the rigid form of control manifested in the collective machine was until our own time confined to great mass enterprises and large-scale operations. This original defect limited the extension of megatechnics until mechanical substitutes for the human operatives could be invented.

The importance of the bureaucratic link between the source of power—the divine king—and the actual human machines that performed the works of construction or destruction can hardly be exaggerated: all the more because it was the bureaucracy that collected the annual taxes and tributes that supported the new social pyramid and forcibly assembled the manpower that formed the new mechanical fabric. The bureaucracy was, in fact, the third type of 'invisible machine'—one might call it a communications-machine—co-existing with the military and labor machines, and an integral part of the final totalitarian structure.

Not the least important qualification of a classic bureaucracy is that it originates nothing: its function is to pass on, without alteration or deviation, the orders that come from above, from central headquarters. No merely local information or human consideration is admitted, to alter this inflexible transmission process. Only corruption or outright rebellion can modify this rigid organization. Such an administrative method ideally requires a studious repression of all the autonomous functions of the personality, and a readiness to perform the daily task with ritual exactitude. Not for the first time, as we have seen, does such a ritual order enter into the process of work: indeed, it seems highly unlikely that submission to colorless repetition could have been achieved at this point without the millennial discipline of religious ritual.

Bureaucratic regimentation was in fact part of the larger regimentation of life, introduced by this power-centered culture. Nothing emerges more clearly from the Pyramid Texts themselves, with their wearisome repetitions of formulae, than a colossal capacity for enduring monotony: a capacity that anticipates the peak of universal boredom achieved in our own day. This verbal compulsiveness is the psychal side of the systematic general compulsion that brought the labor machine into existence. Only those who were sufficiently docile to endure this regimen—or sufficiently infantile to enjoy it—at every stage from command to execution could become efficient units in the human machine.

4: THE MAGNIFICATION OF PERSONALITY

The marks of this cosmic mechanical order can be easily recognized. To begin with, as noted before, there was a change of scale. The habit of 'thinking big' was introduced with the first human machine; for a superhuman scale in the individual structures magnified the sovereign authority. At the same time it tended to reduce the apparent size and importance of all the necessary human components, except the energizing and polarizing central element, the king himself.

Paradoxically, the monopoly of power brought about a monopoly of personality, too, for only the king was endowed with all the attributes of personality, both those incorporated in the communal group and those that, precisely at this period, it would seem, were slowly beginning to emerge in the human soul, which was now pecking through the social shell in which its embryonic existence had been spent.

At this earliest stage, personality and power went together: both centered in the king. For the sovereign alone could make decisions, alter ancient local customs, create structures and perform collective feats that had never been imagined, still less carried out before. In short, he could behave like a responsible person capable of rational choice, released from tribal custom: free to be, when the situation demanded, a non-conformist, and able to introduce by edict and law deviations from the ancestral pattern. Like the king's original monopoly, that of immortality, some of these prerogatives would, under pressure, be passed along eventually to the whole community. But here it is the magnification that must be noted: all the old dimensions were overpassed, just as the physical bounds of the village horizon and the small group were lifted. Now the sky was the limit, and the city was no less than a whole world in itself, closer to Heaven in every dimension.

Both in practice, and even more in fantasy, these magnifications applied to time and space. Kramer notes that in the early dynasties reigns of incredible length are attributed to legendary kings: a total of close to a quarter of a million years for the eight kings before the Flood and a total of twenty-five thousand years for the first two dynasties after the Flood. This tallies with similar periods that Egyptian priests were still assigning to ancient history when Herodotus and Plato visited them. Even in pure fantasy these are big numbers. This new cultural trait reached a climax in the abstract calculations of the Maya, Thompson tells us: "On one stela at the city of Quirigua, accurate computation sweeps back over ninety million years; on another stela at the same site the date reached is some four hundred million years ago."

But this multiplication of years was only the secular side of the more general expansion of power, symbolized in the royal claim to immortality. In the beginning this was, in Egypt, solely the attribute of the divine king, even though—as one notes in Sumer, where the entire court was simultaneously massacred in the interior of the royal tomb in Ur in order presumably to accompany the sovereign to another world—the servants and the ministers of the king might also share this hope for immortality.

In the Sumerian deluge myth Ziusudra, the king (Noah's counterpart), is rewarded by the gods An and Enlil, not by a symbolic rainbow, but by being given eternal "life like a god." The desire for life without limits was part of the general lifting of limits which the first great assemblage of power by means of the megamachine brought about. Human weaknesses, above all the weakness of mortality, were both contested and defied.

But if the biological inevitability of death and disintegration mock the infantile fantasy of absolute power, which the human machine promised to actualize, life mocks at it even more. The notion of 'eternal life,' with neither conception, growth, fruition, nor decay—an existence as fixed, as sterilized, as loveless, as purposeless, as unchanging as that of a royal mummy—is only death in another form. What is this but a return to the state of arrest and fixation exhibited by the stable chemical elements that have not yet combined in sufficiently complex molecules to promote novelty and creativity? From the standpoint of human life, indeed of all organic existence, this assertion of absolute power was a confession of psychological immaturity—a radical failure to understand the natural processes of birth and growth, of maturation and death.

The cult of the old fertility gods never shrank from facing death: it sought no monumental mockery in stone, but promised rebirth and renewal in the rhythmic order of life. What kingship promised was a grandiloquent eternity of death. If the gods of power had not triumphed, if kingship had not found a negative mode of increasing the scope of the human machine and therewith elevating the royal claim to absolute obedience, the whole further course of civilization might have been different.

Along with the desire for eternal life, achieved by material as well as magical agents, kings and their gods nourished other ambitions that carried over the centuries to become part of the vulgar mythology of our own age. Etana, in the Sumerian fable, mounts an eagle to go in search of a curative herb for his sheep when they are stricken with sterility. At this early date, the dream of human flight was born, or at least was recorded, though that dream still seemed so presumptuous that Etana, like Daedalus, was hurled to death as he neared his goal.

Soon, however, kings were guarded by winged bulls; and they had at their command heavenly messengers who conquered space and time in order to bring commands and warnings to their earthly subjects. Rockets and television sets were already beginning secretly to germinate within this royal myth of the machine. The Genii of the Arabian Nights are only popular later continuations of these earlier forms of power-magic.

This power drive, which was the mark of the sky-oriented religions, became in turn an end in itself. Within the span of early 'civilization,' 3000 to 600 B.C., the formative impulse to exercise absolute control over both nature and man shifted back and forth between gods and kings. Joshua commanded the sun to stand still and destroyed the walls of Jericho by martial music: but Jehovah himself, at an earlier moment, anticipated the Nuclear Age by destroying Sodom and Gomorrah with a single visitation of fire and brimstone; and a while later he even resorted to germ warfare in order to demoralize the Egyptians and aid in the escape of the Jews.

In short, none of the destructive fantasies that have taken possession of leaders in our own age, from Kemal Ataturk to Stalin, from the Khans of the Kremlin to the Kahns of the Pentagon, were foreign to the souls of the divinely appointed founders of the first machine civilization. With every increase of effective power, extravagantly sadistic and murderous impulses erupted out of the unconscious. This is the trauma that has distorted the subsequent development of all 'civilized' societies. And it is this fact that punctuates the entire history of mankind with outbursts of collective paranoia and tribal delusions of grandeur, mingled with malevolent suspicions, murderous hatreds, and atrociously inhumane acts.

Paradoxically, despite the promise of an endless after-life, the other great prerogative of this royal technics is speed: all the king's projects must be executed within his own lifetime. Speed itself, in any operation, is a function of effective power and in turn becomes one of the chief means of ostentatiously displaying it. So deeply has this part of the myth of the machine become one of the basic assumptions of our own technology that most of us have lost sight of its point of origin. But royal commands, like urgent commands in the Army, must be performed 'on the double.' The current commitment to supersonic locomotion as a status symbol, already comically exposed in the intercontinental oscillations of the 'jet set' in business and government, has its beginnings here.

Nothing better illustrates this acceleration of pace than the fact that in Egypt, as later in Persia, each new monarch in the Pyramid Age built a new capital for use in his own lifetime. Compare this with the centuries needed to build a medieval cathedral in free cities that lacked royal resources for assembling power. On the practical side, road-building and canal-building, which were the chief means for hastening transportation, have been all through history the favored form of royal public works: a form that reached a brief technological peak in the Iron Age, when the Romans under Nero planned the cutting of the Corinth Canal through ninety-eight feet of dirt and rock: a work that, if it had then been consummated, would have topped all their road and aqueduct construction.

Only an economy of abundance, at a time when there were at most probably only four or five million people in the Nile Valley, could have afforded to drain off the labor of a hundred thousand men annually, and provide them with sufficient food to perform their colossal task; for in relation to the welfare of the community that was the most sterile possible use of manpower. Though many Egyptologists cannot bring themselves to accept the implications, John Maynard Keynes' notion of 'Pyramid Building,' as a necessary device for coping with the surplus labor force in an affluent society whose rulers are averse to social justice and economic equalization, was not an inept metaphor. This was an archetypal example of simulated productivity. Rocket-building is our exact modern equivalent.

5: THE LABORS OF CONSUMPTION

But the most lasting economic contribution of the first myth of the machine was the separation between those who worked and those who lived in idleness on the surplus extracted from the worker by reducing his standard of living to penury. Forced poverty made possible forced labor: in an agricultural society both rested on the royal monopoly of land and the control of the usufruct. According to Akkadian and Babylonian scriptures, the gods created men in order to free themselves from the hard necessity of work. Here, as in so many other places, the gods prefigured in fantasy what kings actually did.

In times of peace, kings and nobles lived by the pleasure principle: eating, drinking, hunting, playing games, copulating, all in ostentatious excess. So at the very period when the myth of the machine was taking form, the problems of an economy of abundance first became visible in the behavior and the fantasies of the ruling classes—here, too, mirroring in advance the processes at work in our own age.

If we note attentively the aberrations of the ruling classes throughout history, we shall see how far most of them were from understanding the limitations of mere physical power, and of a life that centered upon an effortless consumption: the reduced life of the parasite on a tolerant host. The boredom of satiety dogged this economy of surplus power and surplus goods from the very beginning: it led to insensate personal luxury and ever more insensate acts of collective delinquency and destruction. Both were means of establishing the superior status of the ruling minority, whose desires knew no limits and whose very crimes turned into Nietzschean virtues.

One early example of the vexing problems of affluence lies at hand. An Egyptian story, translated by Flinders Petrie, reveals the emptiness of a Pharaoh's life, in which every desire was too easily satisfied, and time hung with unbearable heaviness on his hands. Desperate, he appealed to his counsellors for some relief from his boredom; and one of them put forth a classic suggestion: that he fill a boat with thinly veiled, almost naked girls, who would paddle over the water and sing songs for him. For the hour, the Pharaoh's dreadful tedium, to his great delight, was overcome; for, as Petrie aptly remarks, the vizier had invented the first Musical Revue: that solace of the 'tired businessman' and soldiers on leave.

Too often, however, these passing modes of relief proved insufficient. Among the all-too-scanty literary documents as yet unearthed, two dialogues on suicide significantly remain, one Egyptian and one Mesopotamian. In each case a member of the privileged classes, with every luxury and sensual gratification open to him, finds his life intolerable. His facile dreams are unsalted by reality. The Egyptian debate between a man and his soul dates from the period following the disintegration of the Pyramid Age, and betrays the desperation of an upper-class person who had lost faith in the ritualistic exaltation of death as the ultimate fulfillment of life, which rationalized the irrationalities of high Egyptian society. But the Mesopotamian dialogue between a rich master and his slave, dating from the First Millennium B.C., is even more significant: for the principal finds that no piling up of wealth, power, or sexual pleasure produces a meaningful life. Another seventh-century B.C. 'Dialogue About Human Misery' expands the theme: the fact that it has been called a Babylonian Ecclesiastes indicates the depth of its pessimism—the bitterness of power unrelieved by love, the emptiness of wealth condemned to enjoy only the goods that money can buy.

If this is what the favored few could expect, in justification of thousands of years of arduous collective effort and sacrifice, it is obvious that the cult of power, from the beginning, was based upon a gross fallacy. Ultimately the end product proved as life-defeating for the master classes as the mechanism itself was for the disinherited and socially dismembered workers and slaves.

In short, from its earliest point of development on, under the myth of divine kingship, the demoralizing accompaniments of unlimited power were revealed in both legend and recorded history. But these defects were for long overlaid by the exorbitant hopes the 'invisible machine' awakened. Though a multitude of single inventions for long lay beyond the scope of the collective machine, which could provide only partial and clumsy substitutes, the fundamental animus behind these inventions—the effort to conquer space and time, to speed transportation and communication, to expand human energy through the use of cosmic forces, to vastly increase industrial productivity, to over-stimulate consumption, and to establish a system of absolute centralized control over both nature and man—all had been planted and richly nurtured in the soil of fantasy during the first era of the megamachine.

Some of the seeds shot up at once in riotous growth: others required five thousand years before they were ready to sprout. When that happened, the divine king would appear again in a new form. And the same infantile ambitions would accompany him, inflated beyond any previous dimension, different only because they were at last realizable.

6: THE AGE OF THE BUILDERS

Now, no institution can thrive solely on its self-deceptions and illusions. Even after allowing for its many heavy impositions and flagrant mischiefs, the megamachine must still be counted as one of the greatest of mechanical inventions: it is doubtful indeed whether non-human machines would have been pushed to their present perfection if the elementary lessons in machine-building had not first been made with malleable human units.

Not merely was the megamachine the model for all later complex machines, but it served to bring a necessary order and continuity and predictability into the rough-and-tumble of daily life, once the food supply and the canal system had overpassed the boundaries of the little neolithic village. What is more, the megamachine challenged the capricious uniformities of tribal custom by introducing a more rational method, potentially universal, that made for greater efficiency.

For the mass of men, it is true, the restricted, inhibited, often oppressive specialized mode of life that 'civilization' imposed did not make sense, as compared with that of the village, whose inner compulsions and conformities were of a more humane order. But the whole structure produced by the megamachine had immeasurably greater significance: for it gave the smallest unit a cosmic destiny that transcended mere biological existence or social continuity. In the new cities all the dissevered human parts were brought together, seemingly in a higher unity.

As we shall see, when we make a fuller canvass of the megamachine, the many negative factors that accompanied it from the start became more formidable, instead of diminishing with its success. But before examining these negative traits, one must account for both the practical success and the apparent popularity of this institution, over many ages and in many different cultures.

At the start, the virtues of divine kingship must have bedazzled all men. For this was the 'Age of the Builders': and the new cities that arose were deliberately designed as a simulacrum of Heaven. Never before had so much energy been available for magnificent permanent public works. Soon cities set on man-made mounds rose forty feet above the flood, with great walls twenty, even fifty feet thick, wide enough at the top for two chariots to drive abreast; likewise 'palaces' were built big enough to house five thousand armed men, who fed and drank from the communal kitchens, to say nothing of temples, like those at Sumer, eighty feet high, whose sacred enclosure was surrounded by another wall. Such a *temenos* was big enough to hold most of the population of a city to witness the sacred ceremonies.

Great buildings whose baked clay surfaces were coated with brilliant glazes, even gold, sometimes encrusted with semi-precious stones, embellished at intervals by monumental sculptures of lions or bulls, dominated the new cities of Mesopotamia; and similar constructions, in different forms and materials, appeared everywhere. Such buildings naturally fostered communal pride: vicariously, the meanest drudge in the new ceremonial centers and cities participated in these feats of power, these wonders of art, daily witnessing a life that was entirely beyond the reach of the humble peasant or herdsman. Even for the distant villager these monumental structures served as magnets which periodically, on festal days, would draw people from all over the land to the great capitals: to Abydos or Nippur, as later to Jerusalem or Mecca, to Rome or Moscow.

These great constructive activities served as foundation for a more intense, consciously directed kind of life, in which ritual was transposed into drama, in which conformity was challenged by new practices, new resources, coming from every part of the great valley, in which there was a sharpening of individual minds through constant intercourse with other superior minds: in short, the new life of the city in which every previous aspect of existence was intensified and magnified. This urban life transcended that of the village in every dimension, importing raw materials greater distances, rapidly introducing new techniques, mixing different racial and national types. In 'The City in History' I have paid due tribute to these collective expressions of order and beauty.

Though villages and country towns set the original patterns of settlement, the construction and cultural elevation of a whole city was largely the work of the megamachine. The rapidity of its erection and the enlargement of all its dimensions, particularly its central nucleus, the temple, the palace, and the granary, bear testimony to royal direction. Walls, fortifications, highroads, canals, and cities, remain in every age what they were in the Age of the Builders: supreme acts of the 'sovereign power.' At the beginning this was no constitutional abstraction, but a living person.

Throughout history, this original image of the city called forth human devotion and effort. The great mission of kingship has been to overcome the particularism and isolationism of small communities, to wipe out the often meaningless differences that separate one human group from another and prevent them from interchanging ideas, inventions, and other goods that might, in the end, intensify their individuality.

Under kingship common standards were set up for weights and measures: boundaries were not merely clearly marked but, partly because of the expansion of royal power, widened, bringing more communities into a system of cooperation. Under a common code of law, conduct became more orderly and predictable, and frivolous deviations became less frequent. To an appreciable extent, this gain in law and order laid the foundations for a wider freedom: it opened the door to a world in every part of which any member of the human race might be at home, as if in his own village.

To the extent that kingship promised such helpful uniformity and universality, every community and every member of the community stood to benefit.

In the building of the city and all the special institutions that accompanied it, kingship came to its greatest constructive culmination. Most of the creative activities we associate with 'civilization' can be traced back to this original implosion of social and technical forces. These works created a well-founded confidence in human powers, different from the illusions and naive self-deceptions of magic. Kings demonstrated how much populous communities, once they were collectively organized in great mechanical units, could accomplish. This was an august achievement, and the vision that made it possible may honestly have seemed godlike. Had it not produced distortions in the human psyche the results might have spread beneficently, in time, through all human activities, elevating and enhancing every common function and purpose throughout the planet.

The mighty cultural heroes and kings who fabricated the megamachine and performed these tasks, from Gilgamesh and Imhotep to Sargon and Alexander the Great, roused their contemporaries from a sluggish passive acceptance of cramped, 'natural' limits: they called upon them to 'plan the impossible.' And when the work was done, that which had seemed impossible of human performance had, in fact, been realized. From around 3500 B.C. on, nothing that men could imagine seemed to lie entirely beyond the reach of royal power.

For the first time in man's development, it would seem, the human personality—at least in a few self-elevated but representative figures—transcended the ordinary limits of space and time. By identification and vicarious participation, as a witness if not an active aid, the common man had an exalted sense of human potentiality as expressed in the myths of the gods, the astronomical knowledge of the priests, and the far-reaching decisions and activities of kings. Within a single lifetime, the mind might encompass a higher state of creativity and a richer consciousness of being than had ever been open to any living creatures before. This, and not the widening of trade opportunities or the march of empires, was the most significant part of the so-called urban revolution.

Though this heightening of the sense of human possibilities was the work of an audacious minority, it could not, like the astronomical lore of the priesthood, remain 'secret knowledge,' for it permeated every activity of 'civilization' and gave it an aura of beneficent rationality. People no longer lived from day to day, piously guided only by the past, reliving it in myth and ritual but fearful of any new departure lest everything be lost. Writing and architecture, indeed the city itself, became stable, independent embodiments of mind. Though urban life developed inner tensions and conflicts to which smaller communities were by reason of their like-mindedness immune, the challenges of this more open mode of existence opened up fresh possibilities.

If all the emergent advantages of these large-scale enterprises had been appreciated and the higher functions of urban life had been more widely distributed, then most of the early failings of the megamachine might in time have been corrected, and even its incidental compulsions could have been lightened and eventually erased. But unfortunately the gods went mad. The deities responsible for these advances exhibited failings that effaced the genuine gains: for they battened on human sacrifices, and meanwhile they invented war as the ultimate proof of 'sovereign power' and the supreme art of 'civilization.' While the labor machine largely accounts for the rise of 'civilization,' its counterpart, the military machine, was mainly responsible for the repeating cycles of extermination, destruction, and self-extinction.

CHAPTER TEN

The Burden of 'Civilization'

1 : THE SOCIAL PYRAMID

Kingship deliberately sought by means of the megamachine to bring the powers and glories of Heaven within human reach. And it was so far successful that the immense achievements of this archetypal unit for long surpassed, in technical proficiency and output, the important but modest contributions made by all other contemporary machines.

Whether organized for labor or for war this new collective mechanism imposed the same kind of general regimentation, exercised the same mode of coercion and punishment, and limited the tangible rewards largely to the dominant minority who created and controlled the megamachine. Along with this, it reduced the area of communal autonomy, personal initiative, and self-regulation. Each standardized component, below the top level of command, was only part of a man, condemned to work at only part of a job and live only part of a life. Adam Smith's belated analysis of the division of labor, explaining changes that were taking place in the eighteenth century toward a more inflexible and dehumanized system, with greater productive efficiency, illuminates equally the earliest 'industrial revolution.'

Ideally, the megamachine's personnel should consist of celibates, detached from family responsibilities, communal institutions, and ordinary human affections: such day-to-day celibacy as we actually find in armies, monasteries, and prisons. For the other name for the division of labor, when it reaches the point of solitary confinement at a single task for a whole lifetime, is the dismemberment of man.

The pattern imposed by the centralized megamachine was transmitted eventually to local trades and crafts, in life-constricting servile work; for there is no human virtue left in handicraft when, as in the making of a spur, for example, seven specialists were employed to perform the seven specialized operations to make this simple tool. The sense that all work was degrading to the human spirit spread insensibly from the megamachine to every other manual occupation.

Why this 'civilized' technical complex should have been regarded as an unqualified triumph, and why the human race has endured it so long, will always be one of the puzzles of history.

Henceforth, civilized society was divided roughly into two main classes: a majority condemned for life to hard labor, who worked not just for a sufficient living but to provide a surplus beyond their family or their immediate communal needs, and a 'noble' minority who despised manual work in any form, and whose life was devoted to the elaborate "performance of leisure," to use Thorstein Veblen's sardonic characterization. Part of the surplus went, to be just, to the support of public works that benefited all sections of the community; but far too large a share took the form of private display, luxurious material goods, and the ostentatious command of a large army of servants and retainers, concubines and mistresses. But in most societies perhaps the greatest portion of the surplus was drawn into the feeding, weaponing, and over-all operation of the military megamachine.

The social pyramid established during the Pyramid Age in the Fertile Crescent continued to be the model for every civilized society, long after the building of these geometric tombs ceased to be fashionable. At the top stood a minority, swollen by pride and power, headed by the king and his supporting ministers, nobles, military leaders, and priests. This minority's main social obligation was control of the megamachine, in either its wealth-producing or its illth-producing form. Apart from this, their only burden was the 'duty to consume.' In this respect the oldest rulers were the prototypes of the style-setters and taste-makers of our own over-mechanized mass society.

The historic records begin with this pyramid of civilization with its division of classes and its broad base of workers crushed by the load above, already firmly established. And since this division has continued right down to our own times—and in countries like India has even hardened into inviolable hereditary castes—it has often been taken as the natural order of things. But we must ask ourselves how it occurred, and on what putative basis of reason or justice it has so long persisted, since an inequality of status, once ingrained in law and property, will only by accident coincide with the natural inequality of ability, due to the re-shuffling of the biological inheritance of every generation.

In the fencing between Leonard Woolley and his Russian communist commentators, in 'Prehistory and the Beginnings of Civilization,' the British archeologist was puzzled by their insistence upon correcting his failure to underline a condition so 'normal' (from his standpoint) that he did not bother even to mention it. Even Breasted was not guiltless of the same oversight; for he dated the beginnings of justice and moral sensitiveness from the moment when the 'Plea of the Eloquent Peasant' to be delivered from arbitrary despoilage and manhandling by a covetous landowner are at last listened to 'at court.'

Unfortunately, Breasted overstressed the improvement in law and morality, what he called "the dawn of conscience," because he mentally started with the savage exploitation of power introduced by such early kings as 'Narmer' and 'Scorpion,' and their successors. Doing so, he completely overlooked the amicable, non-predatory practices of the neolithic village, where forbearance and mutual aid prevailed, as they do generally within pre-'civilized' communities. Breasted saw in this famous papyrus the increased ethical sensitiveness of the ruling classes, ready to relieve the poor peasants from the gross bullying and unconscionable robbery practiced too often by their superiors. But he did not ask how a dominant minority had gained the position that enabled them to exercise such arbitrary powers.

The crisis of conscience that Breasted dwells on would have been more meritorious had it not been overdue: a tardy reparation like the surrender of their feudal privileges by the French nobility on the eve of the Revolution of 1789. If the Eloquent Peasant finally obtained justice, as seems indicated at the point where the document breaks off, it was only, we should remember, after he had been teased and tormented, even flogged, by his betters merely to increase their amusement over his delightful impudence in standing up for his rights and answering back. In the one-way system of communication characteristic of all megamachines, such 'speaking up' constituted an unthinkable affront to the ruling officers, and indeed it still remains so under military discipline. In the 'insolence of office' the modern State has retained the bad manners as well as the overwhelming powers of earlier sovereigns.

The underlying assumption of this system is that wealth, leisure, comfort, health, and a long life belonged by right only to the dominant minority; while hard work and constant deprivation and denial, a 'slaves' diet' and an early death, became the lot of the mass of men.

Once this division was established, is it any wonder that the dreams of the working classes throughout history, at least in those relatively happy periods when they dared to tell each other fairy stories, was a desire for idle days and for a surfeit of material goods? These desires were kept from an explosive eruption, perhaps, by the institution of occasional feasts and carnivals. But the dreams of an existence which counterfeited closely that of the ruling classes, as the brummagem jewelry worn by the poor in Victorian England imitated in brass the gold baubles of the upper classes, have remained alive from age to age: indeed they are still an active ingredient in the fantasy of effortless affluence that currently hovers like a pink smog over Megalopolis.

From the beginning, make no doubt, the weight of the megamachine itself was the chief burden of civilization: not merely did it turn daily work into a grievous penalty, but it diminished the psychal rewards that compensate the hunters, farmers, and herdsmen for their sometimes exhausting labors. Never was this burden heavier than at the beginning, when the greatest public activity in Egypt was mainly directed to supporting the claim of the Pharaoh to divinity and immortality.

To give this whole tissue of illusions a semblance of 'credibility' in the twenty-ninth century B.C. "the tomb of Prince Nekura, son of King Khafre of the Fourth Dynasty, was endowed from the prince's private fortune with no less than twelve towns, the income of which went exclusively to the support of the tomb." Such heavy taxation for such empty ostentation still characterized the Sun God (*Le Roi Soleil*) who built Versailles. But why stop there? This trait of kingship crops up at every point in history.

The cost of this effort was noted, in another context, by Frankfort: "Egypt was drained of talent for the benefit of the royal residence. The graves at Qua-el-Kebir—a cemetery in Middle Egypt used through the third millennium—show the scantiest equipment, and that of the poorest quality of craftsmanship, during the flourishing period of the Old Kingdom, when the pyramids were being built." That says everything. The future historians of the great states now busily projecting manned rockets into space will—if our civilization lives long enough to tell the tale—doubtless make exactly parallel observations.

2: THE TRAUMA OF CIVILIZATION

Though the labor machine can be followed fitfully through history more by public works like roads and fortifications than by any detailed descriptions, we possess most exhaustive documentary knowledge of the megamachine from its massive negative applications in war. For, I repeat, it was as a military machine that the whole pattern of labor organization earlier described, in squads and companies and larger units, was transmitted from one culture to another without substantial alteration, except in the perfection of its discipline and its engines of assault.

This brings us face to face with two questions: why did the megamachine persist so long in this negative form, and, even more significantly, what motives and purposes lay beneath the ostensible activities of the military machine? In other words, how was it that war itself became an integral part of 'civilization,' exalted as the supreme manifestation of all 'sovereign power'?

In its original geographic setting, the labor machine is almost selfexplanatory and even self-justifying. By what other means could the socalled hydraulic civilizations have regulated and utilized the water flow needed for the growing of large crops? No small efforts at local cooperation could have coped with this problem. But war itself offers no such justification: as an institution, it reversed all the patient laboriousness of neolithic culture. Those who attempt to impute war to man's biological nature, treating it as a manifestation of the ravenous 'struggle for existence,' or as a carry-over of instinctive animal aggression, show little insight into the difference between the fantastic ritualized massacres of war and other lessorganized varieties of hostility, conflict, and potentially murderous antagonism. Pugnacity and rapacity and slaughter for food are biological traits, at least among the carnivores: but war is a cultural institution.

The chief non-human species that practice war, with organized armies engaging in deadly combat, are certain varieties of ant. Those social insects some sixty million years ago had invented all the major institutions of 'civilization,' including 'kingship' (actually, queenship), military conquest, the division of labor, the segregation of functions and castes, to say nothing of the domestication of other species and even the beginnings of agriculture. Civilized man's chief contribution to this anthill complex was to add the powerful stimulant of irrational fantasy.

In the earlier stages of neolithic culture, there is still not even a hint of armed combat between villages: possibly even the massive walls around ancient towns like Jericho, as Bachofen suspected and Eliade confirms, performed a magico-religious function before they were found to furnish a decided military advantage. What is conspicuous in neolithic diggings is rather the complete absence of weapons, though tools and pots are not lacking. This evidence, though only negative, is widespread. Among such hunting peoples as the Bushmen, the older cave paintings show no representation of deadly fighting, whereas later pictures, contemporary with kingship, do. Again, though ancient Crete was colonized by distinct and therefore potentially hostile groups, Childe pointed out that "they seemed to have lived together peaceably, as no fortifications have been found."

All this should cause no surprise. War, as Grahame Clark well observed in 'Archaeology and Society' is "directly limited by the basis of subsistence, since the conduct of any sustained conflict presupposes a surplus of goods and manpower." Until neolithic society could produce such a surplus, the paleolithic hunter was preoccupied with hunting his own game. This practice does not sustain more than five or ten people at most per square mile. With such sparse numbers, murderous aggression would be difficult, and worse, suicidal. Even the establishment of 'territoriality' between hunting groups, though probable, no more suggests sanguinary conflict than it does among birds.

The heavy yields of neolithic crops in the great valleys of the Fertile Crescent changed this picture, and altered the possibilities of life for both the cultivator and the hunter. For the hazards of cultivation were increased by the dangerous animals—tigers, rhinoceroses, alligators, hippopotamuses —that infested Africa and Asia Minor. These predators, and even the hardly less dangerous creatures like the big cattle (urus) before they were domesticated, took their toll of both human beings and domestic animals, and often trampled on or devoured the crops.

The courage to deal with such animals, the skill to kill them, belonged to the surviving paleolithic hunters, not to the plodding gardeners and farmers, who at best might net fish or snare birds. The farmer, clinging to the hard-won patch of land, inured to regular toil, was the antithesis of the adventurous, roving hunter, and was disabled for aggression, if not paralyzed, by his mild virtues. Not the least of the scandals that provoked the indignation of an exponent of the old social order, when the Pyramid Age violently ended, was the spectacle of "bird-catchers"—mere peasants, not hunters!—becoming army leaders.

In Egypt and Mesopotamia these sedentary habits must have prevailed before the hunter learned to exploit them: the fact that the original cities of Sumer were often less than a dozen miles apart seems to argue for their having been established in a period before such closeness would provoke property encroachment and conflict. What is more to the point, this passiveness, this submissiveness, to say nothing of the lack of weapons, made it easy for small bodies of hunters to draw tribute—in present-day usage 'protection money'—from much larger communities of farmers. Thus the rise of warriors, to speak in paradox, preceded war.

Almost inevitably, this transformation occurred in more than one place; and at this point the evidence for armed conflict, between two independent and politically organized groups—the term that Malinowski properly insisted upon as the criterion of war, as distinguished from mere birdlike territorial threats, robber-raids, or cannibal head-hunting—becomes indisputable. War implies not only aggression but armed collective resistance to aggression: when the latter is absent one may speak of conquest, enslavement, and extermination, but not war.

Now the equipment, organization, and tactics of an army were not achieved overnight: one must allow for a period of transition before a large mass of men could be trained to operate under unified command. Until towns arose and population was sufficiently concentrated, the prelude to war was an organized but one-sided display of power and bellicosity in raiding expeditions for wood, malachite, gold, slaves.

The radical institutional change to war cannot, I submit, be sufficiently explained on either biological or rational economic grounds. Beneath it lies a more significant irrational component that has as yet hardly been explored. Civilized war begins not by the direct conversion of the hunting chief into the war-making king, but in an earlier passage from the animal-hunt to the man-hunt; and the special purpose of that hunt, if we may cautiously carry back indisputable later evidence into the remote past, was the capture of victims for human sacrifice. There is much scattered data, which I have already touched on in discussing domestication, to suggest that local human sacrifice antedated inter-tribal or inter-urban war.

From the beginning, on this hypothesis, war was probably the by-product of a religious ritual whose vital importance to the community far transcended those mundane gains of territory or booty or slaves by which later communities sought to explain their paranoid obsessions and their grisly collective holocausts.

3: THE PATHOLOGY OF POWER

A personal over-concentration upon power as an end in itself is always suspect to the psychologist: he reads into it an attempt to conceal inferiority, impotence, anxiety. When this tendency is combined with inordinate ambitions, uncontrolled hostility and suspicion, and a loss of any sense of the subject's own limitations, leading to 'delusions of grandeur,' this becomes the typical syndrome of paranoia: one of the most difficult psychological states to exorcise.

Now as it happens, early 'civilized' man had reason to be frightened of the forces that he himself, by his series of technological successes, was unleashing. In the Near East, many communities were escaping from the constraints of a subsistence economy, with its circumscribed, domesticated environment, and were facing a world that was expanding in every direction, widening the area of cultivation, drawing, by rowboat and sail, after 3500 B.C., on raw materials from distant regions, coming in frequent contact with other peoples.

Our own age knows how difficult it is to achieve an equilibrium in an economy of abundance: and our tendency to concentrate responsibility for collective action in a president or a dictator is, as Woodrow Wilson pointed out long before dictators became fashionable again, one of the conditions, the easiest if the most dangerous, for effecting this control.

I have already sought to trace the effects of this general situation in the development of kingship: I wish now to stress more specifically its relation to the sacrificial rituals of war. As the community spread farther and became more closely inter-related, its internal balance became less stable, and the possibility of damage and hardship, of starvation and loss of life, became more serious. Under circumstances beyond local control neurotic anxiety probably grew. The magical identification of the divine king with the whole community did not lessen these occasions for anxiety; for despite royal claims to divine favor and immortality, kings were subject to mortal accidents and misfortunes; and if the king rose higher than common men, his downfall could be more shattering to the whole community.

At an early stage, before any written documents are available, dream and fact, myth and hallucination, empiric knowledge and superstitious guesses, religion and science, formed an indistinguishable welter. One lucky change in the weather after a sacrificial ritual might give sanction to further propitiatory slaughters on an even grander scale. There is reason to suspect, from much later evidence in both Africa and America, put together by Frazer, that the king himself, just because he incarnated the community, may once have been slain as a ritual sacrifice.

To save the adored ruler from such an unseemly fate, a commoner might be temporarily inducted into the office to become, in good season, the sacrificial victim; and when even such a sacrifice became locally unpopular—as indicated clearly in the Mayan classic, the 'Popul Vuh'—substitutes would be found by securing captives from *other* communities. The transformation of these raiding expeditions into full-scale wars between kings, as equal 'sovereign powers,' backed by equally bloodthirsty gods, cannot be documented. But this is the only guess that ties all the components of war together and explains in some degree the hold that this institution has kept throughout history.

The conditions favoring organized war, conducted by a military machine of great potency, capable of completely destroying massive walls, wrecking dams, razing cities and temples, were greatly enlarged by the genuine triumphs of the labor machine. But it is highly doubtful if these heroic public works, which demanded an almost superhuman effort and endurance, would have been undertaken for any purely mundane purpose. Communities never exert themselves to the utmost, still less curtail the individual life, except for what they regard as a great religious end. Only prostration before the *mysterium tremendum,* some manifestation of godhead in its awful power and luminous glory, will call forth such excessive collective effort. That magical power far outweighs the lure of economic gain. In those later cases where such efforts and sacrifices seem to be made for purely economic advantages, it will turn out that this secular purpose has itself become a god, a sacred libidinous object, whether identified as Mammon or not.

All too soon the military organization needed for collecting captives had another sacred duty to perform: that of actively protecting the king and the local god against reprisals, by anticipating their enemy's attack. In this development, the extension of military and political power soon became an end in itself, as the ultimate testimony to the powers of the divinities that ruled the community, and to the supreme status of the king.

The cycle of conquest, extermination, and revenge is the chronic condition of all 'civilized' states, and, as Plato observed, war is their 'natural' condition. Here, as was so often to happen later, the invention of the megamachine, as the perfected instrument of royal power, produced the new purposes that it was later supposed to serve. In this sense the invention of the military machine made war 'necessary' and even desirable, just as the invention of the jet plane has made mass tourism 'necessary' and profitable.

What is most notable, as soon as documents become available, is that the spread of war as a permanent fixture of 'civilization' only widened the collective anxiety that the ritual of human sacrifice had sought to appease. And as communal anxiety increased, it could no longer be overcome by symbolic disembowelment at the altar: that token payment had to be replaced by a collective surrender of life on a far wider scale.

Thus anxiety invited appeasement by magical sacrifice: human sacrifice led to man-hunting raids: one-sided raids turned into armed combat and mutual strife between rival powers. So ever larger numbers of people with more effective weapons were drawn into this dreadful ceremony, and what was at first an incidental prelude to a token sacrifice itself became the 'supreme sacrifice,' performed *en masse*. This ideological aberration was the final contribution to the perfection of the military megamachine, for the ability to wage war and to impose collective human sacrifice has remained the identifying mark of all sovereign power throughout history.

By the time the written records tell of war, all the preliminary events in Egypt and Mesopotamia were buried and unrecorded, though they may in fact have been no different from those we have definite later knowledge of among the Maya and the Aztecs. Yet as late as the time of Abraham, the voice of God might command a loving father to offer up his son on a private altar; and the public sacrifice of prisoners captured in war remained one of the standard ceremonies in such 'civilized' states as Rome. The modern historians' glossing over all this evidence shows how necessary it has been for 'civilized' man to repress this evil memory, in order to keep his self-respect as a rational being: that life-saving illusion.

The two poles of civilization, then, are mechanically organized work and mechanically organized destruction and extermination. Roughly the same forces and the same methods of operation were applicable to both areas. To some extent systematic daily work served to keep in check the licentious energies that were now available for turning mere dreams and wanton fantasies into actualities: but among the governing classes no such salutary check operated. Sated with leisure, war gave them 'something to do' and by its incidental hardships, responsibilities, and mortal risks, provided the equivalent of honorable labor. War became not merely the 'health of the state,' as Nietzsche called it: in addition, it was the cheapest form of mock-creativity, for in a few days it could produce visible results that undid the efforts of many lifetimes.

This immense 'negative creativity' constantly nullified the real gains of the machine. The booty brought back from a successful military expedition was, economically speaking, a 'total expropriation.' But it proved to be a poor substitute, as the Romans were later to discover, for a permanent income tax derived annually from a thriving economic organization. As with the later pillage of gold from Peru and Mexico by the Spaniards, this 'easy money' must often have undermined the victor's economy. When such imperial robber-economies became prevalent and preyed on each other, they cancelled out the possibility of one-sided gain. The economic result was as irrational as the military means.

But to compensate for these insensate explosions of hostility, and these disruptions of orderly life-sustaining patterns of behavior, the megamachine introduced a severer mode of internal order than the most custom-bound tribal community had ever achieved. This mechanical order supplemented the ritual of sacrifice; for order of any kind, no matter how stringent, reduces the need for choice and therewith lessens anxiety. As the psychiatrist, Kurt Goldstein, has pointed out, "compulsive patterns of orderliness" become essential even when anxiety is caused by a purely physical injury to the brain.

The rituals of sacrifice and the rituals of compulsion were accordingly unified through the operation of the military machine. And if anxiety was the original motive that brought about the subjective response of sacrifice, war, in the act of widening the area of sacrifice, also restricted the area where normal human choices, based on respect for all the organism's creative potentials, could operate. In a word, *a compulsive collective pattern of orderliness was the central achievement of the negative megamachine.* At the same time, the gain in power that the organization of the megamachine brought was further offset by the marked symptoms of deterioration in the minds of those who customarily exercised this power: they not merely became dehumanized but they chronically lost all sense of reality, like the Sumerian king who extended his conquests so far that when he returned to his own capital he found it in the hands of an enemy.

From the fourth century on, the stelae and monuments of the great kings abound in insensate boasts of power and vain threats against those who might ransack their tombs or deface their inscriptions—events that nevertheless repeatedly took place. Like Marduk in the Akkadian version of the Creation Epic, the new Bronze Age kings mounted their chariots "irresistible and terrifying," "versed in ravage, in destruction skilled . . . wrapped in an armor of terror." With such sick-making sentiments we are still all too familiar: they are mimicked in the nuclear press releases of the Pentagon.

Such constant assertions of power were doubtless efforts to make conquest easy by terrifying the enemy beforehand. But they also testify to an increase of irrationality, almost proportional to the instruments of destruction that were available: something we have seen again in our own time. This paranoia was so methodical that the conqueror, on more than one occasion, would level a city to the ground, only to build it again immediately on the same site, thus demonstrating his ambivalent role as destroyer-creator, or devil-god, in one.

Half a century ago, the data for such historic acts might have seemed questionable: but the United States government followed precisely the same technique in the wholesale destruction and subsequent post-war rebuilding of Germany: capping an atrocious military strategy—extermination bombing—with an equally demoralized political and economic policy that handed back the victory to Hitler's unrepentant supporters.

This ambivalence, this duality between the two types of megamachine, was expressed in the suave, spine-chilling threat at the end of a Sumerian poem quoted by S. N. Kramer:

> The pickaxe and the basket built cities
> The steadfast house the pickaxe builds . . .
> The house which rebels against the king,
> The house which is not submissive to its king,
> The pickaxe makes it submissive to its king.

Once the cult of kingship became firmly established, the demands for augmented power increased rather than diminished: for cities that had once existed peacefully almost within sight of each other, like the original cluster of towns in Sumer, now became potential enemies: each with its own bellicose god, each with its own king, each with its capability of massing armed force and inflicting destruction upon its neighbor. Under these conditions, what began as a neurotic anxiety, demanding collective ceremonial sacrifice, turned too easily into a rational anxiety and well-founded fear, which necessitated taking counter-measures of the same order —or else abjectly surrendering, as the Council of Elders in Erech proposed to do when threatened.

Note what is said as an encomium of one of the earliest exponents of this power system, Sargon of Akkad, in the 'Sargon Chronicle,' "He had neither rival nor opponent. He spread his terror-inspiring glamor over all countries." To maintain that peculiar halo of power which, Oppenheim notes, radiated only from kings, "5,400 soldiers ate daily in his presence," that is, within the citadel, where they protected the treasury and the temple granary, those monopolistic instruments of political and economic control. The wall around the citadel not merely gave extra security in case the outer walls of the city were breached, but likewise was a safeguard against any uprising by the local population. The very existence of a standing army of this order at daily beck and call indicates two conditions: the need for a ready means of coercion to keep order, and the capacity to enforce strict military discipline, since otherwise the army itself might have turned into a dangerous rebellious mob—as too often happened later in Rome.

4 : THE COURSE OF EMPIRE

The solemn original association of kingship with sacred power, human sacrifice, and military organization was central, I take it, to the whole development of 'civilization' that took place between 4000 and 600 B.C. And under thin disguises it remains central today. The 'sovereign state' today is only the magnified abstract counterpart of the divine king; and the institutions of human sacrifice and slavery are still present, equally enlarged and even more imperious in their demands. Universal military service (conscription on the pharaonic model) has grossly multiplied the number of sacrificial victims, while constitutional government by 'consensus' has only made the power of the ruler more absolute, since dissent and criticism are not 'recognized.'

In time, the magical incentives to war took on a more reputable utilitarian disguise. While the search for sacrificial captives might be enlarged into an even more terror-making slaughter of conquered women and children, the victims, if spared, might be turned into slaves and so add to the labor force and economic efficiency of the conqueror. So the secondary products of military effort—slaves, booty, land, tribute, taxes—supplanted and guilefully concealed the once-barefaced irrational motives. Since a general expansion of economic productivity and cultural wealth had accompanied kingship and seemingly offset its destructive tendencies, people were conditioned to accept the evil as the only way of securing the good: besides, unless the megamachine broke down, they had no alternative.

The repeated death of civilizations from internal disintegration and outward assault, massively documented by Arnold Toynbee, underscores the fact that the evil elements in this amalgam largely cancelled the benefits and blessings. The one lasting contribution of the megamachine was the myth of the machine itself: the notion that this machine was, by its very nature, absolutely irresistible—and yet, provided one did not oppose it, ultimately beneficent. That magical spell still enthralls both the controllers and the mass victims of the megamachine today.

As the military machine grew stronger, the authority of the temple became less necessary, and the palace organization, grown rich and self-sufficient in a large territorial state, often overshadowed that of religion. Oppenheim draws this observation from the period after the fall of Sumer: but the shift in the balance of power and authority occurred repeatedly. Too often the priesthood became the compliant creature of the megamachine it had helped originally to sanctify and firmly establish.

The very successes of the megamachine re-enforced dangerous potentialities that had hitherto been kept in check by sheer human weakness. The inherent infirmity of this whole power system lies exposed in the fact that kings, exalted above all other men, were constantly cozened, flattered, and fed with misinformation—zealously protected from any disturbing counterbalancing 'feedback.' So kings never learned from either their own experience or from history the fact that unqualified power is inimical to life: that their methods were self-defeating, their military victories were ephemeral, and their exalted claims were fraudulent and absurd.

From the end of the first great Age of the Builders in Egypt, that of the Sixth Dynasty Pharaoh, Pepi I, comes corroborative evidence of this pervasive irrationality, all the more telling because it issues from the relatively orderly and unbedevilled Egyptians:

> The army returned in safety
> After it had hacked up the land of the Sand Dwellers
> . . . After it had thrown down its enclosures . . .
> After it had cut down its fig trees and vines . . .
> After it had cast fire into all its dwellings . . .
> After it had killed troops in it by many ten-thousand.

That sums up the course of Empire everywhere: the same boastful words, the same vicious acts, the same sordid results, from the earliest Egyptian palette to the latest American newspaper with its reports, at the moment I write, of the mass atrocities coldbloodedly perpetrated with the aid of napalm bombs and defoliating poisons, by the military forces of the United States on the helpless peasant populations of Vietnam: an innocent people, uprooted, terrorized, poisoned and roasted alive in a futile attempt to make the power fantasies of the American military-industrial-scientific élite 'credible.'

Yet in the very act of countenancing destruction and massacre, war in all its disruptive spontaneity temporarily overcame the built-in limitations of the megamachine. Hence the sense of joyful release that so often has accompanied the outbreak of war, when the daily chains were removed and the maimed and dead to come were still to be counted. In the conquest of a country, or the taking of a city, the orderly virtues of civilization were turned upside down. Respect for property gave way to wanton destruction and robbery: sexual repression to officially encouraged rape: popular hatred for the ruling classes was cleverly diverted into a happy occasion to mutilate or kill *foreign* enemies.

In short, the oppressor and the oppressed, instead of fighting it out within the city, directed their aggression toward a common goal—an attack on a rival city. Thus the greater the tensions and the harsher the daily repressions of civilization, the more useful war became as a safety valve. Finally, war performed another function that was even more indispensable, if my hypothetical connection between anxiety, human sacrifice, and war prove defensible. War provided its own justification, by displacing neurotic anxiety with rational fear in the face of real danger. Once war broke out, there was solid reason for apprehension, terror, and compensatory displays of courage.

Patently, a chronic state of war was a heavy price to pay for the boasted benefits of 'civilization.' Permanent improvement could come only by exorcising the myth of divine kingship, demounting its too-powerful megamachine and abating its ruthless exploitation of man-power.

Psychologically healthy people have no need to indulge fantasies of absolute power; nor do they need to come to terms with reality by inflicting self-mutilation and prematurely courting death. But the critical weakness of an over-regimented institutional structure—and almost by definition 'civilization' was over-regimented from the beginning—is that it does not tend to produce psychologically healthy people. The rigid division of labor and the segregation of castes produce unbalanced characters, while the mechanical routine normalizes—and rewards—those compulsive personalities who are afraid to cope with the embarrassing riches of life.

In a word, the obstinate disregard for organic limits and human potentialities undermined those valid contributions both to the ordering of human affairs and the understanding of man's place in the cosmos that the new sky-oriented religions had introduced. The dynamism and expansionism of civilized technics might have proved a vital counterbalance to the fixations and isolations of village culture had its own regimen not been even more life-restricting.

Now, any system based on the assumption of absolute power is vulnerable. Hans Christian Andersen's fairy tale about the emperor who set out in his airship to conquer the earth and was defeated by a tiny gnat getting into his ear and tormenting him, covers a multitude of other mischances. The strongest city gate might be opened by trickery or treachery, as Babylon and Troy found out; and the mere legend of Quetzalcoatl's return prevented Montezuma from taking effective measures to overwhelm Cortez's small band. Even the sternest royal commands might be disobeyed by men who still heeded their own feelings or trusted their own judgement—as did the tender wood-cutter who secretly defied his king and preserved the life of Oedipus.

After the Second Millennium B.C. the use of the colossal labor machine became intermittent: it never again quite attained that apex of efficiency to which the fine dimensions of the Great Pyramids bear witness. Private property and private employment slowly took over many once-public functions as the prospect of profit became more effective than the fear of punishment. The military machine, on the other hand, though it reached an early peak of regimentation in the Sumerian phalanx, furthered many technological improvements in other departments. It is scarcely an exaggeration to say that *mechanical* invention until the thirteenth century A.D. owed a greater debt to warfare than to the arts of peace.

This holds over long stretches of history. The Bronze Age chariot preceded the general use of wagons for transportation, burning oil was used to repel enemies besieging a city before it was employed for powering engines or heating buildings: so, too, inflated life preservers were used by Assyrian armies to cross rivers thousands of years before 'water-wings' were invented for civilian swimming. Metallurgical applications, too, developed more rapidly in the military than in the civilian arts: the scythe was attached to chariots for mowing down men before it was attached to agricultural mowing machines; while Archimedes' knowledge of mechanics and optics was applied to destroying the Roman fleet attacking Syracuse before it was put to any more constructive industrial use. From Greek fire to atom bombs, from ballistas to rockets, warfare was the chief source of those mechanical inventions that demanded a metallurgical and chemical background.

Yet after all these inventions are duly accounted for and appraised, no single one of them, nor even the whole series, proved so great a contribution to technical efficiency and large-scale collective operations as the megamachine itself. In both its constructive and destructive forms, the megamachine established a new pattern of work and a new standard of performance. Some of the discipline and self-sacrifice of the army has proved a necessary ingredient for every great society that raises its sights above the village horizon: some of the orderly accountancy introduced by the temple and the palace into economic affairs is essential to every larger system of practical cooperation and trade.

Finally, the self-operating machine, detached from detailed human supervision if not ultimate control, was implicit in the abstract model of the megamachine. What was once done clumsily, with imperfect human substitutes, always necessarily on a large scale, paved the way for mechanical operations that can now be managed adroitly on a small scale: an automatic hydraulic electric power station can transmit the energy of a hundred thousand horses. Plainly many of the mechanical triumphs of our own age were already latent in the earliest megamachines, and what is more, the gains were fully anticipated in fantasy. But before we become unduly inflated over our own technical progress, let us remember that a single thermonuclear weapon can now easily kill ten million people, and that the minds now in charge of these weapons have already proved as open to practical miscalculations, humanly distorted judgements, corrupt fantasies, and psychotic breakdowns as those of Bronze Age kings.

5: REACTIONS AGAINST THE
MEGAMACHINE

From the beginning, the balance of mechanized power seems to have fallen on the side of destruction. In so far as the megamachine was passed on intact to later civilization, it was in the negative form of the military machine—drilled, standardized, divided into specialized parts—that its continuity was assured. This applies even to details of discipline and organization, such as the early division of labor between shock weapons and long-distance firepower, between bowmen, spearmen, swordsmen, cavalry, and charioteers.

Do not be a soldier, advises an Egyptian scribe from the New Kingdom: as a recruit "he receiveth a burning blow on his body, a ruinous blow on his eye . . . and his pate is cleft with a wound. He is laid down and beaten. . . . He is battered and bruised with flogging." On such soldierly foundations 'glamorous power' was built: the destructive process began with the preparation of the smallest unit. Obviously the 'Prussianism' of the drill sergeant has a long history.

It would be consoling to believe that the constructive and destructive sides of the megamachine cancelled each other out, and left a place for more central human purposes to develop, based on previous achievements in domestication and humanization. In some degree this actually happened, since vast tracts of territory in Asia, Europe, and America were only nominally conquered, if at all, and apart from paying taxes and tribute their inhabitants led a largely isolated and enclosed communal life, sometimes over-elaborating their own provincialities to the point of self-stultification and ruinous triviality. But perhaps the greatest threat to the efficiency of the megamachine came from within: from its rigidity and repression of individual ability, and from a sheer lack of rational purpose.

Apart from the destructive animus of the military machine it had many inherent limitations. The mere increase in actual power had the effect on the ruling classes of releasing the obstreperous fantasies of the unconscious, and giving play to sadistic impulses that had hitherto had no collective outlet. And at the same time the machine itself was dependent for operation upon weak, fallible, stupid, or stubborn human members; so that the apparatus was liable under stress to disintegrate. These mechanized human parts themselves could not be permanently held together without being sustained by a profound magico-religious faith in the system itself, as expressed in the cult of the gods. Thus beneath the smooth imposing surface of the megamachine, even when supported by awe-making symbolic figures, there must have been from the beginning many cracks and fissures.

The fact is, happily, that human society could not be made to correspond exactly to the theoretic structure that the cult of kingship had erected. Too much of common everyday life escapes effective supervision and control, to say nothing of coercive discipline. From the earliest times on there are indications of resentment, defiance, withdrawal, escape: all celebrated in the classic story of the Jewish exodus from Egypt. Even when no collective retreat proved possible, the daily practices of the farm, the workshop, the marketplace, the hold of family ties and regional loyalties, the cults of minor gods, tended to weaken the system of total control.

As noted before, the most massive collapse of the megamachine seems to have occurred at the early period when the Pyramid Age, to judge by its mortuary remains, was at its height. Nothing short of a revolutionary uprising can account for the inter-regnum of some two centuries that separated the 'Old Kingdom' from the 'Middle Kingdom.' And though the archaic power complex was finally restored, it was modified by various important concessions, including the extension of immortality (once a Pharaonic or upper-class privilege) to the population at large. While the actual incidents that brought on and effected this overthrow of the central government are unrecorded, we have, apart from the eloquent testimony of silence, along with the falling off of construction, a vivid account of changes that could only have followed from a violent revolution, as detailed by an adherent of the old order, Ipu-wer. His lament gives an account of the revolution, seen from the inside, as graphic, if no less fictionized, as 'Dr. Zhivago' does of the Bolshevik revolution.

This first revolt against the established order turned the pyramid of authority, on which the megamachine was based, upside down: the wives of great men were forced to become servants and prostitutes, the papyri tell us, and the common people assumed positions of power. "Doorkeepers say: 'Let us go and plunder.' . . . A man regards his son as his enemy. . . . Nobles are in lamentation while poor men have joy. . . . Dirt is throughout the land. There are really none whose clothes are white in these times. . . . They who built the pyramids are become farmers. . . . The [stored] grain supply of Egypt is now on a come-and-get-it basis."

At this point, obviously, reality had broken through the imposing theological wall and toppled the social structure. For a time the cosmic myth and the centralized power dissolved, while feudal chiefs, big landowners, regional governors, town and village councils, restored to the service of their lesser local gods, took over the burden of government. All this could hardly have happened had not the grim impositions of kingship, despite the superb technological achievements of the megamachine, become intolerable.

What was happily proved by this early revolution is something that we perhaps need to be reminded of again today: that neither exact science nor engineering is proof against the irrationality of those that operate the system. Above all, that the strongest and most efficient of megamachines can be overthrown, that human errors are not immortal. The collapse of the Pyramid Age proved that the megamachine exists on a basis of human beliefs, which may crumble, of human decisions, which may prove fallible, and human consent, which, when the magic becomes discredited, may be withheld. The human parts that composed the megamachine were by nature mechanically imperfect: never wholly reliable. Until real machines of wood and metal could be manufactured in sufficient quantity to take the place of most of the human components, the megamachine would remain vulnerable.

I have cited this revolt, whose consequences if not its detailed chain of causes are attested, because it must stand for many other challenges, uprisings, and slave rebellions that have probably been carefully expunged from the official chronicles. Fortunately, we may add the capture and escape of the Jews, whose forced labor for the Egyptian megamachine was recorded; likewise we know of the slaves' uprising that took place in Rome under the aristocratic leadership of the Gracchi. We have reason to suspect that there were many other human revolts, also suppressed without mercy, as Wat Tyler's rebellion and the Paris Commune of 1871 were put down.

But there were more normal forms of expressing both alienation and resistance, if not active reprisal. Some of them were so normal, indeed, that they were nothing more than the healthy development of small-scale economic operations and secular interests. The city itself, though at the outset a major enterprise of kings, was not merely an active rival of the megamachine, but, as it turned out, a more humane and effective alternative, with a better means of organizing economic functions and drawing upon a diversity of human abilities. For the great economic strength of the city lay not in its mechanization of production, but in its assemblage of the greatest possible variety of skills, aptitudes, interests. Instead of ironing out human differences and standardizing human responses to make the megamachine operate more effectively as a single unit, the city recognized and emphasized differences. By continued intercourse and cooperation urban leaders and citizens were able to utilize even their conflicts to draw on unsuspected human potentialities, otherwise suppressed by regimentation and social conformity. Urban cooperation, on a voluntary give-and-take basis, was throughout history a serious rival to mechanical regimentation, and often effectually superseded it.

The city, it is true, never completely escaped the compulsions of the megamachine: how could it, when the citadel, the very expression of the organized coalescence of sacred and temporal power, occupied the center: a visible reminder of the king's unescapable presence? Yet the life of the city favored the many-tongued human dialogue against the tongue-tying monologue of royal power, though these valuable emerging attributes of urban life were certainly never part of the original royal intention, and were often suppressed.

Similarly the city gave encouragement to small groups and associations, on the basis of neighborhood and vocation, whose threatening independence the constituted sovereign authority always eyed with suspicion. The fact remains, as Leo Oppenheim has pointed out, that in Mesopotamia, if not in Egypt, the city alone had sufficient force and self-respect to challenge the state organization. "A small number of old and important cities enjoyed privileges and exemptions with respect to the king and his power. . . . In principle, the inhabitants of the 'free cities' claimed with more or less success, depending upon the political situation, freedom from corvée work, freedom from military service . . . as well as tax exemptions." Or to use the terminology I have introduced, these ancient cities claimed a large measure of freedom from the megamachine.

6: CURBS ON THE MEGAMACHINE

Since the basic institutional transformations that preceded the construction of the megamachine were magical and religious, one should not be surprised to find that the most effective reaction against it drew on the same potent sources. One such possible reaction has been suggested to me by two correspondents: namely that the institution of the Sabbath was, in effect, a way of deliberately bringing the megamachine periodically to a standstill by cutting off its manpower. Once a week the small, intimate basic unit, the family and the Synagogue, took over; reasserting, in effect, the human components that the great power complex suppressed.

Unlike all other religious holidays, the Sabbath spread from Babylon across the world, mainly through three religions; Judaism, Christianity, Islam. But it had a limited local origin, and the hygienic reasons put forward by Karl Sudhoff to justify it, though they are physiologically sound, do not account for its existence. To cut out one whole day from the working week is an expedient that could occur only in an area where there was an economic surplus, a desire to throw off an onerous compulsion, and a necessity to re-assert the more significant concerns of man. The latter possibility would have direct appeal, one might suppose, to an oppressed and exploited group like the Jews. On the Sabbath day alone, the lowliest classes in the community enjoyed a freedom, a leisure, a dignity that only the chosen minority enjoyed on other days.

Such a curb, such a challenge, was obviously not the result of any deliberate appraisal and criticism of the power system: it must have sprung from far deeper and more obscure collective sources: at bottom perhaps a need to control the inner life by ordered ritual as well as by compulsory labor. But the Jews, who espoused the Sabbath and passed it on to other peoples, had certainly more than once become the victims of the megamachine by being taken wholesale into captivity; and during their Babylonian exile they combined the Sabbath with still another by-product of that episode, the institution of the Synagogue.

This unit of organization escaped the restraints of all the older religions tied to territorial gods, a remote priesthood, and a capital city, since it could be transplanted anywhere; while the leader of such a community, the rabbi, was a scholar and a judge, rather than a priest and a dependent upon royal or municipal power. Like the village community, the Synagogue was an I-and-Thou, face-to-face association: it was held together, not just by neighborly nearness, not just by a common ritual and a special day for religious observance, but likewise by regular instruction and discussion in matters of custom, morals, and law. This last intellectual office, derived from the city, was what had been lacking in village culture.

So far as is known, no other religion before 600 B.C. combined these essential attributes, including transportability in small units and universality, though Woolley traces these features back to the household religious practices that Abraham may have acquired at Ur, where even burial took place in a crypt beneath the dwelling house. By means of the Synagogue, the

Jewish community recovered the autonomy and the capacity for self-replication which the village had lost through the growth of larger political organizations.

This fact accounts not alone for the miraculous survival of the Jews through endless centuries of persecution, but for their worldwide distribution. Even more significantly, it shows that this small-scale organization, though as unarmed and open to oppression as a village, could maintain itself as an active nucleus of self-sustaining intellectual culture for over twenty-five hundred years, when every larger mode of organization, based on power alone, had disintegrated. The Synagogue had inner fortitude and persistence that highly organized states and empires, for all their temporarily effective instruments of coercion, always lacked.

Admittedly, the small communal unit, in its Judaic form, had serious weaknesses. For one thing, its underlying premise, the existence of a special relationship established between Abraham and Jehovah, which made the Jews a Chosen People, was as presumptuous as the claims of divine kingship. That unfortunate solecism for long prevented the example of the Synagogue from being more widely imitated, and from serving, until the heresy of Christianity arose, as a means of establishing a more universal community. Jewish exclusiveness outdid even that of the tribe or the village, where at least marriage out of the group was often encouraged. But despite this weakness it seems plain, from the very antagonisms the Jewish community awakened, that both in the Synagogue and in the practice of the strict Sabbath it had found a way of obstructing the megamachine and challenging its inflated claims.

The hostility that the Jews and the early Christians constantly evoked in great States was a gauge of the frustration that mere military power and 'absolute' political authority experienced in dealing with a small community held together by a traditional common faith, inviolable rituals, and rational ideals. For power cannot long prevail unless those upon whom it is imposed have reason to respect it and conform to it. Small, seemingly helpless organizations that have an inner coherence and a mind of their own have in the long run often proved more effective in overcoming arbitrary power than the biggest military units—if only because they are so difficult to pin down and confront. This explains the efforts of sovereign states all through history to curb and suppress such organizations, whether they were mystery cults, friendly societies, churches, guilds, universities, or trade unions. And in turn, that antagonism suggests the way in which the modern megamachine may in future be curbed, and brought under some measure of rational authority and democratic control.

CHAPTER ELEVEN

Invention and the Arts

1 : THE TWO TECHNOLOGIES

Since the megamachine was in essence an invisible organization, the historic record tells us nothing specifically about its existence: what we know is derived from the details that must be pieced together.

Those who designed the machine were of course unconscious that it *was* a machine: for how could they identify it as such when the few existing machines, all far more primitive in design, supplied no clue? But of one thing we may be certain: because the motive power of this machine required a great assembly of human prime movers, it could flourish only in a few prosperous agricultural regions that favored urban civilization, with its facilities for concentrating and coercing a large population. Without this constant flow of manpower, the machine could no more work than a watermill on a dried-up river.

This explains why the megamachine, as a productive mechanism, never took hold in many sparsely populated areas of the world. Smaller communities on a tribal or feudal basis might, once the megamachine was in being, mimic many of the accoutrements of kingship in close detail, from the Shilluks of Africa to the Polynesian kingdoms of the Pacific. But as a working organization, the megamachine lapsed in such areas; and if evidences of its existence nevertheless remain, as is perhaps indicated by the stone statues of Easter Island, and certainly by the cities and roads of the Peruvian and the Mayan Empires, one must assume a more dense population than now exists in those regions.

In short, a miniature megamachine is a self-contradictory, almost comic concept, even apart from the difficulty in small communities of achieving the necessary de-socialization and de-personalization of the individual parts. When the feat of universalizing the megamachine was finally achieved, it was only by translating its attributes into their non-organic equivalents in wood or metal.

Once invented, the original megamachine underwent no further improvement as a whole, though various parts might, by training, reach a higher degree of automatism. But the invisible machine, as a smoothly working apparatus, never exceeded in either output or meticulous workmanship the high standards of the Pyramid Age. The Macedonian phalanx was not more completely 'mechanized' than the Sumerian phalanx had been some two thousand years before: nor was the Roman phalanx more economic of energy than the Macedonian; and two thousand years later, the famous British military square, even when equipped with muskets, was still, as war machine, on the same level as its predecessors.

Along that particular line, invention was arrested at an early stage. But this arrest was partly an indication of the adequate performance of the megamachine when conditions favored it. For the massive achievements in civil engineering, from the construction of the Mesopotamian canal system —traced out recently by Thorkild Jacobsen and his associates—to the building of the Great Wall of China, were uniformly done under royal authority through its local agents and appointees. No small community could engage in such enterprises, even if any custom-bound Council of Elders were sufficiently aspiring to conceive them.

The line of technological advance for long lay outside the province of the megamachine; and it was in large part a continuation of the same kind of small-scale enterprise, based on empirical knowledge, seasoned by broad human experience, which had brought about the domestication of plants and animals and vastly raised the energy potential of the human community. These improvements were far less spectacular than the wholesale constructions and destructions of the megamachine; and most of them, like agriculture itself, were the work of many little people, pooling their experience, maintaining their traditions, and concerning themselves more with the quality and human value of their products than with mere quantitative displays of power or material wealth. The handicraft tradition, like the older inventions of social organization, language and agriculture, was never exclusively in the hands of a self-favoring minority in command of a centralized organization.

Almost from the beginnings of civilization, we can now see, two disparate technologies have existed side by side: one 'democratic' and dispersed, the other totalitarian and centralized. The 'democratic' mode, based on small-scale handicraft operations, was kept alive in a multitude of little villages, in partnership with farming and herding, though spreading into the growing country towns and finally lured into the cities. Craft specialization and interchanges by barter and purchase were necessary to this economy, as they had been, indeed, in paleolithic times: so that though special raw materials, copper or iron for the smith, or minerals for glazing pottery, or special dyes for cotton or linen cloth, might come from the outside, most of the resources and most of the skills to use them were home-grown. Such innovations as took place came in slowly, without disrupting the ancestral patterns.

To make the contrast between democratic and authoritarian technics clear, let me define the term democracy in this context, since I have already characterized the authoritarian system.

'Democracy' is a term now confused and corrupted by indiscriminate use, and often treated with patronizing contempt, when not foolishly worshipped as if it were a panacea for all human ailments. The spinal principle of democracy is the perception that the traits and needs and interests that all men share have a superior claim to those put forward by any special organization, institution, or group. This is not to deny the claims of superior natural endowment, special knowledge, experience, or technical skill: even primitive democratic groups acknowledge some or all of these distinctions. But democracy consists in favoring the whole, rather than the part; and ultimately only living human beings can embody and express the whole, whether acting alone or with the help of others. "An institution is the lengthened shadow of a man." Yes: but only part of a man.

Democracy, in the sense I here use the term, is necessarily most active in small communities and groups, whose members meet face to face, interact freely as equals, and are known to each other as persons: it is in every respect the precise opposite of the anonymous, de-personalized, mainly invisible forms of mass association, mass communication, mass organization. But as soon as large numbers are involved, democracy must either succumb to external control and centralized direction, or embark on the difficult task of delegating authority to a cooperative organization.

The first choice is the easier one; or rather, it is hardly a choice, but what happens automatically when no sufficient effort is made to lift the spontaneous democratic mode of customary control to a higher level of intelligent organization. Historic experience shows that it is much easier to wipe out democracy by an institutional arrangement that gives authority only to those at the apex of the social hierarchy. The latter system, in its first stages, often achieves a high degree of mechanical efficiency: but at a prohibitive human cost.

Unfortunately, the forms and methods of totalitarian technics could not be confined to the megamachine; for wherever the population was concentrated in large cities, where a large-scale organization of a landless and increasingly traditionless proletariat took place, compulsive methods made their way even into the processes of handicraft and progressively 'mechanized' them—mechanized, that is, in the human sense. The large-scale organization of the proletariat in specialized workshops and factories, using what now seem like 'modern' methods, is reasonably well documented for the Hellenistic and Roman world, as Rostovtseff showed; but must have begun at a much earlier date. In this way the original practices of the megamachine began to pervade even the more humane institutions derived from an earlier economy.

Both kinds of technics had their virtues and their disadvantages. Democratic technics had the security that comes from petty operations under the direct control of the participants, following a customary pattern, in a familiar environment: but it was at the mercy of local conditions and could suffer grievously from natural causes, ignorance, or bad management without being able to get help from elsewhere. Authoritarian technics, at home with quantitative organization, capable of handling larger numbers of people and drawing by trade or conquest on other regions, was better capable of producing and distributing surpluses, under rulers with sufficient political intelligence to establish just distribution. But the megamachine wiped out its own gains in efficiency, in the workshop as in the state, by cupidity and sadistic exploitation. Ideally, each mode had something to give the other: but neither established for long any effective cooperation.

If the small agricultural community favored democratic technics, the increased use of metals—first copper, then bronze, and finally iron—which coincided with the rise and spread of kingship, helped to foster the authoritarian form and carry it into other industries, from era to era. The persistence of military operations in itself incited improvements in the arts of metallurgy; and it was in the mines, smelteries, and foundries that the industrial processes exhibited both the harsh coercions and the heroic exertions that had hitherto been the special characteristics of the military regime.

In hunting and agriculture work had been a sacred function, one of collaborating with the forces of nature, and invoking the gods of fertility and organic abundance to countenance with their favor the efforts of the human community: pious exaltation and cosmic wonder mingled with strenuous muscular exercise and meticulous ritual. But for those who were drafted into the megamachine, work ceased to be a sacred function, willingly performed, with many pleasurable rewards in both the act and its fruition: it became a curse.

In the Book of Genesis God associated this curse with Adam's exile from the tropical exuberance of the Garden of Eden, for that exile imposed the necessity, brought in with cereal crops, to dig and delve in the hard soil. Doubtless it was natural for free-moving pastoral nomads, like the Jews, to associate this curse with the unfamiliar rugged duties of agriculture: to down Cain, the farmer, and raise up Abel, the shepherd. But this interpretation conceals the historic fact. Actually, it was mining, mechanization, militarism, and their derivative occupations that took the joy out of daily work and turned it into an implacable, mind-dulling system of drudgery.

Wherever tools and muscle power were freely used, at the command of the workers themselves, their labors were varied, rhythmic, and often deeply satisfying, in the way that any purposeful ritual is satisfying. Increase of skill brought immediate subjective satisfaction, and this sense of mastery was confirmed by the created product. The main reward of the craftsman's working day was not wages but the work itself, performed in a social setting. In this archaic economy there was a time to toil and a time to relax: a time to fast and a time to feast: a time for disciplined effort, and a time for irresponsible play. In identifying himself with his work and seeking to make it perfect, the worker remolded his own character.

All the praise of tool-making and tool-using that has been mistakenly applied to early man's development becomes justified from neolithic times onward, and should even be magnified in evaluating the later achievements of handicraft. The maker and the object made reacted one upon the other. Until modern times, apart from the esoteric knowledge of the priests, philosophers, and astronomers, the greater part of human thought and imagination flowed through the hands.

Under democratic technics, the only occupation that demanded a lifetime's attention was that of becoming a full human being, able to perform his biological role and to take his share in the social life of the community, absorbing and transmitting the human tradition, deliberately bringing the ceremonies he performed, the food he planted, the images he shaped, the utensils he carved or painted to a higher degree of esthetic perfection. Every part of work was life-work. This archaic attitude toward work was widespread; and despite all the efforts Western man has made, since the sixteenth century, to corrupt and destroy this basic culture, it still lingered in peasant communities, as well as in the surviving tribal enclaves that were intact at the beginning of the present century. Franz Boas noted the high regard for craftsmanship among supposedly primitive peoples; while Malinowski emphasized the same attitude among his near-neolithic 'Coral Gardeners.'

Machine culture in its original servile form did not share these life-enhancing propensities: it centered, not on the worker and his life, but on the product, the system of production, and the material or pecuniary gains therefrom. Whether kept in operation by the taskmaster's whip or by the inexorable progression of today's assembly line, the processes derived from the megamachine worked for speed, uniformity, standardization, quantification. What effect these objectives had upon the worker or upon the life that remained to him when the workday was over was no concern of those who commanded these mechanical operations. The compulsions produced by this system were more insidious than outright slavery, but as with slavery, they finally debased the controllers as well as the working force so controlled.

Under household slavery, indeed, personal relations might be established—and sometimes were established—between the slave and his master; and this in turn might result in a recapture of autonomy, since the favored slave, as in Rome, might acquire property, perform outside services for money, and might finally purchase his freedom. Slaves employed in making works of art, a much larger part of industrial production in ancient times than in the present age, achieved an inner freedom, a personal delight, that provided them with a life not essentially different from those voluntarily dedicated to the work: indeed, in fifth-century Greece and elsewhere they worked side by side. But wherever the practices derived from the megamachine were dominant, all work became a curse, even if the worker were legally free, and in many industrial operations, it was a form of punishment, even if the laborer had committed no crime.

The wider use of metals did not lift this curse, though it did in fact provide better and cheaper tools, as well as weapons. For the extraction and breaking up and smelting of ores, along with the further working up of the metals, demanded prolonged physical effort, under far more unhygienic and depressing conditions than those under which the farmer and the craftsman of the more domestic trades labored. In the small workshop, the carpenter, the leatherworker, the potter, the spinner and weaver, though often unduly confined and economically harried, had the benefit of human companionship, more or less on a family pattern, often with family assistance.

But mining underground was from the beginning a dismal, dangerous, back-breaking occupation, particularly when done with the crude tools and apparatus that prevailed up to the sixteenth century of our era, and in many places well into the twentieth. Physical coercion, disease, bodily injury gave mining at every step the features of a battlefield: both the landscape and the miner bore the scars of this operation, even when the latter remained alive. From the earliest times, as Mircea Eliade points out, blood sacrifice had been a ritual accompaniment of metallurgy. The curse of war and the curse of mining are almost interchangeable: united in death.

There is plenty of historic evidence of this association. Though peasants were sometimes conscripted for mining operations, as well as for war, so repellent was this labor that for the greater part of history only slaves or criminals were assigned to the mines: it was "confinement with hard labor," a prison sentence, and no work for free men.

As the cult of power widened its province, the heavier demand for metals in war, notoriously the chief consumer of metals, extended this slavery and these sacrificial rituals over wider areas. And if, as V. Gordon Childe supposed, metal-workers were the first full-time specialists, then that division of labor itself shared and intensified the primal curse of labor, which embittered and actually shortened life. With the 'advance' of civilization, this system of brutalized toil, on the prison-model of the mine and the galley, was eventually carried over into the more commonplace tasks of daily life.

Traced back to its origins, the curse of labor is the curse of the megamachine: a curse extended beyond the period of conscription to a whole lifetime. That curse gave rise to the compensatory dream of a Golden Age, part memory, part myth: the picture of a life when there was no harsh struggle or competition, when the wild animals meant no harm and even man was kind to his fellows. This dream first appears on an Akkadian tablet; and much later was transferred to the future, as an after-life in Heaven, when all work would cease, and everyone would enjoy an existence of sensuous beauty, material amplitude, and endless leisure: a replica in terms of mass consumption of all that actually took place in the great palaces and temples for whose expansive maintenance and further elevation the megamachine was first invented.

With the increased division of labor in many urban trades, the field of activity for the individual worker shrank, and the chance of shifting from one occupation to another, as one does in the seasonal round of farm work, became more remote. At a quite early date the city, once conceived as a representation of Heaven, took on many of the features of a military camp: a place of confinement, daily drill, punishment. To be chained, day after day, year after year, to a single occupation, a single workshop, even finally to a single manual operation, which was only a part of a series of such operations—that was the worker's lot.

Each specialized trade, precisely through its specialization, now acquired its typical 'occupational disabilities'—its lopsided posture, its over-developed muscles, its bleached complexion, its myopic eyes, its enlarged heart, or its silicate-clogged lungs, with their associated diseases and permanent anatomical deformities. Too often these ailments were chronic and persistent: a higher death rate betrayed the lower life rate. Down to our own time, the expectation of life for an English farm laborer, often living in a crowded dwelling, eating coarse food, constantly exposed to wind and rain, remained superior to that of the factory worker, even when the latter was far better provided, not only with higher wages, but with more sanitary domestic facilities.

Under such conditions, the 'curse of labor' was no idle epithet. In the Egyptian summary of the benefits enjoyed by the scribe above all other occupations, these disadvantages of specialization were set forth, as related to one occupation after another: the daily hardships, the filth, the danger, the nightly weariness. The scholars who imagine that this reckoning was grossly exaggerated, and call the document a satire, know too little about the actual condition of the urban working classes in any period.

All the miseries the scribe depicts were worse, of course, in those crafts which were practiced indoors, in underlighted, ill-ventilated quarters—in contrast to the poorest peasant, who is foot-free and mobile. Witness the weaver's lot: "He cannot breathe the [open] air. If he cuts short the day of weaving, he is beaten with fifty thongs. He must give food to the door keeper to let him see the light of day." This passage clearly shows that the discipline of the megamachine had already been extended to the urban workshop, thousands of years before it reached the eighteenth-century factory.

If conditions under the megamachine were oppressive, they remained dismal enough in many ordinary trades all through history, though the picture was never uniformly bad, and in certain periods and in certain cultures—Athens in the fifth century B.C., or Florence in the twelfth century A.D., to take only the most obvious examples—was distinctly bright. Is it strange that out of such depressing circumstances there came not only the feeling that labor itself was inherently a curse, but that the most desirable life possible would be one in which magical mechanisms or robots would perform all the necessary motions under their own power, without human participation of any kind? In short, the idea of the mechanical automaton, which would obey all orders and do all the work.

This dream has haunted civilization throughout history, repeated with magical variations in a hundred fairy stories and popular myths, long before it took form in the modern slogan: 'Let automation abolish all work.' Often this dream was accompanied by another that sought to release mankind from the other curse that the megamachine had imposed upon the underlying population: the curse of poverty. The cornucopia of plenty, the blessed land where an inexhaustible supply of foods and goods came forth at a wave of the hand: in other words, the infantile contemporary heaven of an ever-expanding economy—and its end product, the affluent society.

The curse of work was a real affliction for those who came under the rule of authoritarian technics. But the idea of abolishing all work, of transferring the skill of the hand without the imagination of the mind to a machine—that idea was only a slave's dream, and it revealed a desperate but unimaginative slave's hope; for it ignored the fact that work which is not confined to the muscles, but incorporates all the functions of the mind, is not a curse but a blessing. No one who has ever found his life-work and tasted its reward would entertain such a fantasy, for it would mean suicide.

2: WAS INVENTION ARRESTED?

The authoritarian industrial and military technics that supported civilization reached a plateau in purely mechanical aptitude, most interpreters of technical history have supposed, early in the Iron Age: a period that can be roughly dated as beginning around 1200 B.C. On the direction of that progress and the results of that mastery the testimony of ancient wisdom leaves little doubt. "The Iron Age," said Albius Tibullus, in the first century before Christ, repeating Hesiod's early lament, "sings the praise not of love but of pillage. . . . From it comes blood, from it slaughter, and death draws nearer."

Doubtless most of Tibullus's contemporaries regarded this description as 'hysterical,' but by the fifth century A.D. the grim results could no longer be ignored; and the marks left by the Iron Age, if only because they intensified and extended the capacities of the megamachine, are still with us. But when historians compare the total volume of invention during this long period with what took place in Western Europe since the eighteenth century, they habitually look around for some explanation of what seems to them a curious technological backwardness, which causes them to ask: What arrested invention? In doing so, they close off a line of inquiry I purpose to open up by asking: Was invention actually arrested?

But first let us survey the impact of iron itself. For the manufacture of digging and cutting tools, no less than weapons of war, iron gave a distinct advantage over other metals. To that extent it lightened the burden of work, or at least raised the level of efficient production from the same number of man-hours. In agriculture, the iron hoe was an immense improvement over the bone or stone hoe; and the spade and the shovel and the iron pickaxe gave the farmer tools that could cope with any kind of soil; while the iron axe was so efficient that it should be held partly accountable—along with the goat—for the ruthless destruction of the forest cover over the whole Mediterranean area.

Dr. Fritz Heichelheim suggests further that the use of iron at first must have been a social leveller, by improving the condition of the working classes and extending cultivation with the iron plow to heavier, richer bottomland soils. But the increased production of iron also made it cheaper to equip armies and tempted rulers to embark on more extensive conquests. Note that the same classical scholar observes that the "population of the Mediterranean area decreased between 201 and 31 B.C., owing to the Roman wars of conquest, civil wars, social revolutions, and slave hunting."

Once the use of iron had become common, the present general impression of the technology that prevailed between 100 B.C. and 1500 A.D. is that it had reached a dead center, and that, instead of showing cumulative progress, with an acceleration of inventions, there was a widespread slacking off of technical activity. Even such a competent historian of technics as R. J. Forbes records this judgement, and attributes it, as many others have done, to the prevalence of slavery, which supposedly removed the chief incentive to the production of labor-saving machines. This is on many counts an extremely doubtful explanation. Was it not a labor-saving machine, Eli Whitney's cotton gin, that helped increase the demand for slaves in the cotton states of North America?

Sometimes, on the other hand, this supposed failure of technical invention and initiative is attributed to the upper-class divorce from manual labor, since the ancients held that theoretical studies, free from the taint of manual labor and vulgar use, should alone occupy the liberal mind. Even Archimedes, that prince of technicians, seems to have shared this notion, though war evoked from him a series of ingenious devices to disable the Roman fleet attacking Syracuse.

But the patrician contempt for labor, which extended even to commerce, was far from absolute: the aristocratic youth of Athens learned about life from an old stone-cutter, named Socrates. Nor did it prevent the rise of an active merchant class. There was nothing to keep free workers, self-employed or masters of small workshops, from inventing machines had they felt sufficiently interested. The crafts involved in the Aegean export trade, producing pottery and textiles in a system closer to mass production than to individual piecework for a single customer, must already have subdivided and specialized their operations. With a few more steps, such activities might have been transferred to machines, as happened in Europe between the fifteenth and nineteenth centuries.

Though the upper classes indeed might look with disdain upon 'base mechanic employments' as they called them, it was only in the metallurgical arts and on large-scale public works of engineering and building, that slavery or forced labor prevailed. Even under tyrannous regimes there was plenty of scope for both the engineers and craftsmen to make improvements; and various improvements were indeed made.

What has misled judgement in our own age is that the greatest technological achievements of the ancient world were in the realm of statics, not dynamics: in civil, not mechanical engineering: in buildings, not machines. If the historian finds a lack of invention in earlier cultures, it is because he persists in taking as the main criterion of mechanical progress the special kinds of power-driven machine or automaton to which Western man has now committed himself, while treating as negligible important inventions, like central heating and flush toilets—or even ignorantly attributing the latter to our own 'industrial revolution'!

Doubtless slavery and upper-class contempt both had an insidious effect in undermining respect for the worker as a person, and perhaps in lowering his own interest in his job. Shakespeare's cruel caricature of 'Snout' and 'Starveling,' as if bodily deformity and undernourishment were themes for amusement, echoed a thousand similar epithets. These attitudes and institutional barriers may have discouraged interest in mechanical invention: but that is not the whole story.

The common belief that no important technical progress took place between the perfection of the megamachine in the Iron Age and its resurrection in our own era is also partly due to the fact that modern observers tend to under-rate the productivity of the ancient world. There must have been a reliable surplus, in many departments besides agriculture, to have permitted the costly wars and massive urban destruction that constantly took place; and not a little of this surplus was the result of mechanical inventions.

The chief center of these inventions was Greece: the very place where servile work was regarded as unbecoming a free citizen. But it is hardly an accident that fresh mechanical inventions sprang up here; for this was a culture whose main cities, Athens in particular, challenged and overthrew the institution of kingship at an early date. Even in the Homeric epics, kings were little more than provincial chiefs, living in manor houses much like those of later feudal Europe, not exalted sacred beings exercising divine prerogatives; and Greek myth, though perhaps it had Mesopotamian roots, never seriously harbored the "silly nonsense"—to use Herodotus' scornful characterization—associated with the cult of divine kingship. Even at the height of Hellenic urbanization, the democratic village measure kept its hold; and typically it was a handful of resolute free men rather than a massed army who held the mountain passes or manned the efficient Greek war galleys.

The fact is that most of the components of later complex machines were either invented by the Greeks, between the seventh and the first centuries B.C. or were manufactured with the aid of machines and mechanical parts the Greeks first invented. Witness two key inventions: the screw and the lathe.

The invention of the screw by the Greeks, possibly in the seventh century B.C., made feasible a whole series of other inventions. Archimedes applied the principle of the screw to the raising of water; and this opened up new territories for agriculture all over North Africa and the Middle East. Later irrigation machines, once called typically Oriental, were in fact, Heichelheim reminds us, invented during the third century B.C. as a consequence of Hellenistic progress in mathematics. Later than Archimedes, Ctesibius invented suction and force pumps, which soon came into use; while Archytas, supposedly the inventor of the screw, applied geometry to mechanics, as other geometers had done to architecture. This is neither the first nor the last example of the interplay between the exact sciences and the machine.

The invention of the lathe was an advance of comparable importance, since accurately turned and bored cylinders and wheels lie at the core of every rotating machine. Though neither the time nor the place of this invention can be definitely assigned—some authorities give it an early Mesopotamian origin—it seems likely that machine-turned spindles antedated the screw. At all events, without the lathe the passage from the human to the non-human machine could hardly have been accomplished.

Though the perfection of the lathe was a slow process, it was from the beginning as great a labor-saving device as the wheeled vehicle or the sailboat, and because of its many applications, equally important. The most direct use of the product of the lathe was in lifting devices, pulleys, winches, and derricks, of use in loading goods and hoisting sails; but in the classic Greek tragedy this facility also had a part to play: the God who intervened in human affairs at a critical point was called 'the God from the Machine' because he descended from above by means of a real machine. Does not the fact that a Greek audience found nothing incongruous in this device suggest an acceptance of the machine as a supernatural agent?

While the screw and the lathe were outstanding inventions, many other contributions joined them. The stamping of metals to make coins was a seventh-century Greek invention, which revolutionized economic transactions, though the further application of the stamping process to printing had to wait many centuries. As for the immense skill shown by the Greeks in the casting of bronze statues by the 'lost wax' method—there the supposed technical indifference or incompetence of the Greeks falls fantastically wide of the truth. Anyone who recalls Cellini's account of the difficulty in the casting of his Perseus, a relatively small figure, will have some insight into the superb technical mastery that made possible the casting of many far bigger bronze statues in Greece.

So, too, in our admiration for the final form of the Greek temple, architectural critics often forget the resourceful engineering necessary to transport the heavy stones for the Parthenon up the steep slopes of the Acropolis. No less remarkable was the shaping and fitting together of the massive foundation stones employed for the temple of Apollo at Delphi: these smooth-faced but completely irregular blocks of stone, fitted together like a jigsaw puzzle, without cement, were a sound device for withstanding earthquakes: but anyone who contemplates the scribing and exact fitting of these stones will not be tempted to under-rate Greek engineering.

One must grant that these brilliant technical innovations were not always immediately applied, any more than was Hero of Alexandria's aeropile; and let us not forget either that parallel inventions, quite as original, were made in China, India, and Persia—inventions that help account for their dazzling performances in sculpture and architecture. But it is doubtful if the series of mechanical improvements that were introduced into Western Europe after the eleventh century could have been developed, or even conceived, without this long series of preliminary efforts.

I have kept to the last the most revolutionary of all these mechanical inventions, also seemingly Greek—that of the watermill. The original model for this invention might conceivably have come, by way of Alexander's armies, from India, where toy water wheels were used for the mass production of Buddhist prayers. But again, it is hardly an accident that the watermill as a practical invention, not a magical toy, came from Greece, the culture that had doggedly retained the democratic technics of the archaic village and had never submitted supinely to the totalitarian ideology of kingship, as revived by Alexander the Great and later Hellenistic 'savior kings.'

In further confirmation one notes that the Athenians had not, either, accepted that other necessary component of the megamachine, a trained permanent bureaucracy. The Athenians retained, as a proud mark of citizenship, the administrative functions that must otherwise have been delegated to specialized functionaries; and instead of making administration a lifetime function, they rotated offices. Thus the mechanical prime mover, in its pure form, not using even animal power, was a Greek invention: the first successful effort to displace the collective human machine as a source of energy for productive work.

On the present evidence, it seems likely that the earliest type of water-mill was a small horizontal one, now called 'Norse,' good only for local domestic use, though thus all the more available on small streams. This may well have been the contribution of the Greek mountain village; for the earliest literary citation of the watermill is in a first-century verse by Antipater of Thessalonica (Salonika), as follows: "Cease from grinding, ye women who toil at the mill; sleep late even if the crowing cocks announce the dawn. For Demeter has ordered the Nymphs to perform the work of your hands, and they, leaping down on the top of the wheel, turn its axle, which with its revolving spokes, turns the heavy concave Nysirian millstones. We taste again the joys of primitive life, learning to feast on products of Demeter, without labor."

This distinct reference to an overshot water wheel, though it has been repudiated without explanation by an historian of technics, would in fact indicate a much earlier date of invention, since the undershot wheel is supposedly the older, less efficient, type. On the most prudent estimate, one must allow one or two centuries before such an invention would come to the attention of a poet, even a local poet, and awaken such lyric praise over what must have been a well-demonstrated success. Probably the smaller simpler, type of mill was invented still earlier, and it remained in existence in the Hebrides Isles right through the nineteenth century.

What is important to realize is that with this invention the unavoidable drudgery of domestic flour-grinding was, at least in principle, over, though hand-querns continued in use. The potentialities of waterpower as a labor-saving device were appreciated. If this invention did not spread rapidly over the Mediterranean area, geographical conditions, rather than human inertia, may have been largely responsible; for in Greece the mountain streams in the summer often dry up to a trickle, or worse, and even at other times of the year cannot be relied upon anywhere without a mill-dam and a mill-race.

Though the range of the watermill was necessarily limited, its spread and large-scale utilization, where possible, is now well attested. The discovery of sixteen watermills, set on eight symmetrical floors, at Barbegal near Arles, dating from A.D. 308–316, proves, as Bertrand Gille points out, that under Diocletian and Constantine a scarcity of slave labor had led to a large-scale introduction of power machinery, which displaced both the slave and the free domestic system by one based upon mechanical prime movers. Perhaps this is the first historic evidence of wholly mechanized mass production, though another poet, Ausonius of Bordeaux, only slightly later, records the use of watermills for sawing limestone in the Moselle valley. There is no reason to suppose, when the documentary evidence for the watermill's use on a great scale reappears in the eleventh century, that this was a fresh re-invention.

Though I have stressed the three key Greek inventions, because they have been under-rated, I must add various others derived from them, such as the auger, the pulley, the winch, and the screw-press for mechanically squeezing oil and wine. These indicate that the common appraisal of the whole period as backward in technology because of slavery reflects only a stereotyped academic judgement, which unfortunately was crystallized before the contrary evidence came to light.

Now what applies to the Greeks applies no less to a whole series of other inventions in other countries and later centuries. These were long lost to view, some doubtless forever, because they were not recorded in books or preserved as relics. In the industrial museum at Doylestown, Pennsylvania, there is a whole array of clever early American inventions, mostly in wood, to simplify domestic labor and facilitate farm work. But like the machine for chopping turnips for fodder, most of these contraptions have disappeared with changes in cultivation and feeding, and but for such museums would be completely forgotten.

Just as our machine-infatuated age has failed to rate the breeding of vegetables and fruits on a par with mechanical invention, so it has also grudged to give credit for the processing of foods, by salting, smoking, cooking, brewing, distilling, as another realm of invention. The lively records from both Mesopotamia and Egypt, hailing improvements in the taste of beer, call attention to similar efforts in other departments. Though we cannot date the first pressing of olive oil, or the fabrication of the first sausage, both are attested in classic Greek literature. The sausage itself is such an admirable device for preserving meat in a handy form that it has continued in existence down to our own day, without any further improvement until the dubious one of plastic casing was introduced. None of these concrete improvements should be dismissed because they must be evaluated in other terms than mechanical ingenuity and productivity.

So, too, in our preoccupation with large-scale industrial uses, we have forgotten technological innovations in other departments. A great variety of Roman surgical instruments, finely specialized, reminds us that invention did not cease here; and Herodotus' account of the series of hysterectomies performed on the King of Lydia's concubines seems surely to indicate the discovery of effective anesthetics—a professional secret shared by the priesthood at Delphi, who removed a cataract on the eye after putting their visitor to sleep. But unfortunately the secret was so well guarded in both cases it seems not to have been transmitted to later generations.

3 : THE BROADER RECORD

Before I round out the picture of early technical accomplishments our own age has ungratefully assimilated without even noticing, let me point out the one massive cause for both industrial and social retardation that the technical historians have so far slighted: a far more serious cause than slavery; and this is the repeated, indeed chronic, devastations and decimations of war.

Against the many positive advances, one must place this colossal negation. In the burning of villages and the wrecking of cities, something more than buildings and workshops was repeatedly wiped out: namely the traditions of craftsmanship, the trade secrets, the new inventions, along with the secure sense of the future that makes men willing to sacrifice valuable days of their life for a later fruition. Only a part of the technical tradition could be sustained and transmitted under these constant assaults, even if the workers were enslaved instead of being killed. For with slavery came a loss of initiative and in all likelihood a certain amount of vengeful sabotage—"the conscientious withdrawal of efficiency."

Since most craft knowledge was carried in the head, and was effective only when appropriate materials were at hand, and when the process was passed on by imitation and verbal instruction, the losses through war must have been great—incalculably great. To omit war as a cause for general technological retardation, and to center on its by-products—slavery and ruling class contempt for the utilitarian arts—is to treat secondary factors as primary ones.

Perhaps the largest failure to acknowledge substantial technological achievement has been in the realm of the domestic arts: the slow but constant improvement in utensils and equipment that serve domestic convenience and increase comfort. I refer to a wide range of inventions from textiles and cutlery to chairs and beds. If the forms of some of these objects, in pottery for example, have remained stable, it is because no improvement was possible. The degraded or silly fantasies in furniture, pottery, and cutlery that now seek to command contemporary attention by their hideous novelty furnish ample proof, by contrast, of this earlier success.

If we turn to the furnishing of houses and the equipment of kitchens, we shall find that the whole horizon of invention widened out at an early period. The dovetail joint, for example, was an old Egyptian invention that kept sliding drawers—themselves another useful invention—from falling apart. The wicker chair, in form like the modern Chinese peeled chair, was also an Egyptian invention, while the Etruscans, before the Roman conquest, had fabricated both chairs and bedsteads of bronze, thousands of years before the iron bedstead had been boastfully advertised as a singular mark of Victorian mechanical progress.

Many ingenious inventions for infant care, personal hygiene, or agricultural efficiency date from the seemingly uninventive inter-regnum after the horse had been introduced and iron had become the chief industrial metal. Even a partial list of inventions or popular adaptations would include the folding camp stool, the potty stool for infants (Greek), the bath tub, the shower bath, piped hot-air heating and piped water, water closets, piped sewage disposal, the flail, the reaping machine, the barrel, the churn, the pump, the horseshoe, the stirrup, the hod, the wheelbarrow, and not least, paper. Note that only a few of these inventions can be called machines. Many of them, as one would expect from their agricultural or domestic setting, are utensils or utilities: derivatives from the primarily neolithic art of containers on which I dwelt earlier.

As for the city—itself a complex social invention with many differentiated parts—it was the scene of many other inventions, both functional and, what is at least equally important, meaningful. The public bath, the gymnasium, the theater, the park, were all true inventions: not less useful because they lie outside the mechanical realm. In our modern commitment to dynamism, industrial turnover, rapid transportation, we have overlooked the fact that a life without stable containers would fall to pieces, as in fact our own life is now rapidly doing. All over the world the city has been wantonly sacrificed to the private motorcar—although mono-transportation by motorcar is the most inefficient possible substitute for the complex transportation network needed to serve—and save—the city.

Shortly I will review this long, supposedly stagnant, inter-regnum from quite another angle than that of productive efficiency. So far I have been concerned only to show that the actual productivity in both inventions and their applications has been under-estimated because of the current Western fixation on tools and machines. After such an accounting, one is nevertheless left with certain areas in which the absence of technical

progress defies explanation on any plausible rational grounds: glass for
example. The first glass beads date back to around 4000 B.C.; and the
culture that invented the pottery kiln and the ore-furnace could easily have
carried glass-making forward; for its chief material, sand, is far easier to
procure than metal-bearing stones. But apart from beads, the earliest glass
objects known date from 2500 B.C., and only some thousand years later
have the first glass vessels been dated.

At the end of the first century B.C. we indeed have evidence of glass
blowing; and less than a century later Seneca reports as recent "the use of
glass windows which let in the full brilliance of the day through a transparent pane." But though there were no serious technical obstacles to the
manufacture of window glass—and though there was need for better in-
door light in offices, scriptoria, and workshops—transparent window glass
seems to have remained a rarity even in Rome and continued so in
European dwelling houses down to the sixteenth century and later.

This tardiness is all the more mysterious since only recently (1965) a
huge slab of glass, estimated to weigh over eight tons, was discovered in
a cave near Haifa, dated between A.D. 400 and 700. Was this a unique
technological stunt, like the column of chemically pure iron found in India,
or was it a sound project that was nullified by another outbreak of armed
violence? At all events, the employment of glass for a wide variety of
purposes, from vases and chemical alembics to spectacles and mirrors,
dates only from the thirteenth century—itself still often erroneously supposed to be a period of empty theological disputation and technical stagna-
tion, despite Lynn Thorndike's ample documentation to the contrary.

Certainly, then, there was an unexplainable backwardness in more than
one province where technical improvement was easily conceivable without
any violation of existing social arrangements or craft traditions. But some
of this backwardness may be explained on the same theory I have applied
to backwardness in tool-making before late paleolithic culture: there was
a major concentration of interest in other departments—religious ritual,
magic, literature, the plastic and graphic arts. Once the fundamental crafts
and the simple machines were established, advances within technology
came mainly through increased skill, elaboration of form, refinement of
detail. To sacrifice esthetic invention or functional 'rightness' in order to
double the output, or even to hurry the process of production, was foreign
to the whole scheme of pre-mechanized civilization, whether democratic or
authoritarian.

Not that quantity was altogether neglected: this came in with capitalism
and long-distance trade. Even in symbolic objects quantity and cheapness
might play a part. By reducing the size of a fabricated article, as in a

Tanagra figurine, it became possible to put more of such products on the market at lower prices. On the whole, the concern for quality for long served as a brake upon production; but if it lessened the output and decreased the circle of possible consumers, it also evened up the account by decreasing the pace of obsolescence and eliminating a large source of waste. When these contradictions between authoritarian and democratic technics are reckoned with, they help to give a more accurate picture of technical development than one based upon a gross comparison between ancient technology and that of the present age.

4 : THE PRIMACY OF ART

Now, the whole picture of 'backwardness' changes as soon as we cease to judge earlier technologies by the provincial standards of our own power-centered culture, with its worship of the machine, its respect for the uniform, the mass-produced, the mass-consumed, and with its disregard for individuality, variety and choice, except in strict conformity to the demands of the megamachine.

In terms of today's criteria all past cultures up to our own were indeed uninventive. But once we realize that under handicraft production, even when subservient to the megamachine, the great realm of invention was in the arts, the position of the two technologies is reversed. In esthetic and symbolic terms, it is our present culture that has become painfully uninventive, ever since the handicrafts and the folk-arts that went with them lost their lifeblood in the nineteenth century. The end products in painting and sculpture, at least those most exploitable commercially, are now deliberately degraded to a level far below the earliest paleolithic carvings.

While utilitarian inventions made slow and intermittent advances up to the nineteenth century, esthetic inventions, marked by a proliferation of styles, patterns, and constructional forms have characterized and differentiated every culture, even the humblest. As in man's earliest departures from his dumb animal forebears, it is into the arts of expression and communication that the most intense human energies have until now always poured: here, and not in manufacture or engineering, was the major realm of invention.

To make even a cursory inventory of the esthetic inventions between 3000 B.C. and A.D. 1800, would be a more formidable task than to write a history of technics proper. For such a canvass would be nothing less than a massive encyclopedia of all the arts, those of the folk as well as those of the palace: not merely as forms in space, but as the languages of the human spirit, comparable in their richness and subtlety to the spoken language itself.

Esthetic invention played fully as large a part as practical needs in man's effort to build a meaningful world; and because of the demands it made it was also a major stimulus to technics. The greatest technical conquests of ancient civilization, once past the neolithic processes of domestication, were in the domains of building and the domestic arts. From the earliest Sumerian ziggurat onward, architecture was the scene of a succession of major inventions: each building, by its combinations of volume, mass, color, ornamental pattern, texture, was a new invention, expressing and modulating an idea about human and cosmic relationships. The pyramid, the obelisk, the tower, the arch, the dome, the steeple, the groined vault, the flying buttress, the stained-glass window—all these are examples of untrammelled technical audacity, brought into play, not by the satisfaction of physical needs or the desire for material wealth, but by the more fundamental pursuit of significance.

Though architecture, with its utilization of many arts and its organic complexity, may well serve as the central exhibit of esthetic inventiveness, the same endless flow of designs characterized the other arts and touched even their commonplace pots and textiles. No article, even of vulgar daily use, was regarded as finished, unless it bore some unmistakable stamp, by its painting or modelling or shaping, of the human spirit. This mass of esthetic inventions compares favorably with the total mass of mechanical inventions during the last few centuries. But so far from suppressing technics, as our current economy suppresses art, these two modes of invention interacted.

The current separation of art and technics, then, is a modern solecism. Until the machine monopolized our attention there was a continual interplay between quantitative order and working efficiency on one hand, and the qualitative values and purposes that reflected the human personality on the other. To deny the name of invention to the creative expression of subjective forms is to deny the unity of the organism itself and the impress of the human personality.

The fabrication of musical instruments, beginning with Pan's pipes, the drum, and the harp, covers at least as wide a span of time as the art of weaving. Perhaps it is no accident that one of the earliest observations in mathematical physics was the discovery by Pythagoras of the relation between the length of a vibrating string and the accompanying musical note.

So far from being backward, the subjective arts not only produced fresh modes and styles of their own, but in turn stimulated mechanical invention. Thus Hero of Alexandria designed a windmill to work an organ, and later, steam was generated to work an organ bellows, long before either force was used to pump out a mine.

The reciprocal relation between art and technics was maintained, to their common advantage, through all the ages of small-scale handicraft production. The violin whose perfection was such a superb contribution, not only to baroque music, but to all later orchestral composition, was itself an extraordinary invention: for that deceptively simple-looking instrument, as fabricated in Cremona, had as many as seventy separate parts, each of wood specially selected, seasoned, and shaped for its function; while the musical compositions it made possible were as much inventions as the instruments that played them.

Even the most cursory historic survey of the arts reveals a fertility of invention in design unsurpassed by any utilitarian equivalent in engineering until the nineteenth century; and the working out of new esthetic forms imposed further demands upon technical ingenuity, as in the series of textile inventions, starting historically with the Damascus weavers, going on through the tapestries of the Middle Ages, and finally in the elaborate ornamental patterns that called forth the Jacquard loom. The latter, incidentally, drives home the point I have been making, for the complex punch-card instructions of the Jacquard loom served as model for the later inventions of sorting and calculating machines.

In short it was in the decorative, the symbolic, and the expressive arts that progress was maintained, even in ages that, in retrospect, otherwise seem stagnant. Well before the steam engine and the power loom, the first large gains in quantitative production were made here: not merely by means of the printing press, but in the arts of engraving, etching, and lithographing, which made pictures in quantity, often by the greatest artists, available at reasonable prices for use in private homes.

Thus not merely esthetic invention, but mechanical invention for the sake of achieving or perfecting the purely esthetic or symbolic results, characterized a large part of pre-automated production. This contribution has been under-rated, even in its technical implications, by those who would reduce technics to the conquest of time and space and energy. That craft tradition, handed on mainly by word of mouth and personal example, was not easily lost or destroyed, since it had a worldwide distribution. If China had ever forgotten how to make glazed pots, Japan or Italy could have supplied the missing art. If all the workshops of a town were burned down, the individual craftsmen, if they escaped with their tools, could if necessary have replaced them. War might retard further improvements but neolithic technology, with its worldwide distribution, could not be completely suppressed until megatechnics itself became equally universal.

To destroy such a well-diffused technical tradition, one must entirely wipe out the supporting culture and the individual personality. And eventually that result was brought about, after the sixteenth century, by the 'invention of invention,' which gave to the machine the primacy that had once belonged to the craftsman-artist, and which reduced the personality to just those numbered parts that could be transferred to the machine.

Ironically, but tragically, that happened just at the point when democratic technics, centered in the small workshop, had at last at its command sufficient mechanical power to rival the performances of the megamachine. With the introduction of small-scale power machinery, which could have increased quantitative production without destroying esthetic sensibility or undermining personal creativity, the flowering of the arts that took place in Europe from the thirteenth century on might have gone on steadily. A genuine polytechnics was in the making, capable of reconciling the order and efficiency of the megamachine with the creative initiative and individuality of the artist. But within a few centuries the whole system was undermined by the new impersonal market economy and the resurrection in a new form of the totalitarian megamachine.

There were many processes in handicraft that could well have been shortened, simplified, or perfected by the machine, just as pottery-making had been originally perfected by the invention of the potter's wheel. Anyone who has had the pleasure I once had, of seeing an old-fashioned woodturner in the Chiltern Hills in England split a seasoned log with an axe into equal billets, and turn out on his power lathe, with swiftness and precision, a chair leg, will know that there was no necessary enmity between handicraft and the machine itself. Just the contrary; under personal control, the machine or the machine tool was a boon to the free worker.

Two thinkers during the last century were quick to realize the advantages of an advanced technology—utilizing small machines with efficient and cheap electric power—for restoring the intimate human scale and with it the communal cooperations of the face-to-face community without forfeiting the benefits of rapid communication and transportation: Peter Kropotkin and Patrick Geddes. In his 'Fields, Factories, and Workshops,' Kropotkin outlined this potential new economy. Curiously, Dr. Norbert Wiener, whose own scientific work furthered automation, re-discovered these possibilities two generations later, without being conscious of any prior analyses by Kropotkin, Geddes, and myself. But the dominant forces of the nineteenth century, including the authoritarian communism of Karl Marx, remained on the side of big organizations, centralized direction, and mass production: with no thought for the worker except as a unit in the megamachine. So it has been only in the contemporary American home, with its array of automatic heating and cooling apparatus, washing machines, mixers, grinders, beaters, polishers, cleaners, that these possibilities have been even partly explored.

Up to now the handicraftsman's freedom could not survive an authoritarian economic system, based on the organization of complex machinery that no single worker could buy or control, and promising 'security' and 'abundance' only in return for submission. The philosopher, A. N. Whitehead, has seen the importance of this culminating period in Western handicraft better than most historians, and his words are worth quoting. "So far as their individual freedom is concerned, there was more diffused freedom in the City of London in the year 1633 . . . than there is today in any industrial city of the world. It is impossible to understand the social history of our ancestors unless we remember the surging freedom which then existed within the cities of England, of Flanders, or the Rhine Valley, and of Northern Italy. Under our present industrial system, this type of freedom is being lost. This loss means the fading from human life of values infinitely precious to it. The divergent use of individual temperaments can no longer find their various satisfactions in serious activities. There only remain ironbound conditions of employment and trivial amusements for leisure."

Apart from Whitehead's choice of the seventeenth century as a highpoint, perhaps true for England but far too late for the rest of Europe, his characterization brings us to the great divide in Western history, the point at which democratic technics was so overwhelmed by the authority, the power, and the massive success, on its own narrow terms, of megatechnics. But before confronting that history and attempting to explain its results, we must reckon with a counter-force that had been at work for some two thousand years: that of the Axial religions and philosophies, the diverse yet kindred systems of value that challenged and sought to lift the heavy burden of 'civilization,' by directing all change toward the transformation, not of the environment, but of the individual soul.

5: THE MORALIZATION OF POWER

With the spread of urban civilization, an immense amount of technical facility and material wealth was accumulated: in many areas, the life in these centers of power offered incentives, opportunities, and fulfillments that were beyond the scope of the archaic village. Yet the mass of mankind, until the present era, has never lived in cities, nor altogether been willing to accept, as an ultimate gift, the kind of life offered there. The ruling classes themselves shared some of this discontent with the imputed advantages of civilization, as the 'Dialogues on Suicide' I have cited show: they either maintained estates in the countryside, where they would reside periodically, or, when the whole complex political apparatus broke down, would take refuge there permanently, partly compensating for the vanished goods of 'civilization' by returning to the older pursuits of hunting, fishing, planting, breeding.

As for the mass of urban workers, they must have viewed their dismal lot, if they were conscious at all, with a feeling of galling disappointment. In accepting the division of labor they had lost their own individual wholeness, without having it restored, by fellowship and cooperation, on a higher communal level. The spectacles of power, provided by the megamachine, might divert them or exalt them; but vicarious living is no better than vicarious eating: at best, the worker was forced to starve in the midst of plenty, and had reason to feel cheated. This sense of disillusion with what life had to offer is evident in early Mesopotamian literature and was constantly reiterated. Vanity of vanity, all is vanity, saith the Preacher. And the sum of that vanity is that "they have sought out many inventions." On these terms 'civilization' stank.

So, among the various conditions that help account for the slowness in enlarging the province of the megamachine, once the original spurt of constructive activity had reached an apex, something more than the negations of war must be taken into account: there was a recurrent disillusion with power and material wealth themselves when alienated from the purposeful and significant life-course of the community. This disillusion in time touched the exploiters as well as the exploited.

The ruling classes constantly were enervated by the surfeit of goods and pleasures they had so ruthlessly monopolized for themselves. They had lapsed, too many of these insolent rulers and their agents, from a human to a distinctly simian level: like the apes, they snatched food for themselves, instead of sharing it with the group: like them, too, the more powerful claimed more than their share of women: like them, again, they were in a constant state of nettled aggression toward possible rivals. In short, they had alienated themselves from their distinctly human potentialities and in that sense, the real gains in power and wealth had led to a dead end: they produced no equivalent wealth of mind.

Between 3500 and 600 B.C. the physical shell of civilization had thickened: but the creature within it, he who had fabricated the shell, felt increasingly pinched and constricted, if not immediately threatened. The rewards of large-scale organization and mechanization were small in proportion to the sacrifices demanded. Only this increasing sense of disillusion can explain the popular revolt that began slowly between the ninth and the sixth century B.C.: a revolt of the inner man against the outer man, of the spirit against the shell. Because this revolt did not depend upon physical weapons, it could not be put down by whips, truncheons, or shackles; and it quietly threatened to shatter the whole power system based on land monopoly, slavery, and the life-time division of labor.

The first scholar to describe this simultaneous movement and understand its significance was an almost forgotten Scotsman, J. Stuart Glennie, who also called attention to a five-hundred-year cycle in culture: and both Karl Jaspers and I have independently called these new religions and philosophies 'Axial'—a deliberately ambivalent term which includes both the idea of 'value,' as in the science of Axiology, and centrality, that is the convergence of all separate institutions and functions upon the human personality, around which they revolve.

This revolt began in the mind, and it proceeded quietly to deny the materialistic assumptions that equated human welfare and the will of the gods with centralized political power, military dominance, and increasing economic exploitation—symbolized as these were in the walls, towers, palaces, temples of the great urban centers. All over Europe, the Middle East and Asia—and notably out of the villages rather than the cities—new voices arose, those of an Amos, a Hesiod, a Lao-tzu, deriding the cult of power, pronouncing it iniquitous, futile, and anti-human, and proclaiming a new set of values, the antithesis of those upon which the myth of the megamachine had been built. Not power but righteousness, these prophets said, was the basis of human society: not snatching, seizing and fighting, but sharing, cooperating, even loving: not pride, but humility: not limitless wealth, but a noble self-restricting poverty and chastity.

By the sixth century B.C. this challenge had spread everywhere: the same general attitude toward life, the same contempt for the goods of civilization, the same scorn for those leaders in the court, the camp, the temple, and the marketplace who, as William Blake said, would "forever depress mental and prolong corporeal war." Above all, the same espousal of the poor and lowly, hitherto the easy victims of power.

From India and Persia, clear across to Palestine, Greece, and eventually Rome, the new spirit ignited: a seemingly spontaneous combustion. And in each place a new kind of personality sprang up and fathered a succession of similar personalities. This was a popular movement, not an upper-class fad. No longer was the ideal man a hero, a being of extraordinary bodily dimensions and muscular prowess, like Gilgamesh, Herakles, or Samson: no longer a king who boasted of the number of lions he had killed, or the number of rival kings whose gods he had captured and whose persons he had humiliated or mutilated: nor would this ideal figure boast of the number of concubines he had engaged in sexual intercourse in a single night.

The new prophets were men of a modest humane disposition: they brought life back to the village scale and the normal human dimensions; and out of this weakness they made a new kind of strength, not recognized in the palace or the marketplace. These meek, withdrawn, low-keyed, outwardly humble men appeared alone, or with a handful of equally humble followers, unarmed, unprotected. They did not look for institutional support: on the contrary, they dared to condemn and defy those in established positions, even predicting their downfall if they continued their established practices: "Mene, mene, tekel upharsin." "Thou art weighed in the balances and art found wanting."

Even more intransigently than kings, the Axial prophets dared depart from customary usages and traditions, not only those of civilization, but the sexual cults, with their orgies and sacrifices that derived from neolithic practices. For them, nothing was sacred that did not lead to a higher life; and by higher they meant emancipated from both materialistic display and animal urgencies. Against the personified corporate power of kingship they stood for the precise opposite: the power of personality in each living soul.

The humble manual vocations, rather than the high offices of the scribe or the civil servant, sustained these new prophets. "Work," said the author of 'Works and Days,' "is no disgrace; it is idleness which is a disgrace. . . . Whatever be your lot, work is best for you." Amos was a shepherd, Hesiod a farmer, Socrates a stone-cutter, Jesus of Nazareth a carpenter, Paul a tent-maker. True, Siddartha, the Buddha, was a prince; but he left his palace and his family to find a new vision of life in the solitude of the forest; while Confucius, though a scholar and a gentleman, was one of the chronically unemployed: not welcome at court, despite his courtly punctilio.

What is important to note is that this new movement rejected the obvious goods and achievements of 'civilization' no less than its patent failures and evils. This was not merely a revolt against the system of regimentation that had elevated the ambitious and ruthless and depressed the cooperative and amiable: it was a revolt against all the pomps and vanities of worldly success, against ancient rituals that had become empty— "vain repetitions such as the heathen use"—against gigantic images, imposing buildings, gluttonous feasts, promiscuous sexuality, human sacrifices: all that degraded humanity and shrivelled the spirit. These new figures persuaded rather than commanded: they did not seek to be rulers but teachers, each a 'teacher of righteousness,' inciting his followers to return to their own centers and be guided by their hidden lights.

By withdrawal from ordinary duties, by fasting and meditation, the new leaders had found within themselves the possibility of living a new kind of life which reversed the previous scale of values, even that of archaic agricultural society, with its over-emphasis of sexuality, its exclusive concern with kinsmen and neighbors; but even more emphatically they rejected the standards of civilization. Facing the heavily armored personalities produced by kingship, these prophets were spiritually naked and physically unarmed; so many Davids confronting the brass-bound Goliaths of the megamachine. The new leaders were bold enough to present this stripped personality as a model for imitation. For Confucius, one of the most influential of these new prophets, only those who sought to perfect the personality, with the aid of music, ritual, and learning, could be called 'complete men.'

The age ushered in by these prophets and their universal religions or philosophies was a new age: so definitely so that one of the greatest of the prophets supplanted Caesar's name on the calendar by which most peoples still reckon time. In seeking fellowship with others of their kind, without regard to the claims of local gods or to territorial and linguistic boundaries, they established the human personality as more important than its physical and institutional agents.

In relying upon direct human contact by means of word and example, in disciplining and re-directing the natural appetites, in focussing present activities upon distant future ends, these prophets had severally returned to the human center and the special artifacts of the mind. They picked up the thread dropped when the pressures of increasing numbers turned the neolithic cultures of the Near East, even before the onset of civilization, to the one-sided exploitation of the environment.

Among the older type of kings, with their boasts of divine power, not one left a permanent impression upon later generations by reason of any change of character he effected: indeed, the deliberate imitation of the king's personality would have been an unpardonable affront, if not a sacrilege. In the very act of admiring their splendid tombs and their monuments, we must smile, as we smile over the boastful inscription to posterity left by Ozymandias, at their colossal vanity and their childish ambition. What deep inferiority required such inordinate compensations? What mental disturbance prompted such exhibitions? But it is another matter with the new kind of spiritual leaders, "who waged contention with their time's decay, and of the past are all that will not pass away." Isaiah, Buddha, Confucius, Solon, Socrates, Plato, Jesus, Máhomet—these and their like are still in one degree or another alive, more solid and intact than any physical monument, still identifiable in the face and posture of their lingering descendants, as if the change effected had been enregistered in the genes.

To imagine that a transformation as profound and as widespread as this could have no effect upon technology would be possible only to those who think that society has always been divided into watertight compartments. This new way of life, by discrediting alike the aims and the means of 'civilization,' diverted human energy from its service more by withdrawal and abstention than by any overt struggle with the ruling classes. By going back to the original instruments of man's development, seeking to re-orient his mind, pulling it out of its ever-deepening institutional ruts, they had seemingly opened up a way to further development; though in fact, as I showed in 'The Transformations of Man,' they were all-too-soon absorbed back into the institutions they had challenged.

Kingship itself became affected by these new spiritual considerations: in the person of the Buddha, to begin with, then Asoka in India, and Marcus Aurelius in Rome. Nor was it the Jewish prophets alone that dared to admonish kings and bid them conform to a higher morality. Dio Chrysostom (A.D. 40–115) in his first discourse on kingship, did not hesitate to point the lesson. "The good king," he wrote, "also believes it to be due to his position to have the larger portion, not of wealth, but of painstaking care and anxieties; hence he is actually more fond of toil than many are of pleasure or wealth. For he knows that pleasure, in addition to the general harm it does to those who constantly indulge there, also quickly renders them incapable of pleasure, whereas toil besides conferring other benefits, continually increases a man's capacity for toil."

This struck a new note, as stridently contradictory to the original postulates of kingship as Christianity itself. An ancient king would hardly have believed his ears had anyone uttered such words in his presence, for he never counted his exertions in battle as anything but pleasure—unless defeated. But Marcus Aurelius would soon seek to live his life on those terms.

Human life, as thus conceived, was no longer cheap but infinitely precious: not to be squandered on the pursuit of ephemeral goods. This new faith in the central role of the personality shifted the emphasis from mechanical organization to human association and mutual aid; and this, as Kropotkin demonstrated, had its effect upon technics. From the twelfth century on this change can be documented by the practices in Western

Europe of the medieval guilds: for their work was attached to acts of charity and succor and fellowship—aid for widows and orphans, decent offices of burial for the dead, participation in brotherly feasts and ceremonies, the performance of mystery dramas and spectacles.

This new Axial religious and ethical reorientation was fated to have a profound influence upon technics: for one thing, it helped to alleviate the lot of the slave and then brought about the gradual renunciation of slavery itself. In peaceful occupations, if not in war, that source of power for the megamachine was disconnected and abandoned; and this reform hastened the pace of inventing alternative non-human power systems and machines. No one can doubt that this was a positive advance.

Unfortunately, in Europe, the main ecclesiastical organization that had radically challenged the old materialistic values of 'civilization' had itself risen to power by taking over the bureaucratic administrative organization of the Roman state. In time the Papacy would even command armies of its own, along with still later means of coercion practiced by the Inquisition, equipped with ingenious mechanisms in graduated torture, hardly surpassed in our time by Nazi inquisitors and their ugly military counterparts in other countries. In the rebuilding of the Roman state on Axial lines, the Roman Catholic Church itself became, paradoxically, a kind of etherealized megamachine, working for the glory of God and the salvation of souls under a Divine King. And again, to complete this likeness, this took place under a direct representative of Godhead, the Pope, whose pronouncements on matters of faith and morals, supported by the priesthood, were deemed absolute.

But by the time the new values of the Axial prophets had been incorporated in social institutions and embodied in new architectural forms and works of art, sketches and working drawings for a more potent type of megamachine were already, so to say, on the drawing boards. After centuries of erosion, the old megamachine was due for a complete overhauling, even in the army, where the tradition, though not unbroken, had been most faithfully preserved.

To re-build the megamachine on modern lines, both the old myths and the old theology needed to be translated into a more universal language, which would allow the king in person to be overthrown and removed, only to return again in more gigantic and dehumanized form as the Sovereign State, endowed with absolute but far from divine powers. But before this could happen, a long period of preparation was needed, in which the main Axial faiths, Buddhism, Confucianism, Christianity, and Islam, all had an active though largely unconscious part to play.

CHAPTER TWELVE

Pioneers in Mechanization

1: THE BENEDICTINE BLESSING

Now we come to one of the curious paradoxes of history: the fact that certain missing components, necessary to widen the province of the machine, to augment its efficiency, and to make it ultimately acceptable to the workers as well as to the rulers and controllers, were actually supplied by the other-worldly, transcendental religions: in particular by Christianity.

Some of these components had been perfected through the Axial philosophies. Confucianism, with its emphasis on ritual, filial duty, moderation, learning, laid the ground for the exemplary bureaucratic organization of Imperial China, not based on status and privilege alone, but recruited by examination from every class. The first comprehensive effort to reconstitute the machine on a new basis, with greater emphasis on lifelike mechanisms as such, and less on the machine-molding of the human parts, took place in the Christian Church. This was largely responsible for the fact that Western civilization caught up with, and then surpassed, the technical inventiveness of China, Korea, Persia, and India.

Christianity not merely reconstituted the original forces that were combined in the megamachine, but added precisely the one element that was lacking: a commitment to moral values and social purposes that transcended the established forms of civilization. By theoretically renouncing power achieved mainly by the coercion of men, it augmented power in the form that could be more widely distributed and more effectively controlled in machines.

The results of this mutation have become fully visible only since the seventeenth century; but the place where the change first transpired was, it would seem, the Benedictine monastery. In this new institution, all that the machine had hitherto been able to do only by making extravagant claims to divine mandate backed by large-scale military and paramilitary organizations, was now done on a small scale, by small companies of men, recruited on a voluntary basis, who accepted work—indeed the whole technological order—not as a slave's curse but as part of a free man's moral commitment.

This change proceeded from the fact that from the third century after Christ in Western Europe there had been a steady withdrawal of interest from the goods and practices of 'civilization,' accompanied by a wholesale retreat from the great urban centers of power, like Rome, Antioch, and Alexandria. Little groups of mild, peaceable, humble, God-fearing men and women, from all classes, withdrew from the noisy tumult and violence of the secular world, to establish a new mode of life, dedicated to their soul's salvation. When organized as communities, these groups introduced into the daily routine a new ritual of ordered activity, a new regularity of performance, and a measure of accountable and predictable behavior hitherto unattainable.

The Benedictine Order, instituted by Benedict of Nursia in the sixth century, distinguished itself from many similar monastic organizations by imposing a special obligation beyond the usual one of constant prayer, obedience to their superiors, the acceptance of poverty, and the daily scrutiny of each other's conduct. To all these duties they added a new one: the performance of daily work as a Christian duty. Manual labor was prescribed for no less than five hours a day; and as in the organization of the original human machine, a squad of ten monks was under the supervision of a dean.

In its organization as a self-governing economic and religious society, the Benedictine monastery laid down a basis of order as strict as that which held together the earlier megamachines: the difference lay in its modest size, its voluntary constitution, and in the fact that its sternest discipline was self-imposed. Of the seventy-two chapters comprising the Benedictine rule, twenty-nine are concerned with discipline and the penal code, while ten refer to internal administration: more than half in all.

By consent, the monk's renunciation of his own will matched that imposed upon its human parts by the earlier megamachine. Authority, submission, subordination to superior orders were an integral part of this etherealized and moralized megamachine. The Benedictine Order even anticipated a later phase of mechanization, by being on a twenty-four hour basis; for not merely were lights burned in the dormitory during the night, but the monks, like soldiers in combat, slept in daytime clothes, so as to be ready at once for canonical duties that broke into their sleep. In some ways this order was more strict and far-reaching than that of any army, for no periodic letdowns or sprees were permitted. These systematic privations and renunciations, along with regularity and regimentation, passed into the discipline of later capitalist society.

Probably the practical necessity to become self-supporting, in an era when the old urban economy was collapsing, and when self-help and agricultural productivity was the only alternative to helpless starvation or abject submission to slavery and serfdom, dictated Benedict's original insistence upon the obligation to perform manual labor. But whatever the immediate reason, the ultimate effect was to supply something that had been missing alike among the favored classes and the depressed workers in earlier urban cultures: a balanced life, a kind of life that had been preserved, though at a low intellectual level, only in the basic village culture. The privations and abstentions imposed by monasticism were for the sake of enhancing spiritual devotion, not to put more goods or power at the disposal of the ruling classes.

Physical work no longer occupied the entire day: it alternated with emotional communion through prayer and plainsong. Here the slave's working day, from sunrise to nightfall, gave way to the five-hour day: with a plenitude of leisure, be it noted, that owed nothing *in the first instance* to any labor-saving machinery. And this new scheme of living was esthetically enhanced through the creation of spacious buildings, well-tended gardens, thrifty fields. This regimen, in turn, was balanced by intellectual effort in reading, writing, discussion, not least in the planning of the varied agricultural and industrial activities of the monastic community. Shared work had the benefit of shared mind.

The order and regularity that was introduced into the monks' day—with every office performed in due succession at stated intervals, the seven 'canonical hours'—was timed and paced by the hour-glass, the sun-dial, and eventually, the clock. From the monastery, this time-keeping habit spread back to the marketplace, where in the classic era it had perhaps originated: so that from the fourteenth century on a whole town would time its activities to the ringing of the tower clock's bells.

Thus the Benedictine monastery had within its own confines taken over the discipline and order that the great collective labor machine had originally introduced as an attribute of assertive temporal power. But at the same time the monastery had rationalized and humanized this discipline; for the monastery itself had not merely kept to the human scale—only twelve members were required to form one—but it had abandoned the once tightly organized complex of civilization: the small-scale division of labor:

class exploitation: segregation: mass coercion and slavery: fixation for a lifetime in a single occupation or role: centralized control.

Each able-bodied member of the monastery had an equal duty to work; each received an equal share of the rewards of work, though the surplus was largely devoted to buildings and equipment. Such equality, such justice, had rarely before characterized any civilized community, though it is a commonplace among primitive or archaic cultures. Each member had an equal share of the goods and food: and received medical care and nursing, plus extra privileges, such as a meat diet in old age. Thus the monastery was an early model of the 'welfare state.'

In moving from one occupation to another in the course of a day, this regimen overcame one of the worst and most persistent defects of orthodox 'civilization': a lifetime's confinement to a single type of work, and a whole day concentrated upon work alone, to the point of utter exhaustion. Such moderation, such equalization of effort, such promotion of variety, had been possible before only in small, traditional, unambitious communities, that forwent the advantages of any richer intellectual and spiritual development. Now it became a model for cooperative effort on the highest cultural plane.

Through its regularity and efficiency the monastery laid a groundwork for both capitalist organization and further mechanization: even more significantly, it affixed a moral value to the whole process of work, quite apart from its eventual rewards. Admittedly, monasticism had achieved these admirable results by oversimplifying the human problem. Above all, it had left out the prime form of human cooperation—that between the sexes—and it had not allowed for the fact that full-blooded men and women, who must necessarily be open to the lusts of the flesh if they are to reproduce, could not make full use of the monastic pattern. Other ideal communities, with equally remarkable economic and technical achievements, like the Shaker colonies in the United States, would break up later on that same rock.

Unfortunately, the sexual one-sidedness of monastic organization made its own wry contribution to mechanization: in later developments the divorce between the factory and the office on one side, and the home on the other, became as marked as that between the earliest archetypal bachelor armies for war and labor and the mixed farming communities from which they were drawn. The lesson of the ant-hill, that specialized work can best be done by sexual neuters, was increasingly applied to human communities, and the machine itself thus tended to become an agent of emasculation and defeminization. That anti-sexualism left its mark on both capitalism and technics. Current projects for artificial insemination and extra-uterine pregnancy reflect it. These same natural drives broke through the surface of monastic order, sooner or later: both the lust for power and the power of lust proved difficult to control.

Yet the Benedictine system demonstrated how efficiently the daily work could be done, when it was collectively planned and ordered, when cooperation replaced coercion, and when the whole man—sexuality aside—was employed: above all, when the kind and quantity of work done was regulated by the higher needs of human development. By their own example, the Benedictines denied the slavish premise that all labor was a curse and that manual labor was a particular degradation. They proved, in fact, that it contributed, without the aid of any special gymnastics, such as the Greeks had introduced, to both physical fitness and mental balance. In moralizing the whole process of work, the monastery had raised its productivity; and the term, "le travail Bénédictin" became a byword for zealous efficiency and formal perfection.

Manual work thus ceased to be identified with mindless drudgery; and the exercises of the mind ceased, by the same token, to be disembodied 'headwork' utilizing only a minimum of organic aptitudes, a mere game performed with abstract counters, lacking the touch of the senses, and the continued testing of abstract thought by relevant concrete experience and deliberate action. In the act of accepting work as a common daily burden, monastic order lifted its weight: work, study, and prayer went together. And if the Benedictine motto was "to labor is to pray," this meant that the offices of ritual and work had at last become transferrable and interchangeable: yet each part of life was directed toward a more exalted destination.

2: THE MULTIPLICATION OF MACHINES

From the organization of the tasks of the day, each at its appointed hour, the Benedictines took a further step: the multiplication and assemblage of machines. In emancipating themselves from the dehumanized routines of the collective machine, the old labor army or war army, the monks discovered the true uses and advantages of the machine. For this new type of machine was no longer a massive man-eating megamachine, like that of the Pharaohs, but a labor-saving device that partly dispensed with human muscle. This was not the least triumph of the new discipline.

Though the Benedictines had helped to lift the curse of exhausting manual labor and had equalized its burdens to a larger degree than ever before—at least within their own community—they were not under the illusion that all forms of work are equally a blessing. In the practice of their own rule, they must have discovered what Emerson learned by personal experience—and I myself a century after Emerson: namely, that even one of the most rewarding forms of manual labor, that in a garden, if performed more than a short time, dulls the mind. For while a whole day spent in gardening is the best of anodynes and the most benign of soporifics, the higher functions of the mind go to sleep after all heavy work: indeed physical fatigue has done more to prevent rebellions against the harsh exactions of daily work than either strong drink or brute pressure. Even the benign Emerson could say shrewdly about the immigrant labor gangs which built the first railroads, toiling at starvation pay for fifteen or sixteen hours a day: "It served better than the police to preserve order."

The true solution for this difficulty, the monks had found by the eleventh century, was the invention and wider use of labor-saving machines. This began with the systematic adoption of prime movers like the horse-powered treadmill, the watermill, and later the windmill. The invention of machines and their organization into large work-units went on together. The main features of this process of rationalization were elegantly embodied in the original building plans for the monastery of St. Gall, which have survived the dilapidation and destruction of the original complex of buildings. With the centralized administrative system introduced in the twelfth century by the Cistercians, it is noteworthy that new monasteries were built according to a standard plan.

Monastic mechanization was itself part of an over-all rationalization, which embraced the entire technological process, and only in recent times has it been appraised at its full value. The changeover to free industry, based not on tools and craft processes alone, but greatly aided by labor-freeing machines, began around the tenth century, and was first marked by a steady increase in the number of watermills in Europe. As early as 1066, when William the Conqueror seized England, there were 8,000 watermills, serving less than one million people. At the very modest estimate of 2.5 horsepower per mill, this was twice the energy that was available through the assemblage of the 100,000 men who built the Great Pyramid, and probably more than twenty times in relation to the population of their respective countries.

Though sufficient data are lacking to enable one to speak here except in the most tentative manner, we can now perhaps understand why the first effective labor-saving machines came, not from the technically advanced centers of empire, but from the barbarian peoples on the fringes, who had never wholly succumbed to the sacred myths of divine kingship: that is, from Greece and Gaul, or from Rome itself *after* the breakdown of imperial power.

André Varagnac pointed out that both the Celtic and the Germanic tribes had stubbornly kept to the customs of democracy and resisted the Roman attempts to impose the impersonal forms of their 'mechanized' civilization. He made the additional point that these 'barbarians' were technically inventive during the supposedly Dark Ages: indeed, as the megamachine crumbled away again, new specialized machines and specialized crafts began to proliferate; and for lack of surplus manpower in Western Europe, horsepower and waterpower had an increasingly important role to play.

In the 'eotechnic phase,' as I called it in 'Technics and Civilization,' this diffusion of free energy was a far higher technological contribution than the Pharaonic mode of concentration in human masses. Wherever water flowed rapidly or wind blew, it was possible to install prime movers and to turn solar energy and the earth's rotation to human advantage. The smallest village or monastery had as much use for these new machines as the biggest town; and the progressive employment of such agents was marked. These innovations contributed directly both to the rise and the subsequent prosperity of the free towns, where free labor now was able to establish itself in corporations and guilds that were largely independent of either feudal or royal establishments.

But the monastery, through its very other-worldliness, had a special incentive to develop mechanization. The monks sought, as Bertrand Gille has pointed out, to avoid unnecessary labor in order to have more time and energy available for meditation and prayer; and possibly their willing immersion in ritual predisposed them to mechanical (repetitive and standardized) solutions. Though they themselves were disciplined to regular work, they readily turned over to machinery those operations that could be performed without benefit of mind. Rewarding work they kept for themselves: manuscript copying, illumination, carving. Unrewarding work they turned over to the machine: grinding, pounding, sawing. In that original discrimination they showed their intellectual superiority to many of our own contemporaries, who seek to transfer both forms of work to the machine, even if the resultant life prove to be mindless and meaningless.

3: MACHINES FOR LEISURE

Lest there be any doubt as to the extent to which mechanization entered the Cistercian monastery, let me quote at length from Bertrand Gille, who in turn quotes from Migne's account of Saint Bernard in his 'Patrologia Latina.'

"The river enters the abbey as much as the wall acting as a check allows. It gushes first into the corn-mill where it is very actively employed in grinding the grain under the weight of the wheels and in shaking the fine sieve which separates the flour from the bran. Thence it flows into the next building, and fills the boiler in which it is heated to prepare beer for the monks' drinking, should the vine's fruitfulness not reward the vintner's labour. But the river has not yet finished its work, for it is now drawn into the fulling-machines following the corn-mill. In the mill it has prepared the brothers' food and its duty is now to serve in making their clothing. This the river does not withhold, nor does it refuse any task asked of it. Thus it raises and lowers alternately the heavy hammers and mallets, or to be more exact, the wooden feet of the fulling machines. When by swirling at great speed it has made all these wheels revolve swiftly it issues foaming and looking as if it had ground itself. Now the river enters the tannery where it devotes much care and labour to preparing the necessary materials for the monks' footwear; then it divides into many small branches and, in its busy course, passes through the various departments, seeking everywhere for those who require its services for any purpose whatever, whether for cooking, rotating, crushing, watering, washing or grinding, always offering its help and never refusing. At last, to earn full thanks and to leave nothing undone, it carries away the refuse and leaves all clean."

This is not, as Gille is careful to point out, an isolated showpiece of medieval technology: "most of the early abbeys had an extensive water-system of this type," and the "abbey of Fontenay in Burgundy still has its factory, a huge structure with four rooms built at the end of the twelfth century." I know no better description of the effective use of a power technics, applied to just those laborious operations that deplete human energy by their monotony and lower the tone of the whole organism. Thus in the time of Bernard of Clairvaux, well before the twelfth-century resurgence of urban life throughout Europe, a whole series of technological advances had been instituted by the Benedictine monasteries which released labor for other purposes and immensely added to the total productivity of the handicrafts themselves.

How great this release was can be discovered by the number of holidays the medieval worker enjoyed. Even in backward mining communities, as late as the sixteenth century more than half the recorded days were holidays; while for Europe as a whole, the total number of holidays, including Sunday, came to 189, a number even greater than those enjoyed by Imperial Rome. Nothing more clearly indicates a surplus of food and human energy, if not material goods. Modern labor-saving devices have as yet done no better.

With the importation of the Persian invention of the windmill in the twelfth century, the supply of power greatly increased in areas that could count on this source of energy: by the fifteenth century, batteries of windmills surrounded every progressive town. This development must be coupled with the great moral and political feat that accompanied it: the progressive diminution of both human slavery and serfdom and finally, in every advanced industrial country in Europe, its total abolition.

The original step taken by Christianity and Islam, of accepting the slave as an equal member of the spiritual community, was now crowned for the first time in the history of civilization by the progressive abolition of slavery itself. Largely as a result of the technological advance first promoted by monasticism in the pursuit of a holy life the "impossible" conditions for the abolition of slavery laid down by Aristotle in a famous passage of the 'Politics' had at last been met. "For if every instrument, at command, or from foreknowledge of its master's will, could accomplish its special work (as the story goes of the Statues of Daedalus), or what the poet tells us of the tripods of Vulcan, how 'self-taught they moved into the godlike course'; if the shuttle thus would weave and the lyre play of itself, then neither would the chief workman want assistants nor the master slaves." That consummation now approached.

As a result of this combination of an ordered routine of life and technical mastery, the Benedictine monasteries prospered: they traded their surplus products with other houses of the order all over Europe, and in addition, invested such a large part of their capital in more august Abbey churches and other buildings, that they fell under the condemnation of more sensitive Christian souls, who saw that the private property which the monks renounced at their induction was more than offset by their communal opulence through collective proprietorship—which in time was coupled with a richer diet and more copious or more refined drinks, including distilled liquors like brandy and the very cordials that still bear the name of the Benedictine and Carthusian orders.

To handle these economic enterprises, the kind of order that first regularized devotions proved applicable in every form of record-keeping and exact measurement. By the twelfth century the efficient rationalization that had been achieved in the monastery was ready to be transferred to secular occupations. For the Benedictines had proved what the English evangelist, John Wesley, was to point out many centuries later: that Christian thrift, sobriety, and regularity would inevitably lead to worldly success. Most of the habits that Max Weber erroneously treated as the special property of sixteenth-century Calvinist Protestantism were in effective operation in the medieval Cistercian monastery.

4: THE MEDIEVAL EQUILIBRIUM

To sum up. The Benedictine commitment to 'labor and prayer' had done more than take the ancient curse off work. For the productivity of this system established, likewise, the economic value of a methodically ordered life; and that moral was not lost on contemporary craftsmen and traders. The Venetian merchant, Louis Cornaro, in his classic essay on attaining longevity, took this regularity and abstemiousness to be the guarantee not only of a fruitful life but of financial prosperity. These 'Protestant' virtues long antedated Calvinism.

What the monastery began, the medieval guilds carried through; for they not merely laid down a fresh basis for association in craft and trade, but they restored to work the esthetic and moral values, conditioned by religion, that governed the rest of their life. They, too, were autonomous corporate bodies, which established a common discipline for the performance of work and the regulation of wages and prices. With the gradual retreat of both slavery and serfdom, abetted and hastened by the scarcity of labor in the fourteenth century, after the Black Death, the status of the worker improved and the demand for machines widened. The restoration of European productivity within a century, after the loss of between a third and a half of the total population, bears witness to both the abundant human and mechanical energies available.

This transformation was decisive: so there is no need to exaggerate it. The integration of work with moral regulations, esthetic expression, and social security, was never completed in the guilds, any more than in the monasteries. As wealth accumulated, particularly in the wholesale trades and in mining and shipping ventures, the economic gap between the poor and the wealthy guilds grew. In an effort to protect their own particular craft from outside competition and to feather their individual family nests, each guild not merely restricted membership, but too often turned its back upon the technical advances that were being made outside the legal protection of urban centers; and as happens today with the trades unions, they took no account of the growing body of casual workers, handicapped by poverty and lack of training. Admittedly, then, the gains in both productivity and creativity were uneven: yet the total result down to the sixteenth century, which marks a turning point, remains impressive.

Thanks to this emerging economy, which combined handicraft skills with mechanization and power machinery, something like a balance had been brought about, more favorable to a diversified and humane life than anything that earlier technics had ever before achieved, since in regions like the Low Countries, it introduced many improvements in transportation, agriculture, and industry. The technical pace of the previous three thousand years was being quickened without any loss of the esthetic aptitudes that had once been too assiduously developed for the benefit of the upper classes alone. By the sixteenth century the printing press had eliminated the class monopoly of knowledge, and the reproductive processes in printing, etching, engraving had once more democratized the making of images; while in one department after another the sort of material goods that had once been restricted to a small caste were now becoming available to a larger population: indeed, the power machine promised to extend all these advantages to their theoretic limits.

In many departments, until the seventeenth century a nice balance was established between the rural and the urban, between the organic and the mechanical, between the static and the dynamic components. What this regime lacked in power, it made up for in time: for even its common products were fabricated for durability; while its great architectural works were not only built over the centuries but planned to outlast the centuries: so well that many of them remained erect, at the end of the Second World War, amid the rubble of destroyed modern buildings all around them.

Unlike the continuity achieved in the art and architecture of early Egypt after the Pyramid Age, medieval continuity was preserved in the midst of constant change of both contents and form; and its effectiveness over the centuries was in radical contrast to the one-generation forced labor economy of the Pharaohs, or that of the absolute monarchs, like Louis XIV and Peter the Great, who claimed similar powers in the seventeenth century.

But the process of moralizing work and integrating it with every other human activity was never fully accomplished. For the one universal institution of the medieval period in Western Europe, the Christian Church, at a critical moment in the fourteenth century threw its authority on the side of the forces specializing in power—absolutism, militarism, and capitalism—divorced from the social commitments of the monastery, the guild, the free city. Between them, however unintentionally, these institutions laid the basis for a dehumanized technology, and in the end for something that proved even more fateful, a new myth of the machine. Let us examine the beginnings of this process.

5 : THE MECHANIZATION OF MAMMON

If the development of automatic sources of power was one of the essential contributions of the monastic regime, the re-shaping of capitalist enterprise in its systematic modern form was the other, as G. C. Coulton, that knowledgeable medieval historian, held. But whereas monasticism was originally dedicated to a single object, the pursuit of individual salvation, capitalism in its orthodox form was dedicated to the glorification of Mammon and the achievement of salvation of a more tangible sort, by expanding the opportunities for profit and capital accumulation and ostentatious consumption.

In concentrating upon that pursuit capitalism necessarily addressed itself to overthrowing the restrictive, continent practices of all the Axial religions. That the original monastic thesis of renunciation and self-abnegation should have produced its capitalist antithesis, avarice and acquisitiveness, might not have surprised Karl Marx, but it remains one of the wry turns of history.

Capitalism is not of course a modern phenomenon. One means here by capitalism the translation of all goods, services, and energies, into abstract pecuniary terms, with an intensified application of human energy to money and trade, for the sake of gains that accrue primarily to the owners of property, who ideally are prepared to risk their savings on new enterprises as well as to live off the income of established industrial and commercial organizations. Defined in these broad terms, capitalism makes its first appearance, in primitive mercantile form, soon after kingship; and as capital investment grows, takes on an increasingly corporate form. Though the possibilities for gain first came through the control of land and the exaction of rents, capitalist enterprise naturally took hold of areas like ship-building, maritime trading, mining, and smelting, which demanded large investments, provided that these enterprises were too small or too complex to be handled economically by the clumsy bureaucratic organization of the state.

As scholars dig more deeply into the early Mesopotamian and Egyptian records, it seems likely that state capitalism, with the merchant as an official of the state, may have preceded private capitalism if not private barter; and if capitalism, from the thirteenth century on, took over much of the discipline of monastic organization, this was only following earlier habits of regimentation established by the primeval megamachine. The capitalist, as landlord, as merchant, as speculator, in the early stages, might be likened, a little unkindly, to jackals who feasted on the less attractive offal left over from the royal lion's kill.

For long, indeed, trade and industry and banking remained at the mercy of the sovereign power. Their profits and privileges were constantly depleted in times of war by the destruction of cities and by the pillage of their temples, treasuries, and rich inhabitants, and in times of peace by wholesale extortions, levies, and unduly burdensome taxes, often unevenly assessed by corrupt tax farmers.

To flourish at all, mercantile capitalism had to operate over as wide an area as any empire, and take risks of losses larger than any small-scale merchant dared. Capitalist operators needed a special kind of acuteness, agility, inventiveness, and enterprise to offset these disabilities; and it is hardly surprising that from early times on the alphabet, coined money, and Arabic numbers come from people who were pre-eminently long-distance traders and colonial exploiters. Marco Polo was neither the first nor the last of such adventurers: Jacob Fugger in one age, Rothschild in another, and John D. Rockefeller in a third incarnated this institution.

The classic theory of capitalist accumulation was first set forth in the Middle Ages, not by economists, but by the schoolmen in their purely theological doctrine of the Treasury of Salvation: the piling up of earthly merits through continence and sacrifice for the sake of an immense future reward in Heaven. One of these schoolmen, Vincent of Beauvais, in the thirteenth century admonished people to work, not just for a living, but for the sake of accumulation, which would lead to the further production of wealth. Scholars who go on repeating Max Weber's anachronistic equation of the spirit of capitalism with Protestantism should find a way to conceal the abundant medieval data that contradict it.

Protestantism, as it first appeared in the doctrines of the heretic merchant, Peter Waldo, in the twelfth century, was in fact a vehement protest against the new capitalism and a repentant effort to go back to the way of life of the early Christians, who despised worldly goods and the insidious temptations of trade. The social outlook of the Waldensians, the Wycliffites, the Lollards, the Beguins, the later Anabaptists, was first and last militantly anti-capitalist; and so for that matter were the autarchic economic principles and the anti-usury polemics of Martin Luther.

When Francis of Assisi, a century after Waldo, made a similar effort to restore by humble daily labor the central doctrines of early Christianity, the continued pressure of capitalistic expansion doomed him to failure: poverty did not serve capital accumulation, and voluntary service for the good of the community could not be anything but disruptive in the new wage system which replaced serfdom. The very papal authority that cannily incorporated the Franciscan order into the Church, promptly proclaimed, under John XXII, that the current belief that early Christians actually practiced communism was a damnable heresy.

Now, the desire for money, Thomas Aquinas pointed out, knows no limits, whereas all natural wealth, represented in the concrete form of food, clothing, furniture, houses, gardens, fields, has definite limits of production and consumption, fixed by the nature of the commodity and the organic needs and capacities of the user. The idea that there should be no limits upon any human function is absurd: all life exists within very narrow limits of temperature, air, water, food; and the notion that money alone, or power to command the services of other men, should be free of such definite limits is an aberration of the mind.

The desire for limitless quantities of money has as little relevance to the welfare of the human organism as the stimulation of the 'pleasure center' that scientific experimenters have recently found in the brain. This stimulus is subjectively so rewarding, apparently, that animals under observation willingly forgo every other need or activity, to the point of starvation, in order to enjoy it. When capitalists become aware of the nature of such pecuniary over-stimulation, once called the curse of Midas, they either commit suicide or turn repentantly to public service and philanthropy.

In the ideal capitalist ego, the miserly hoarding of money blended with the zealous acquisition of illimitable riches, just as the abstemious habits of the monk combined with the adventurous activity of the soldier. To speak in Freudian terms, it appealed to both 'anal' and 'oral' personality types. The new capitalists deserved in large measure the title later bestowed on them, 'Merchant Adventurers'; and at an early period these conflicting yet complementary strands of inheritance came together in the order of Knights Templars, those warrior-bankers of the Middle Ages. So, too, it was in no defiance of the new capitalist spirit that the trading posts of the great Hansa towns were in fact run as monastic enclaves, under a strict military discipline.

This combination of traits was in due course transmitted to the scientific ideology of the seventeenth century: a readiness to entertain daring hypotheses, a willingness to dismember organic complexities, while subjecting every new theoretic insight to cautious observation and experimental test. Despite their different origins and their seemingly incompatible aims, the monk, the soldier, the merchant, and the new natural philosophers or experimental scientists were more closely united than they realized. Like John Gabriel Borkman, the Ibsen hero who summed up the capitalist spirit of the nineteenth century, each was ready to forgo love and to sacrifice life, in order to exercise power—however sublimated and transmuted that power might seem to be.

But at the same time capitalism, in satisfying its insatiable desire for pecuniary riches, took over and translated into its own special terms the economy of abundance that had originally been the work—and the mark— of divine kingship. The actual increase in productivity brought an often happy release from the nagging constraints of natural poverty and economic backwardness; and it prompted a steadily growing revolt against the ascetic inhibitions of orthodox Christianity, which had been easy to popularize in a Time of Troubles when no tempting alternatives were available, but now seemed gratuitous and needlessly life-denying.

Within a few centuries, the new capitalist spirit challenged the basic Christian ethic: the boundless ego of Sir Giles Overreach and his fellows in the marketplace had no room for charity or love in any of their ancient senses. The capitalist scheme of values in fact transformed five of the seven deadly sins of Christianity—pride, envy, greed, avarice, and lust—into positive social virtues, treating them as necessary incentives to all economic enterprise; while the cardinal virtues, beginning with love and humility, were rejected as 'bad for business,' except in the degree that they made the working class more docile and more amenable to cold-blooded exploitation.

6: PECUNIARY INCENTIVES TO DYNAMISM

Werner Sombart observed that if he were pressed to give a date for the inauguration of capitalism, he would say that the publication of Leonardo Pisano's 'Liber Abbaci,' the first popular treatise on arithmetic, would be that date, A.D. 1202. Any such single starting point would be challengeable; one might cite a score of equally critical moments. But one of the most important traits of the new capitalism, its concentration on abstract quantities, was indeed furthered by such instruction.

The new form of universal accountancy isolated from the tissue of events just those factors that could be judged on an impersonal, quantitative scale. Counting numbers began here and in the end numbers alone counted. This was ultimately a more significant contribution of capitalism than any of the actual goods the merchant bought and sold. For only when the habit of using mathematical abstractions became ingrained in a dominant part of the community could the physical sciences resume the place they had first occupied in the great trading cities of Ionic Greece. Again, this connection was not accidental. Thales, the archetypal scientist, was a speculative philosopher in more than one sense, for according to Diogenes Laërtius, he had made a fortune by cannily cornering oil presses one season when the olives were specially abundant.

Wherever the capitalist spirit took hold, people became familiar with the abstractions of the counting house: timing, weighing, and measuring, in ever more exact amounts, became the mark of this whole regime. The change was not spontaneous, but the result of deliberate intention and persistent indoctrination. From the thirteenth century on, the grammar school, with its fundamental courses in reading, writing, and arithmetic, inculcated the elementary symbols for long-distance buying and selling, for making contracts, for book-keeping and bill-rendering. The need for reliable information and careful forecasts, in order to trade in commodities not seen till delivered, furthered the appreciation of quantitative appraisals in every department: not merely just weights and measures, but accurate astronomical observations in navigation.

The impersonal, bureaucratic order of the counting house vied with monastic and military order in laying the foundations for the inflexible discipline and impersonal regularity that has now gradually extended itself to every aspect of institutional life in Western civilization. This order has been smoothly translated into automatic machines and computers, even more incapable of exercising humane judgement and discretion than a trained clerk. The new bureaucracy devoted to managerial organization and coordination again became a necessary adjunct to all large-scale, long-distance enterprises: book-keeping and record-keeping set the pace, in standardized uniformity, for all the other parts of the machine. The failure to reckon with this mathematical aspect of mechanization, as a prelude to industrial inventions, has resulted in a warped and one-sided picture of modern technics. This account gives to specific tools and machines by themselves the priority in effecting changes that first took place in the human mind and were translated later into institutions and mechanisms.

During the centuries when capitalism and mechanism were being shaped, their ultimate tendencies were largely concealed; for they were both curbed by the stubborn rivalry and the formidable inertia of many other institutions. As late as the sixteenth century the theologians of the University of Paris denounced the opening of State Banks on the ground that usury (lending money at interest) was a sin in Christian theology; and the humane protection offered by the guilds to their own members was still so effective in the eighteenth century that new enterprises, using cheaper methods of production, were forced, as Adam Smith pointed out, to establish themselves in the countryside, or in nearby unincorporated suburbs, bootlegging their products into the town.

Other-worldly Church doctrines and restrictive feudal customs, such as tethered, non-purchasable landed property, guild regulations, high standards of craftsmanship, family interests, all slowed down the pace of capitalism's conquests. The desire for qualities, for long conflicted with the demand for quantities. Even as late as the sixteenth century, when the great Augsburg financier and industrialist, Jacob Fugger the Elder, offered to take a brother of his into his highly profitable business, the latter refused, on the ground that such a sinful enterprise would endanger his soul's chances for salvation. At that period there was still an open choice.

In its early spread in Western Europe, furthermore, urban capitalism was subject to the same perversions as had arisen under kingship. The leading mercantile cities resorted to armed force in order to destroy rival economic power in other cities and to establish a completer economic monopoly. These conflicts were more costly, destructive, and ultimately even more futile than those between the merchant classes and the feudal orders. Cities like Florence, which wantonly attacked other prosperous communities like Lucca and Siena, undermined both their productivity and their own relative freedom from such atrocious attacks. When capitalism spread overseas, its agents treated the natives they encountered in the same savage fashion that it treated their own nearer rivals.

In sum, where capitalism prospered, it established three main canons for successful economic enterprise: the calculation of quantity, the observation and regimentation of time ('Time is Money'), and the concentration on abstract pecuniary rewards. Its ultimate values—Power, Profit, Prestige—derive from these sources and all of them can be traced back, under the flimsiest of disguises, to the Pyramid Age. The first produced the universal accountancy of profit and loss; the second ensured productive efficiency in men as well as machines; the third introduced a driving motive into daily life, equivalent on its own base level to the monk's search for an eternal reward in Heaven. The pursuit of money became a passion and an obsession: the end to which all other ends were means.

With this shift from the contemplative life of the religious to the active life of merchants, sailors, financiers, industrial enterprisers, these canons took on the form of moral imperatives, if not neurotic compulsions. Yet so well established was the older system of values that even into the nineteenth century the ambition to retire from active business in the prime of life with a 'competence' still seemed to many merchants more attractive than the piling up of more money by incessant application to business.

But it was in science that the abstractions of capitalism came in the long run to play an even greater role and to bring an even greater reward. When the Royal Society was founded in London in the middle of the seventeenth century, merchants and bankers took a leading part in it, not merely as providers of funds, but as active experimenters in the new science. The notion that every item of exchange must be accounted for and that 'the books must balance' preceded by centuries Robert von Mayer's doctrine of the conservation of energy.

Thus account-keeping and time-keeping had not merely been secularized by the sixteenth century, but they had ensured that such sacrifices as this regimen exacted were attached to the promise of a tangible reward. Under kingship, rewards for the privileged classes had not flowed directly from their services, but were dependent upon the caprices of the ruler and were often ill-proportioned to either the effort expended or the value of the result. But under the new accountancy of capitalism, failure was directly penalized by loss and even more significantly success, tied to efficiency and foresight, was abundantly rewarded.

Capitalism, in other words, relied on the method of conditioning used successfully by animal trainers to ensure obedience to orders, and to secure the performance of difficult feats. And whereas kingship had emphasized punishment, a method that has a definite limit in the death of the individual too severely punished, there was no limit under early capitalism to the possibility of reward. Moreover, this new motive did not appeal only to a single class: it theoretically held out a promise and a hope to the humblest individual who would strictly apply himself to business. From small beginnings, by thrift, acumen, and concentration, great fortunes might be made, as 'Poor Richard' later insisted. Any Dick Whittington might, theoretically, become Lord Mayor of London.

Edward Thorndike, whose psychological experiments established the value of rewards rather than punishments for effective conditioning, was conscious of the contrast between the punishment method, traditionally used by political government, and the contrary method employed by business. "The shift from the status of feudalism," he observed, "to the contract of the modern world was to some extent a shift from use and wont supported by threats and punishment to experimentation supported by hopes and rewards. Business was especially permeated by the reward principle." What was peculiar to the capitalist economy however, was the fact that immediate rewards were mainly in the abstract form of money, and any further sharing of rewards by the worker and consumer was put off until the investing and managerial groups were fully satisfied—though in principle no limit was placed on their demand for ever larger gain. In the teleology of business enterprise, profit was the final end of life. Compared to this, the ancient pharaonic system, which sought "Life, Health, Prosperity" was better grounded in organic realities.

In short capitalism exploited and universalized a powerful positive motivation that—for decent human reasons—had never been tapped in more primitive societies. For centuries, it is true, the capitalist continued to use the negative form, of penalization rather than reward, to ensure compliance by the worker, while self-righteously reserving rewards for himself, his managerial colleagues, and the investors.

Money, as the nexus in all human relations and as the main motivation in all social effort, replaced the reciprocal obligations and duties of families, neighbors, citizens, friends. And as other moral and esthetic considerations diminished, the dynamics of money power increased. Money was the only form of power which, through its very abstraction from all other realities, knew no limits—though finally this indifference to concrete realities would meet its nemesis in the progressive inflations of an 'expanding economy.'

7 : ENTER THE SORCERER'S APPRENTICE

Though capitalism had begun to establish a new style of thinking by the sixteenth century, it did not operate alone: indeed it could hardly have made headway so rapidly without gaining the support of other institutions and interests, some of them newly awakened to life by the first successes in creating power machines and automatons.

Since the thirteenth century, new inventions had been stirring everywhere in Europe, to mingle with far more ancient ones that had never entirely disappeared. For long these fantasies of power, power that transcended man's natural limitations, including his biological mortality, had floated from mind to mind: first of all, perhaps, man's envy of the birds and his desire to conquer the air. That dream, expressed early in Mesopotamian myth, took more realistic form in the story of Daedalus among the Greeks, and then spread everywhere, even to Peru, in the figure of Ayar Katsi, the flying man—to say nothing of the flying carpet in the 'Arabian Nights.'

Similarly, the old wish for the cornucopia of endless plenty called adventurers to distant lands, along with the search for the elixir of life, the panacea—nowadays called the wonder drug—that would cure all human ailments. And the alchemists, many centuries before their like-minded scientific successors, such as Herman Muller and F. H. C. Crick, played with the idea of creating a living homunculus in a test tube.

Now the passage of a myth from the unconscious to actual life is an obscure and devious one. Until it begins to be supported by changes in daily existence, and in turn confirms them, one can hardly do more than guess at its existence, for it remains at best an elusive impulse, a seemingly empty wish, often too outrageous when expressed publicly to be taken seriously, certainly too deeply buried to affect the surface of life.

Yet the conception of new kinds of power-machines, which could be put together and made to work without magical hocus-pocus, fascinated various minds from the thirteenth century on, notably Albertus Magnus, Roger Bacon, and Campanella—all, be it noted, monks. Dreams of horseless carriages, flying machines, apparatus for effecting instantaneous communication, or transmuting the elements, multiplied. These fantasies were no doubt incited by such rudimentary machines as were already in operation: for there must have been a moment when the first windmill or the first automaton moving by clockwork seemed as marvellous as the first dynamo or the first 'talking machine' less than a century ago.

One remarkable accompaniment of these dreams should parenthetically be noted, for it was soon to be repressed or peremptorily dismissed. In their initial announcement, such dreams did not always promise a happy ending: significantly, they were mixed with forebodings.

The ancient Norse legends contain premonitions of a massive disaster, that of Ragnarok, the Twilight of the Gods, when the world would go up in flames and the brutal giants and demons would triumph over all that was benignly human and godlike. In another vein, the deliberate suppression, by the Christian Church, of the anatomical dissection of human cadavers, even for the improvement of medical and surgical practice, testified to a fear that such a violation of nature for the sake of detailed, exact knowledge alone might be no less inimical to human salvation than the naked pursuit of power—though the culture that inhibited such research stultified itself by inventing the ingenious machines for torture used by the Inquisition.

These mythical promptings and yearnings in time produced a happy effect that nullified the premonitions of disaster. By the end of the fifteenth century certainly, the more alert minds of Europe were aware that a great cyclical change, long in preparation, was about to take place. Poliziano prophetically interpreted the discovery of the New World by Columbus as heralding a beneficent change in the whole life of man; and his contemporary, the Calabrian monk, Campanella, eagerly anticipated the technical achievements that would accompany this re-colonization of the planet by the militant powers of Europe. Campanella, in his utopia, 'The City of the Sun,' pictured boats which would "go over the waters without rowers or the force of the wind, but by some marvellous contrivance"; and at the end of the account the Grand Master who rules this commonwealth says "Oh, if you knew what our astrologers say of the coming age, that it has more history within a hundred years than all the world has in four thousand years before."

Others were stirred by the same intuitions at the same time. Mankind, or at least an awakening minority in Western Europe, was already headed for a new world; and if they did not find it as Sir Thomas More did at the other side of the globe, they believed, as he also did, that they might install it at home, with the aid of a beneficent monarch, by ordaining uniform ordinances and rational regulations, or introducing new mechanical inventions, like the chicken incubator in 'Utopia.' Not least, they might deliberately create more humane social institutions than as yet existed.

Though a whole literature of utopias soon followed in the wake of More's picture of an ideal commonwealth, it is significant that the only one whose direct effects can be traced was the mere fragment of a utopia left behind by Francis Bacon: for it was his 'The New Atlantis' that first canvassed the possibility of a joint series of operations that would combine a new system of scientific investigation with a new technology. At a moment when the bitter struggle within Christianity between contentious doctrines and sects had come to a stalemate, the machine itself seemed to offer an alternative way of reaching Heaven. The promise of material abundance on earth, through exploration, organized conquest, and invention, offered a common objective to all classes.

8: THE RADICAL INVENTIONS

Now it was not solely in fantasy, as was noted, that the first efforts to install machines and expand human control were initiated. Though the medieval innovations of the windmill and the watermill made the great advances of the eighteenth century possible centuries before the steam engine came into common use, the central inventions, upon which everything else depended, were all made in Europe before the sixteenth century; and these inventions profoundly changed the space-time framework of the civilized world—and modified both the external environment and man's internal character.

The first set of inventions rested upon the improvement of glass manufacture, which made it possible, thanks to the increased scientific knowledge of optics, recorded in Roger Bacon, to supply pure glass for spectacles, by which defects of eyesight, particularly those brought on by old age, could be corrected. The invention of spectacles prolonged and enriched the mental life of mature people by an average of fifteen years, if sixty be taken as their expectation of life at forty-five; and in many cases, where nearsightedness began earlier, it added an even longer period of mental activity. Amid all the factors that have been uncovered to account for the 'revival of learning' the effect of spectacles was surely not the least.

But the immediate effects of this invention did not end with the prolongation of reading ability through corrective optical diagnosis; for it was the knowledge so learned that led first to the simple magnifying glass, and then to the discovery of the extraordinary magnification possible through using compound lenses. The invention of the microscope and the telescope in the seventeenth century altered all the dimensions of the world: that which had been invisible heretofore, because it was either too small or too distant became visible under closer scanning. Thus these inventions opened up the new world both of micro-organisms and of distant stars and galaxies: a far greater New World than Columbus or Magellan explored.

For the first time, to use a now overworked cliché, it was possible to see both the cosmos and the organic environment *in depth*. Without moving a foot from the microscope or the astronomical observatory, modern man could take into consciousness potentialities that had hitherto not been touched even in his most audacious dreams. This first transformation of spatial dimensions owed nothing to machines for instantaneous communication and rapid transportation, which came much later: the whole vast change was achieved by the glassmakers and lens grinders and optical scientists, with the aid of the simplest tools and utensils. Again 'finding' preceded 'making' and static inventions set the pace for dynamic transformations.

The importance of spectacles was enormously advanced by the other great invention that came a few centuries later: the printing press and its perfection through the invention of movable type for setting up a printed page. This transformed the slow hand process of manuscript copying, which itself had already become standardized, accurate, and elegantly stylized, into a machine process. That final perfection of this art was the outcome of a series of inventions that swept across the world from China and Korea, through Persia and Turkey, until the final steps in the invention were taken, almost simultaneously, in Haarlem and Mainz, with Gutenberg and Johann Fust putting on the finishing touch of casting movable type. This stands as the first example of applying mass production through molding to a dynamic process, with standardized, interchangeable, replaceable parts. The printing press in its own history typifies the changeover from the mechanization of the worker to the mechanization of the work process itself. (For a fuller discussion, see my 'Art and Technics.')

But apart from the direct effects of the printing press upon the invention of later machines, it had a social result that was perhaps even more important: for almost at a stroke, the cheap and rapid production of books broke down the ancient class monopoly of knowledge, particularly of the kind of accurate, abstract knowledge, of mathematical operations and physical events, that had long been the monopoly of a small professional class. The printed book made all knowledge progressively available to all those who learned to read even if poor: and one of the results of this democratization was that knowledge itself, as contrasted with legend, dogmatic tradition, or poetic fantasy, became a subject of intense independent interest, spreading by means of the printed book into every department of life, and immensely increasing the number of minds, past, present, and future, having intercourse with each other.

The enrichment of the collective human mind, through the printing and circulation of books, is comparable only to that linking together of individual brains and experiences through the invention of discursive language. The increase of the scope of scientific discovery and the tempo of mechanical invention can both be largely attributed to the printed book, and from the seventeenth century on, to the printed scientific paper and review. Changes that might have taken centuries to achieve through the circulation of a limited number of manuscripts took place almost overnight through the agency of print.

The third key invention, the clock, became the source of a whole line of other inventions which completed in the realm of time and motion what the magnifying glass had done in space. The mechanical clock dates from the fourteenth century, though parts of the mechanism, and the process of time-measurement itself had come in with the earlier water-clock and the astronomical armillary, which followed the movement of the planets and the seasons. The machine that mechanized time did more than regulate the activities of the day: it synchronized human reactions, not with the rising and setting sun, but with the indicated movements of the clock's hands: so it brought exact measurement and temporal control into every activity, by setting an independent standard whereby the whole day could be laid out and subdivided.

In the sixteenth century the tower clock in the late medieval market-place, which struck the hours, moved into the upper-class home on the mantel shelf, and by the nineteenth century, reduced to the size of a watch it became part of the human costume: exposed or pocketed. Punctuality, ceasing to be "the courtesy of kings" became a necessity in daily affairs in those countries where mechanization was taking command. The measurement of space and time became an integral part of the system of control that Western man spread over the planet.

Karl Marx was one of the first to understand the place of the clock as the archetypal model for all later machines: in a letter to Friedrich Engels in 1863 he observed that "the clock is the first automatic machine applied to practical purposes; the whole theory of *production and regular motion* was developed through it." The italics are his and he did not exaggerate; but the influence of the clock went far beyond the factory, for not only were some of the most important mechanical problems in transmitting and governing motion worked out in clockworks, but the clock, by its increasing success in achieving accuracy, crowned by the invention of the ship's chronometer in the eighteenth century, made it the model for all instruments of precision.

The clock, in fact, is the paragon of automatons: almost all that we can achieve and all that we can expect in automatons was first worked out in the clock. In the progress from the great cathedral clocks of the sixteenth century to the tiny 'self-winding' wrist watch, with its calendar and alarm, one finds, too, the earliest example of the process of miniaturization, over which electronic technology currently, and with reason, is so proud. The automation of time, in the clock, is the pattern of all larger systems of automation.

Between the twelfth and the sixteenth century, then, the key inventions upon which a whole order of new machines would be built, as the first step in creating a new kind of megamachine, had been assembled: the watermill, the windmill, the magnifying glass, the printing press, the mechanical clock. On those inventions all the later technical advances, different in kind as well as in power from those of earlier industrial cultures, largely depended. It was this new technical assemblage that gave the scientists of the seventeenth century the agents they needed to achieve what was to become a world revolution, strangely similar, in all its main assumptions and goals, to those of the Pyramid Age.

9: THE PREMONITIONS
OF LEONARDO DA VINCI

In the mind of Leonardo da Vinci (1452–1519), one of the greatest intellects of a great age, a multitude of practical inventions accompanied his ideal projections. He and other contemporary artist-engineers demonstrated, as early as the sixteenth century, how many of the technical achievements of our own time had already been sampled in fantasy and even tested in actual or pictured models.

By now everyone is familiar with Leonardo's many daring but remarkably practical constructions, and his equally practical anticipations: likewise with his unsuccessful Great Bird. The latter was actually a glider, with wings which could not move, a failure for reasons that his near contemporary, Borelli, was soon to explain by his remarkable researches on the locomotion of animals, and in particular on the anatomy of birds. For even if Leonardo's wings had been feather-light, they would have required enormous pectoral muscles on the scale of a bird's breast to flap them.

Yet in doing justice to Leonardo, the inventor and engineer, scholars have tended to overlook how disturbed he was by his own mechanical fantasies. Like Roger Bacon, he too had foreseen in his usual enigmatic way (labelled a dream) that "men shall walk without moving [motorcar], they shall speak with those absent [telephone], they shall hear those who do not speak [phonograph]." But in another fantasy, written in the form of a letter, Leonardo conjures up the image of a hideous monster that would attack and destroy mankind. Though Leonardo gave the monster a tangible, gigantic, sub-human form, his actual performances come all too close to the hideous scientifically engineered exterminations our own age has witnessed. The monster's imperviousness to attack only completes resemblance to the airborne atomic, bacterial, and chemical weapons that now have it in their power to wipe out all of mankind. Leonardo's description, printed in MacCurdy's translation of the Notebooks under 'Tales,' demands direct quotation.

"Alas, how many attacks were made upon this raging fiend; to him every onslaught was as nothing. O wretched folk, for you there availed not the impregnable fortresses, nor the lofty walls of your cities, nor the being together in great numbers, nor your houses or palaces! There remained not any place unless it were the tiny holes and subterranean caves where after the manner of crabs and crickets and creatures like these you might find safety and a means of escape. Oh, how many wretched mothers and fathers were deprived of their children! How many unhappy women were deprived of their companions. In truth, my dear Benedetto, I do not believe that ever since the world was created there has been witnessed such lamentation and wailing of people, accompanied by so great terror. In truth the human species in such a plight has need to envy every other race of creatures . . . for us wretched mortals there avails not any flight, since this monster when advancing slowly far exceeds the speed of the swiftest courser.

"I know not what to say or do, for everywhere I seem to find myself swimming with bent head within the mighty throat and remaining indistinguishable in death, buried within the huge belly."

There is no way of proving that this nightmare was the reverse side of Leonardo's hopeful anticipations of the future: but those who have lived during the last half century have experienced both the mechanical triumphs and the human terror they have generated, and we know, even better than Leonardo, by what a large factor his anticipated evils have been multiplied.

Like his successors who actually promoted the myth of the machine and caused it to gain practical ascendancy, Leonardo could have had no conscious foreboding that he was both prefiguring and serving a myth. On the contrary, like them, he probably felt that he was creating a more sensible rational order, in which his acute intelligence, with more adequate methods and agents than man had ever possessed before, would bring all natural phenomena under the sway of the human mind. These technical premises seemed so simple, their aim so rational, their methods so open to general imitation, that Leonardo never saw the need to put the question we now must ask: Is the intelligence alone, however purified and decontaminated, an adequate agent for doing justice to the needs and purposes of life?

Yet some insight into this limitation had already lurked under the surface of Leonardo's conscious interests and tainted his otherwise favorable picture of what rational invention could do for man. He was, intellectually speaking, too large a personality to fit into any of the standard categories of engineer, inventor, artist, or scientist; though like his near contemporaries, Michelangelo and Dürer, and many earlier and later figures, he ranged freely over a wide territory, from geology to human anatomy. But he realized the limitations of mechanical invention alone. In one of his notes he wrote: "Would that it might please our Creator that I were able to reveal the nature of man and his customs even as I describe his figure."

Leonardo had at least a glimpse of what was missing from the mechanical world picture. He knew that the man he dissected and accurately depicted was not the whole man. What neither the eye nor the scalpel could reveal was equally essential to the description of any living creature. Without an insight into man's history, his culture, his hopes and prospects, the very essence of his being was not accounted for. Thus he knew the limitations of his own anatomical descriptions and mechanical inventions: the visible world represented in his paintings was but an eviscerated mummy; and he demonstrated in his own experience that the suppressed part of his unconscious world would at last erupt in the same nightmares that now haunt all of humanity.

Unfortunately, Leonardo's talents, as happens to so many of the best scientists and technicians today, were at war with his conscience. Seeking for fuller command of the machine, he was ready, like so many of our present-day scientists, to sell his services to the Duke of Milan, one of the leading despots of his day, provided he got an opportunity to exercise his inventive talents. Yet because the new ideological framework had not yet been put together, Leonardo retained an intellectual freedom and a moral discipline that only rarely could be achieved after the eighteenth century. Although Leonardo, for example, invented the submarine, he deliberately suppressed this invention "on account of the evil nature of men, who would practice assassination at the bottom of the sea." That reservation marks a moral sensitiveness equal to his inventive abilities: only a relative handful of scientists, like the late Norbert Wiener or Leo Szilard in our day, have shown any parallel concern and self-control.

Leonardo's consistent concern with moral problems, with the kind of human being he was himself becoming and was in turn helping to create, sets him off from those who confined their attention to observations, experiments, and equations without the faintest sense of responsibility for their consequences. In all likelihood his sensitiveness to the social outcome of invention created an inner conflict that curbed his success: but so strong were the pressures of both mechanization and war that he nevertheless was driven by his mechanical demon, not merely to invent submarines, but land tanks and rapid-firing guns, and many kindred devices. Yet if Leonardo's imaginative anticipations and internal conflicts had been general, the whole tempo of later mechanization might have been slower.

Leonardo was proud of his status as an engineer: he even listed half a dozen engineers of classical times, from Callius of Rhodes to Callimachus in Athens, he who was skilled in making great bronze castings, as if to establish his own place among his ancient peers. With a sense of history later engineers lost, he ransacked the annals of antiquity for suggestive hints from Greek or Persian engineers. He even cited the fact—to our present astonishment—that the Egyptians, the Ethiopians, and the Arabs had used the old Assyrian method of inflating wineskins to buoy up camels and soldiers in fording rivers; and he advocated building unsinkable boats for transporting troops, also after an ancient Assyrian model.

In his military preoccupations Leonardo did not stand alone: he was but one of a large group of highly inventive minds in Italy, France, and Germany, all devoted to military engineering, finding immediate service, if not a full use for their inventive powers, in the train of absolute rulers who reproduced, in miniature, the powers and the ambitions of more ancient monarchs. They designed canals with canal-locks, and fortifications; they invented the paddle-wheel boat, the diving bell, the wind-turbine. Even before Leonardo, Fontana had invented the velocipede and the military tank (1420), and Konrad Keyeser von Eichstadt had invented both the diving suit (1405) and the infernal machine.

One need not be surprised that the demand for such inventions did not come from either agriculture or handicraft industry: the stimulus to invention, if not the immediate practical support, came from the same socio-technical power complex that had produced the earliest megamachines: absolutism and war.

Similarly, Leonardo was familiar with the early German method of producing poison gas (from feathers, realgar, and sulphur) to asphyxiate a garrison: a grisly fifteenth-century invention that anticipated its first twentieth-century application by the same nation. Like other military engineers of his time, he played with the possibility of armored tanks, propelled by hand-operated cranks, to say nothing of revolving scythes, advancing in front of a horse-propelled vehicle, to mow down the enemy.

One begins to understand how deeply the old myth of unlimited power had begun to stir again in the modern mind, when one observes how Leonardo, a generous, humane spirit—so tender that he bought caged birds in the marketplace in order to release them—deserted his paintings and spent so much of his energies in both military inventions and fantasies of destruction. Had he concentrated his superb technical skill upon agriculture, he might have effected a mechanical revolution there comparable to that he actually began with his device of the flying shuttle for an automatic loom.

Unlike the unduly sanguine prophets of the nineteenth century, who equated mechanical invention with human improvement, Leonardo's dreams were colored by his consciousness of the spectacle of the human savagery and the murderous malice that some of his own proposed military instruments were designed to serve. These horrors mingled in his dreams with prospective marvels, as in the following prophecy: "It shall seem to men that they see new destructions in the sky, and the flames descending therefrom shall seem to have taken flight and to flee away in terror; they shall hear creatures of every kind speaking human language; they shall run in a moment, in person, to divers parts of the world without movement; amidst the darkness, they shall see the most radiant splendors. O marvel of mankind! What frenzy has thus impelled you!"

The vague, ambiguous prophecies of Leonardo's contemporary, Nostradamus, may easily be dismissed: but Leonardo himself committed to paper even more remarkable forebodings of the world that science and mechanization would eventually bring into existence. In his notes on necromancy, he unsparingly criticized people who were then proclaiming the reality of fantastic powers possessed by "invisible beings" for transforming the modern world. Many of these fantasies were nothing but early unconscious projections of natural forces that later took concrete form; and no one described the consequences of such forces more incisively than Leonardo, even in the act of denying their possibility.

Should the claim of the necromancers be established, Leonardo wrote, "there is nothing on earth that would have so much power either to harm or benefit man. . . . If it were true . . . that by such an art one had the power to disturb the tranquil clearness of the air, and transform it into the hue of night, to create corruscations and tempests with dreadful thunder-claps and lightning flashes rushing through the darkness, and with impetuous storms to overthrow high buildings and uproot forests, and with these to encounter armies and break and overthrow them, and—more important than this—to make devastating tempests, and thereby to rob the husbandmen of the rewards of their labors. For what method of warfare can there be which can inflict such damage upon the enemy as the exercise of the power to deprive him of his crops? What naval combat could there be which should compare with that which he would wage who has command of the winds and can create ruinous tempests that would submerge every fleet whatsoever? In truth, whoever has control of such irresistible forces would be lord over all nations, and no human skill will be able to resist his destructive power. The buried treasure, the jewels that lie in the body of the earth, will become manifest to him; no lock, no fortress, however impregnable, will avail to save anyone against the will of such a necromancer. He will cause himself to be carried through the air from East to West, and through all the uttermost parts of the universe. But why do I thus go on adding instance to instance? What is there which could not be brought to pass by a mechanician such as this? Almost nothing, except the escaping from death."

In the light of history, which, can one say today, is the more re-markable?—these pure fantasies themselves, welling forth out of the un-conscious without any check from history or current experience, or Le-onardo's interpretations of what the social consequences would be, if the necromancers' assertions actually proved true? The first response clearly anticipated in dream what centuries later has become a formidable reality: control over the forces of nature sufficient to bring about total destruction. To Leonardo's credit, he realized in advance—*almost five centuries in ad-vance*—the implications of these terrible dreams. He foresaw what total power would become in the hands of unawakened and unregenerate men, as clearly as Henry Adams did on the eve of its achievement.

In passing judgement upon that necromantic dream, Leonardo made only one mistake: he believed that the dream was baseless "because there are no such incorporeal beings as necromancy assumes." He could not anticipate as a probability what in his day seemed so remote from being even a possibility—namely that science in a few centuries would discover these invisible "incorporeal beings" in the heart of an equally invisible atom. Once that discovery was made, every link in Leonardo's chain of reasoning proved sound.

I am not alone in this interpretation of Leonardo's ominous prophecies; nor was Leonardo himself alone, as Sir Kenneth Clark has pointed out. Clark sees in Leonardo's drawings of deluges a foreboding of cosmic disas-ter, which he connects with other apocalyptic speculations that were current around the year 1500, and which led Dürer to dream of a similar cosmic disaster and record his dream in a drawing dated 1525. Those dreams have proved even more significant than the deformed images and blasted emptiness of many modern paintings: for the latter, so far from being prophetic anticipations, are little better than immediate transcriptions of observable physical ruins and disrupted mental states. Both Leonardo's projects and his anxieties throw a light on what followed.

During the next four centuries, the possibilities of terror that Leonardo exposed in his intimate notes were seemingly laid to rest: they were over-laid by the large apparent increase of orderly scientific interpretation and constructive technical achievement. It was possible, at least for the more prosperous manufacturing classes, themselves growing in number and influence, as against the old feudal and clerical estates, to believe that the benefits of science and mechanization would far outweigh their disabilities. And certainly, a thousand fresh inventions and tangible improvements confirmed many of these hopes.

When scrutinized more closely, the social results were, however, more disturbing than the prophets of mechanical progress were willing to admit: from the beginning in the fifteenth century blasted landscapes, befouled streams, polluted air, congested filthy slums, epidemics of avoidable disease, the ruthless extirpation of old crafts, the destruction of valuable monuments of architecture and history—all these losses counterbalanced the gains. Many of these evils were already noted defensively in Agricola's treatise on mining, 'De Re Metallica.' In the heyday of nineteenth-century industry, John Stuart Mill, no enemy of mechanical progress, could still declare in his 'Principles of Economics,' that it was doubtful if all the machinery then available had yet lightened the day's labor of a single human being. Even so, many of the gains were real: some of them would deservedly become part of the permanent heritage of mankind.

While the goods promised by mechanical invention and capitalist organization were naturally more easy to anticipate than the evils, there was one evil, more mountainous than all the rest put together, which for lack of sufficient historic information at that period it was impossible to perceive in advance or to forfend. This was the resurrection of the megamachine. Through the coalition of all the institutions and forces we have just been examining, the way had been prepared for the introduction of the megamachine on a scale that not even Chephren or Cheops, Naram-Sin, Assurbanipal, or Alexander, could have deemed possible. For the accumulation of mechanical facilities had at last made it possible vastly to enlarge the scope of the megamachine, by progressively replacing the recalcitrant and uncertain human components with specialized mechanisms of precision made of metal, glass, or plastics, designed as no human organism had ever been designed, to perform their specialized functions with unswerving fidelity and accuracy.

At last a megamachine had become possible that needed, once organized, a minimum amount of detailed human participation and coordination. From the sixteenth century on the secret of the megamachine was slowly re-discovered. In a series of empirical fumblings and improvisations, with little sense of the ultimate end toward which society was moving, that great mechanical Leviathan was fished up out of the depths of history. The expansion of the megamachine—its kingdom, its power, its glory—became progressively the chief end, or at least the fixed obsession, of Western Man.

The machine, 'advanced' thinkers began to hold, not merely served as the ideal model for explaining and eventually controlling all organic activities, but its wholesale fabrication and its continued improvement were what alone could give meaning to human existence. Within a century or two, the ideological fabric that supported the ancient megamachine had been reconstructed on a new and improved model. Power, speed, motion, standardization, mass production, quantification, regimentation, precision, uniformity, astronomical regularity, control, above all control—these became the passwords of modern society in the new Western style.

Only one thing was needed to assemble and polarize all the new components of the megamachine: the birth of the Sun God. And in the sixteenth century, with Kepler, Tycho Brahe, and Copernicus officiating as accoucheurs, the new Sun God was born.

BIBLIOGRAPHY

INDEX

BIBLIOGRAPHY

This bibliography seeks to be representative, as a generalist's sources must be, but it does not pretend to be exhaustive enough to meet a specialist's requirements in any individual department. Though I have repeated certain necessary titles already cited in *Technics and Civilization,* that book does fuller justice to medieval and renascence technics. For certain themes in the last two chapters, the bibliography in *The Condition of Man* may prove helpful.

Agricola, Georgius (Dr. Georg Bauer). *De Re Metallica.* First edition: 1546. Translated by Herbert C. and Lou H. Hoover. Dover edition. New York: 1950.
  Classic treatise on advanced mining practices.

Aldred, Cyril. *The Egyptians.* New York: 1961.
  In the Ancient Peoples and Places Series.

Aldrich, Charles Roberts. *The Primitive Mind and Modern Civilization.* New York: 1931.
  Attempt, a dozen years after the English translation of Freud's *Totem and Taboo,* to expand and correct his thesis from a Jungian point of view. Unsatisfactory, yet with valuable insights.

Al-Jazari. See Coomaraswamy.

Ames, Adelbert. See Cantrill, Hadley.

Ames, Oakes. *Economic Annuals and Human Cultures.* Cambridge: 1939.
  Significant analysis of the domestication of food plants, which shows the need to revise the present picture of a sudden agricultural revolution. Ames offers strong evidence for believing that a large part of this domestication, the selection and testing of food plants, must have taken place in paleolithic times. See also Sauer's *Agricultural Origins,* with its complementary data on the part played by tubers. Likewise Edgar Anderson.

Anderson, Edgar. *Plants, Man and Life.* Boston: 1952.
  Penetrating interpretations of plant and human domestication. But see Oakes Ames and Carl Sauer.

  The Evolution of Domestication. See Tax, Sol (editor).

Anshen, Ruth (editor). *Language: An Enquiry into Its Meaning and Function.* New York: 1957.
  See especially Roman Jacobson, Jean P. De Menasce, and Margaret Naumburg.

Ardrey, Robert. *African Genesis: A Personal Investigation into the Animal Origins and Nature of Man.* New York: 1961.
  A Hollywood mystery thriller, full of the usual homicidal violence, shot on African location, and so parading as an authentic scientific documentary.

Armstrong, E. A. *Bird Display and Behaviour*. London: 1947.
The author, a clergyman, interprets bird language in terms of ritual rather than human language.

Atkinson, R. J. C. *Neolithic Engineering*. In Antiquity: December 1961.

Bacon, Roger. *The Opus Major*. 2 vols. Translated by Robert B. Burke. Philadelphia: 1928.

Bartholomew, George A., Jr., and J. B. Birdsell. *Ecology and the Protohominids*. See Howells, William (editor).

Bates, Marston. *Where Winter Never Comes: A Study of Man and Nature in the Tropics*. New York: 1952.
Like Wallace's early observations, a necessary counterbalance to the over-emphasis of the 'temperate' zone in man's development.

The Forest and the Sea: A Look at the Economy of Nature and the Economy of Man. New York: 1960.

Beckmann, J. *History of Inventions, Discoveries, and Origins*. 2 vols. London: 1846. German title: *Beiträge zur Geschichte der Erfundungen*. 5 vols. Leipzig: 1783–1788.

Benedict, Ruth. *Patterns of Culture*. Boston: 1934.
Like Henri Frankfort in archeology, Ruth Benedict's contribution was to appreciate the individuality of each culture, regional or tribal: its dominant theme, its special interwoven pattern, as contrasted to standardized and universal traits spread through large populations.

Bergounioux, R. P. *La Préhistoire et ses Problèmes*. Paris: 1958.

Berndt, Ronald M., and Catherine H. *The World of the First Australians: An Introduction to the Traditional Life of the Australian Aborigines*. New York: 1954.
Comprehensive and judicious.

Bibby, Geoffrey. *The Testimony of the Spade*. New York: 1956.
Admirable historical account of archeological investigation.

Biderman, Albert D., and Herbert Zimmer (editors). *The Manipulation of Human Behavior*. New York: 1961.

Birket-Smith, Kaj. *Primitive Man and His Ways: Patterns of Life in Some Native Societies*. London: 1960.

Bloch, Marc. *Feudal Society*. 2 vols. Chicago: 1961.
Classic.

Boas, Franz. *The Mind of Primitive Man*. New York: 1911.
Critical review of factors in primitive culture, race, environment, mentality, language, in an effort to remove the misapprehensions and condescensions of Western man.

Primitive Art. First edition. New York: 1927. Dover edition: 1955.
Still admirable.

Bowra, C. M. (Sir Maurice). *Primitive Song.* New York: 1962.
A brilliant effort to understand the prehistoric beginnings of song and poesy by a circumspect use of evidence from a handful of surviving primitives in scattered areas whose mode of life, gathering, fishing, hunting, is similar to that followed by paleolithic peoples. Though these deductions and speculations are outside the realm of scientific proof in the strict sense—like my own about early man—the author is so prudently sensitive to these limitations that he meets every reasonable scholarly criterion.

Brach, Jacques. *Conscience et Connaissance: étude sur les êtres artificiels, les animaux et les humains.* Paris: 1957.

Braidwood, Robert J. *Prehistoric Men.* Fifth edition. Chicago: 1961.
Concise and well-illustrated summary: 'popular' but authoritative.

Braidwood, Robert J., and Gordon R. Willey (editors). *Courses Toward Urban Life: Archaeological Considerations of Some Cultural Alternatives.* Chicago: 1962.
Valuable.

Bramson, Leon, and George W. Goethals (editors). *War: Studies from Psychology, Sociology, Anthropology.* New York: 1964.

Breasted, James Henry. *Development of Religion and Thought in Ancient Egypt.* New York: 1912.
Excellent.

A History of Egypt from the Earliest Times to the Persian Conquest. New York: 1905.
See also Hayes and Wilson.

The Dawn of Conscience. New York: 1933.
Important interpretation of Egyptian moral development by a great archeologist, with a wide historical horizon.

Breuil, Henri, and Raymond Lantier. *Les Hommes de la Pierre Ancienne (paléolithique et mésolithique).* Paris: 1951.
An authoritative survey, with all of the Abbé Breuil's vast knowledge and judicious interpretation. But see Leroi-Gourhan and Laming-Emperaire.

Breuil, H., *et al. The Art of the Stone Age: Forty Thousand Years of Rock Art.* New York: 1961.
Remarkable color reproductions accompany the interpretive essays. But see also Giedion and above all Leroi-Gourhan.

Brumbaugh, R. S. *Ancient Greek Gadgets and Machines.* New York: 1966.

Bücher, Karl. *Arbeit und Rhythmus.* Sixth improved and enlarged edition. Leipzig: 1924.
A classic in relation to both technics and art. See Bowra.

Bushnell, G. H. S. *Ancient Arts of the Americas.* New York: 1965.

Butzer, Karl W. *Environment and Archaeology: An Introduction to Pleistocene Geography.* London: 1964.

Caillois, Roger. *Man and the Sacred.* Glencoe, Ill.: 1959.
Fresh interpretations of anthropology's central mysteries: the sacred and the sacrificial.

Cantrill, Hadley. *The Morning Notes of Adelbert Ames, Jr., including a correspondence with John Dewey.* New Brunswick, N. J.: 1960.

Cassirer, Ernst. *Language and Myth.* New York: 1946.
Translation of *Sprache und Mythos,* which contains the essence of his more massive work, *The Philosophy of Symbolic Forms.*

The Philosophy of Symbolic Forms. Vol. I: Language. New Haven: 1953.
A seminal book; but some of its ideas have been carried further by Susanne Langer.

Catlin, George. Episodes from *Life Among the Indians* and *Last Rambles.* Edited by Marvin C. Ross. Norman, Okla.: 1959.

Chapuis, Alfred. *Les Automates dans les oeuvres d'imagination.* Neuchâtel: 1947.

Chapuis, Alfred, and Edmond Droz. *Les Automates: figures artificielles d'hommes et d'animaux: histoire et technique.* Neuchâtel: 1948.

Childe, V. Gordon. *What Happened in History.* Harmondsworth: 1941.
Both an extension and a condensation of *Man Makes Himself* (1936). Childe handles the materials of archeology with great professional competence, only slightly biassed by his original Marxist slant. With Frankfort's *The Birth of Civilization in the Near East,* this is perhaps the most satisfactory general account available.

Social Evolution. New York: 1951.
As late as this Childe still clung to Lewis Morgan's stages: savagery, barbarism, civilization.

The Dawn of European Civilization. New York: 1958.
This pioneer work, first published in 1925, benefited by five revisions.

Clark, Grahame. *Prehistoric Europe: The Economic Basis.* London: 1952.
The best all-round reconstruction of primitive technology with reference, not merely to the surviving artifacts, but to the material and institutional equipment that gave them meaning.

Archaeology and Society. London: 1939. Revised: 1957.

World Prehistory: An Outline. Cambridge: 1961.
Excellent summary of our present knowledge, with a minimum of speculation.

Prehistory and Human Behavior. In Proceedings of the American Philosophical Society. Philadelphia: April 22, 1966.

The First Half-Million Years. See Piggott, Stuart (editor).

Clark, Grahame, and Stuart Piggott. *Prehistoric Societies.* New York: 1965.
Covers Asia and the New World as well as Europe.

Clark, Wilfred Edward Le Gros. *Antecedents of Man.* London: 1960.

Clarkson, J. D., and T. C. Cochran. *War as a Social Institution.* New York: 1941.
Good. Note Malinowski's paper.

Cole, Sonia. *The Neolithic Revolution.* London: 1963.

The Prehistory of East Africa. New York: 1963.
Summary to date of African finds which indicate possibly the long-sought missing link between the apes and man in the intermediary ape-man form of the Australopithecines, identified as tool users. See Dart and Leakey.

Cook, James. *The Journals of Captain James Cook.* 4 vols. Edited by J. C. Beaglehole. London: 1955.
Three of the earliest and greatest of scientific expeditions under a commander worthy of his trust.

Coomaraswamy, Ananda K. *The Treatise of Al-Jazari on Automata.* Museum of Fine Arts. Boston: 1924.
Thirteenth-century Arabic sketches of automata.

Cornaro, Luigi. *Discourse on a Sober and Temperate Life.* New York: 1916.

Coulton, G. C. *Medieval Panorama: The English Scene from Conquest to Reformation.* Cambridge: 1939.

Count, E. W. *Myth as World View: A Biosocial Synthesis.* See Diamond, Stanley (editor).

Cressman, L. S. *Man in the New World.* See Shapiro, Harry L. (editor).

Critchley, Macdonald. *The Evolution of Man's Capacity for Language.* See Tax, Sol (editor).

Dart, Raymond A. (with Dennis Craig). *Adventures with the Missing Link.* New York: 1959.
Dart had the wits to realize the importance of the Taung fossils discovered by his student, Josephine Salmons, and to follow the clue further. But his anthropological extrapolations were as careless and naive as those of his disciple, Ardrey. The imaginary pictures that accompany the text would vitiate even a sounder exposition. Robert Broom and L. S. B. Leakey, who followed his trail, have done better in exposing various proto-hominids that may be in the general line that led to man. But the identification of these little creatures as "early men" on the basis of tool-making alone is premature, not to say suspiciously over-eager.

Darwin, Charles. *The Voyage of the Beagle.* London: 1845.

The Expression of Emotion in Animals. London: 1872.

Daumas, Maurice (director). *Histoire Générale des Techniques. Tome I. Les origines de la civilisation technique.* Paris: 1962.
Shorter French equivalent of the five-volume English survey.

Davis, Emma Lou, and Sylvia Winslow. *Giant Ground Figures of the Prehistoric Deserts.* In Proceedings of the American Philosophical Society: February 18, 1965.

De Beer, Gavin. *Genetics and Prehistory*. Cambridge: 1965.
Relevant data on the effect of blood types, food habits, and climate on supposedly genetic differences and variations. Brief but informative.

De Morgan, Jacques. *Prehistoric Man: A General Outline of Prehistory*. New York: 1925.
Despite its date, still useful in dealing with material culture, with 190 well-chosen line illustrations.

Derry, T. M., and Trevor I. Williams. *A Short History of Technology: From the Earliest Times to A.D. 1900.*
A useful condensation and re-statement of material in the five-volume Singer work.

De Terra, Helmut. *Humboldt: The Life and Times of Alexander von Humboldt: 1769–1859*. New York: 1955.

Diamond, A. S. *The History and Origin of Language*. New York: 1959.
Suggestive at various points, but curiously limited by the author's special contribution: the part played by the forceful use of the arms in words like strike and break, as essential root formations. Overlooks the problem of symbolic substitution, except in relation to the uses of metaphor.

Diamond, Stanley (editor). *Culture in History: Essays in Honor of Paul Radin*. New York: 1960.
A rich treasury, with some notable essays included in this bibliography.

Dobzhansky, Theodosius. *Mankind Evolving: The Evolution of the Human Species*. New Haven: 1962.

Donovan, J. *The Festal Origin of Human Speech*. In Mind: Oct. 1891–July 1892.
Often referred to in bibliographies, but insufficiently appreciated. One of the first approaches that pointed to the social prerequisites of speech by way of music and ritual. Though my own development of this theme was independent of Donovan's hypothesis, this earlier work on the same lines is all the more welcome, since verification is denied to all such speculations.

Driver, Harold E. *Indians of North America*. Chicago: 1961.
Comprehensive and reasonably detailed.

Dubos, René. *The Torch of Life. Continuity in Living Experience*. New York: 1962.

Du Brul, E. Lloyd. *Evolution of the Speech Apparatus*. Springfield: 1958.
An exhaustive comparative analysis of the anatomical and physiological preconditions for speech.

Eccles, J. *The Neurophysiological Basis of Mind*. Oxford: 1953.
Scientifically rigorous, but philosophically open-minded on the unstated and almost unstatable problem of the relation of brain to mind: how neural messages become colors, shapes, or meanings.

Eco, Umberto, and G. B. Zorzol. *The Picture History of Invention: From Plough to Polaris*. Milan: 1961. New York: 1963.

Edwards, I. E. S. *The Pyramids of Egypt*. Harmondsworth: 1947.
Answers most of the questions so far answerable about the nature and construction of these tombs.

Eiseley, Loren C. *Fossil Man and Human Evolution*. In Thomas, William L., Jr., *Yearbook of Anthropology*. New York: 1955.
Exposes the weakness of defining man as a tool-using animal and points in passing to the importance of man's dream life. After developing my own ideas on these lines I realized that Eiseley, more than anyone else, had sparked this approach, though of course he is not responsible for my interpretations.

Eliade, Mircea. *Patterns in Comparative Religion*. New York: 1958.
More circumspect than Frazer and free from his common-sensible explanations of the irrational and mythical but with the flair of the earlier master: an ability to associate relevant phenomena—virtues disparaged by anthropological isolationists who treat cultures as closed systems. See especially the chapters on Sky Gods and Sun-Worship. The bibliographies in all his works are copious, with full use of Italian and German as well as French texts.

The Sacred and the Profane: The Nature of Religion. New York: 1959.

The Forge and the Crucible. New York: 1962.
Appeared in French in 1956 under the title *Forgerons et Alchimistes*. Examines myths and legends associated with metallurgy, relating them, like alchemy itself, to the earlier neolithic matrix of generation.

The Prestige of the Cosmogonic Myth. In Diogenes: Fall 1958.

Elkin, A. P. *The Australian Aborigines*. New York: 1964.

Erman, Adolf. *Life in Ancient Egypt*. Translation. New York: 1894.
Still useful.

The Ancient Egyptians: A Sourcebook of Their Writings. London: 1927.
Valuable: some texts not in Pritchard.

Etiemble. *The Written Word*. London: 1961.
Illustrative material from many sources.

Etkin, William (editor). *Social Behavior and Organization Among Vertebrates*. Chicago: 1964.
Summary of present knowledge. The editor's own papers and Tinbergen's are specially relevant.

Evans-Pritchard, E. E. *The Institutions of Primitive Society*. Oxford: 1954.

Feldhaus, Franz Maria. *Ruehmesblaetter der Technik*. 2 vols. Leipzig: 1926.
A pioneer chronicle: still useful.

Technik der Antike und des Mittelalters. Potsdam: 1931.

Finley, M. I. *Between Slavery and Freedom*. In Comparative Studies in Society and History. The Hague: April 1964.

Finley, M. I. (editor). *Slavery in Antiquity: Views and Controversies*. London: 1948.

Fleure, Herbert John, and Harold Peake. *Times and Places*. Oxford: 1956.
A summation of their ten-volume series on human origins.

Forbes, R. J. *Studies in Ancient Technology*. 5 vols. Leiden: 1955.
Valuable monographs in areas where much new material—such as the early use of
horseshoes, 2nd century B.C.—has been uncovered and much more work needs to be
done.

Forde, C. Daryll. *Habitat, Economy, and Society: A Geographical Introduction
to Ethnology*. London: 1945.
Indispensable.

Frankfort, Henri. *Kingship and the Gods: A Study of Ancient Near Eastern
Religion as the Integration of Society and Nature*. Chicago: 1948.
Important original contribution, vital for any full appraisal of ancient civilization and
technics: but especially for my own theory of the origin of the machine. In differen-
tiating between the roles of the king in Egypt and Mesopotamia, Frankfort tends to
over-emphasize the 'democracy' of the latter and to underestimate the social consequences of divine absolutism.

The Birth of Civilization in the Near East. Bloomington, Ind.: 1954.
Brilliant summary, with many important sidelights upon technical development in early
Egyptian and Mesopotamian civilization.

Frankfort, Henri, John Wilson, Thorkild Jacobsen, *et al. The Intellectual Ad-
venture of Ancient Man: An Essay on Speculative Thought in the Ancient
Near East*. Chicago: 1946.
One of the best general introductions to the mind of ancient civilizations.

Frazer, James George. *The Golden Bough: A Study in Magic and Religion*.
1 vol. abridged edition. New York: 1942.
Originally a vast twelve volume compendium of travellers' tales and folk lore: rough
ore gathered from every available mine, smelted down into impure ingots cast into a
two-part mold, one of classical scholarship, the other of Victorian rationalism. The
weaknesses are now obvious: yet the ore itself, with all these impurities, is valuable;
and anthropologists who cast it completely aside have missed many precious clues to
better observed customs by failing to see them in the worldwide setting on which Frazer
drew.

Magical Origin of Kings. London: 1920.

Freud, Sigmund. *Totem and Taboo: Resemblances Between the Psychic Lives of
Savages and Neurotics*. New York: 1918.
A pioneer work, but written at too early a point in the development of both anthropology and psychoanalysis to remain more than a stimulus to sounder speculation a
generation later. But the underlying themes point to persistent irrational factors I have
treated in this book. My debt to Freud's *Interpretation of Dreams* is more fundamental.

Frobenius, Leo. *Kulturgeschichte Afrikas: Prolegomena zu Einer Historischen
Gestaltlehre*. Zurich: 1933.

Fromm, Erich. *The Forgotten Language: An Introduction to the Understanding
of Dreams, Fairy Tales, and Myths*. New York: 1951.
Fromm's theory of dreams rectifies the distortion of both Freudian and Jungian systems
of interpretation; unfortunately, the second part of the book does not fulfill the promise
of the introductory chapters.

Fuhrmann, Ernst. *Grundformen des Lebens: Biologisch-Philosophische Schriften.* Heidelberg: 1962.

Gantner, Joseph. *Leonardos Visionen von der Sintflut vom Untergang der Welt.* Bern: 1958.

Garrod, D. A. E. *Environment, Tools and Man.* Cambridge: 1946.
Inaugural lectures with needed correction of over-stratified classifications of archeological materials without realistic appraisal of working realities that included materials that have long disappeared.

Geddes, Patrick. *An Analysis of the Principles of Economics. Part I.* London: 1885.
Application of biological observations to the division of labor and its consequences.

Gennep, Arnold van. *Les Rites de Passage.* Paris: 1909.
Now classic description of the 'rites of transition' in their physical and social forms: door, threshold, approach of strangers, as well as pregnancy, birth, courtship, marriage, death.

Gerard, Ralph W. *Brains and Behavior.* See Spuhler, J. N. (editor).

Gesell, Arnold. *Wolf Child and Human Child: being a narrative interpretation of the life history of Kamala, the Wolf Girl.* New York: 1940.
One of a number of attested historic cases that establish the positive role of learning in acquiring species character and the traumatic effects of social deprivation.

Giedion, Sigfried. *The Eternal Present: A Contribution on Constancy and Change. The Beginnings of Art. Part I, Vol. 6.* New York: 1962.

The Beginnings of Architecture. Part II, Vol. 6. New York: 1964.
An exhaustive and stimulating study of ancient art and architecture, considered as a "psychic record." Challenging in its sometimes contentious esthetic assumptions, and indispensable for its rich illustrations alone.

Gille, Bertrand. *Esprit et Civilisation Techniques au Moyen Age.* Paris: 1952.
Invaluable. Discusses medieval advances in technics and suggests further areas of research. See also his excellent contributions to Singer.

The Renaissance Engineers. London: 1965.

Girardeau, Emile. *Le Progrès Technique et la Personnalité Humaine.* Paris: 1955.

Glennie, J. Stuart. *The Application of General Historical Laws to Contemporary Events.* In Sociological Society. Sociological Papers. Vol. II. London: 1905.
The latest and most available exposition of a thesis first set forth in the seventies, on periodicity in history. Glennie discerned five-hundred-year cycles, and was the first to point out the contemporaneity of the Axial religions and philosophies, and the significance of the ethical transformation they introduced.

Goldstein, Kurt. *Human Nature in the Light of Psychopathology.* Cambridge: 1940.
Admirable holist presentation, all the more significant because the work of a neurologist who specialized in brain injuries.

Language and Language Disturbance. New York: 1948.

Hahn, Eduard. *Das Alter der Wirtschaftlichen Kultur der Menschheit: Ein Rückblick und ein Ausblick.* Heidelberg: 1905.
A vigorous, original work, largely neglected till Carl Sauer pointed out Hahn's important insights on plant and animal domestication.

Die Haustiere und ihre Beziehung zum Menschen. Leipzig: 1896.

Die Entstehung der Pflugkultur. Heidelberg: 1909.

Haldane, J. B. S. *Animal Ritual and Human Knowledge.* In Diogenes: Autumn 1953.

Hallowell, A. Irving. *Ojibway Ontology, Behavior, and World View.* See Diamond, Stanley (editor).

Self, Society, and Culture in Phylogenetic Perspective. See Tax, Sol (editor).

Harlow, H. F. and M. K., R. O. Dodsworth, and G. L. Arling. *Maternal Behavior of Rhesus Monkeys Deprived of Mothering and Peer Associations in Infancy.* In Proceedings of the American Philosophical Society: February 18, 1966.

Harrison, H. S. *Pots and Pans.* London: 1923.

The Evolution of the Domestic Arts. London: 1925.

War and Chase. London: 1929.
An excellent series of introductions.

Harrison, Jane. *Ancient Art and Ritual.* London: 1913.
Brief but penetrating, in fact outstanding.

Haskins, Caryl P. *Of Societies and Men.* New York: 1951.

Hawkes, Jacquetta. *Man on Earth.* New York: 1955.
A wise and beautiful statement: but those who need it most will probably appreciate it least.

Hawkes, Jacquetta, and Leonard Woolley. *Pre-History and the Beginnings of Civilization.* Vol. I. In History of Mankind series. New York: 1963.

Hayes, William. *Most Ancient Egypt.* Chicago: 1965.

Heichelheim, Fritz M. *An Ancient Economic History: From the Paleolithic Age to the Migrations of the Germanic, Slavic, and Arabic Nations. Vol. 1.* Leiden: 1958.

Henderson, Lawrence J. *The Fitness of the Environment: An Inquiry Into the Biological Significance of the Properties of Matter.* New York: 1927.
A work whose importance is still insufficiently appreciated: perhaps because its impeccable reasoning shatters the current notion that life is an accidental process that takes place in an equally accidental if not hostile environment.

Herrick, C. Judson. *The Evolution of Human Nature.* Austin: 1956.

Herskovits, Melville J. *Man and His Works: The Science of Cultural Anthropology*. New York: 1952.

Hesiod. *The Works and Days*. (Loeb Classical Library.) Cambridge, Mass.: 1936.

Hoagland, Hudson, and Ralph W. Burhoe (editors). *Evolution and Man's Progress*. In Daedalus: Summer 1961.

Hobhouse, Leonard T. *Development and Purpose: An Essay Towards a Philosophy of Evolution*. London: 1913.

Hocart, A. M. *The Progress of Man*. London: 1933.

Kings and Councillors: An Essay in the Comparative Anatomy of Human Society. Cairo: 1936.

Social Origins. London: 1954.
As a comparative anthropologist, Hocart's original perceptions were based on data drawn from many parts of the world. His work was neglected in his time by the exponents of cultural autonomy in pursuit of other aims, valid, but limited to single cultures or sub-cultures.

Hodges, Henry. *Artifacts: An Introduction to Early Materials and Technology*. London: 1964.

Hoijer, Harry. *Language and Writing*. See Shapiro, Harry L. (editor).

Holsti, Rudolph. *The Relation of War to the Origin of the State*. Helsingfors: 1913.
An unfortunately neglected contribution that now needs more ample re-statement and amplification than I have been able to make.

Hooke, S. H. (editor). *Myth, Ritual, and Kingship: Essays on the Theory and Practice of Kingship in the Ancient Near East and in Israel*. Oxford: 1958.
Useful as counterpoise to Frankfort's overemphasis of differences between ancient cultures and his neglect of well-established similarities, when studying a particular institution.

Hough, Walter. *Fire as an Agent in Human Culture*. United States National Museum. Bulletin 139. Washington: 1926.
As of its date, a full conspectus. But see Thomas, William L., Jr. (editor).

Howells, William. *Back of History: The Story of Our Own Origins*. Revised edition. Garden City: 1963.

Mankind in the Making: The Story of Human Evolution. New York: 1959.

Howells, William (editor). *Ideas on Human Evolution: Selected Essays 1949–1961*. Cambridge, Mass.: 1962.
Invaluable in following the transition from animal to human.

Hubbard, Henry D. *The Ancient Egyptians Had Hardened Steel*. In American Machinist: April 16, 1931.

Hubert, Henri, and Marcel Mauss. *Sacrifice: Its Nature and Function*. Chicago: 1964.
First published in the Année Sociologique in 1898, this is still probably the most adequate explanation of rites of sacrifice, though it leaves the primary motivation, in the case of human sacrifices, with its many grades from finger joints to whole bodies, unexplained, and perhaps, like other irrationalities, unexplainable.

Hudson, W. H. *The Naturalist in La Plata*. New York: n.d.
The rehabilitation of the holist studies of animal behavior by Lorenz, Tinbergen, Portmann and others, should also restore Hudson's first-hand reports, which incidentally correct Darwin's over-emphasis on the role of sexual selection.

Huizinga, J. *Homo Ludens: A Study of the Play-Element in Culture*. London: 1949.
The subtitle as translated unfortunately contradicts the author's explicit intention of ascertaining how far "culture in itself bears the character of play." A teasing book which, by its over-extension, promises more than it can perform, yet is full of rare flashes of insight. See Jane Harrison's *Ancient Art and Ritual*. See also Caillois' *Man and the Sacred* for criticism.

Humboldt, Alexander von. *Cosmos: A Sketch of a Physical Description of the Universe*. 2 vols. Berlin: 1844. London: 1949.
The great work of a great mind: more profound in its many implications than the *Origin of Species*, and with fewer ideological errors to correct. Humboldt's observation that "man is man by virtue of language alone" is both earlier and sounder than the belief that tool-making shaped his whole development.

Huxley, Julian S. *Man Stands Alone*. New York: 1927.
Returns with abundant evidence to the earlier belief in the uniqueness of man, which had been undermined by Darwin's overstress on continuities and his underestimation of the role of symbolic communication.

The Uniqueness of Man. London: 1941.
Deservedly popular papers by a biologist who is also a humanist.

Evolution, Cultural and Biological. See Thomas, William L., Jr. (editor), *Yearbook of Anthropology: 1955*.
Huxley has helped close the gap between biological and cultural evolution as Leslie White has done in reverse.

Isaac, Erich. *Myths, Cults, and Livestock Breeding*. In Diogenes: Spring 1963.
Significant exposition, following up with further data the work of Hahn (*q.v.*) on the religious contributions to domestication.

Jacobsen, Thorkild. *Primitive Democracy in Ancient Mesopotamia*. In Journal of Near East Studies: 1943.

James, E. O. *Myth and Ritual in the Ancient East: An Archaeological and Documentary Study*. London: 1958.
Comprehensive, but without the deep bite of observation in Robertson Smith.

James, William. *The Will-to-Believe*. New York: 1903.
See 'Reflex Action and Theism.'

Jaspers, Karl. *The Origin and Goal of History*. London: 1953.
Note chapter on the Axial period. See Glennie, J. Stuart.

Jespersen, Otto. *Language: Its Nature, Development and Origin.* New York: 1922.
Classic exposition by an immensely learned but genial scholar, full of human as well as linguistic insight. Not less endearing because he broke the professional taboo against speculating on origins.

Jolly, Alison. *Two Social Lemurs.* Chicago: 1966.

Kenyon, Kathleen M. *Digging up Jericho.* London: 1957.
Description of the earliest diggings that changed the then-accepted dates for urban settlements. See Braidwood and Mellaart.

Archaeology in the Holy Land. New York: 1966.

Klemm, Friedrich. *A History of Western Technology.* London: 1959.
Misleading title for a series of well-chosen readings from sources.

Klima, Bohuslav. *Coal in the Ice Age: The Excavation of a Paleolithic Settlement.* In Antiquity: September 1956.
Evidence of a paleolithic settlement, dated 30,000 years ago, with a building identified as a workshop, and evidence that nearby coal was used as fuel.

Koffka, Kurt. *The Growth of the Mind: An Introduction to Child-Psychology.* New York: 1925.

Kraeling, Carl H., and Robert M. Adams (editors). *City Invincible: A Symposium on Urbanization and Cultural Development in the Ancient Near East.* Chicago: 1960.
Though not devoted to technics, by its critical examination of the early context throws many valuable sidelights on its development.

Kramer, Samuel Noah. *The Sumerians: Their History, Culture, and Character.* Chicago: 1963.
Comprehensive summary by a redoubtable scholar who incorporates many recently translated tablets.

Kroeber, A. L. *Anthropology: Race, Language, Culture, Psychology, Prehistory.* New York: 1923. Revised Edition: 1948.
Masterly.

Kroeber, A. L., *et al. Anthropology Today: An Encyclopedic Inventory.* Chicago: 1953.
The first of a useful series, with many contributors.

Kubzansky, Philip E. *The Effects of Reduced Environmental Stimulation on Human Behavior.* See Biderman and Zimmer (editors).

Kuehn, Emil. *The Rock Pictures of Europe.* Fair Lawn, N. J.: 1956.

Laming-Emperaire, Annette. *La Signification de l'art Rupestre Paléolithique.* Paris: 1962.
Judicious critique of past interpretations of paleolithic art, showing the insufficiency of any one theory to account for all the evidence now at hand.

Laming-Emperaire, A. (editor). *La Découverte du Passé: progrès recents et techniques nouvelles en préhistorie et en archéologie.* Paris: 1952.

Lang, Andrew. *Myth, Ritual, and Religion*. 2 vols. Revised edition. London: 1899.
An early essay in comparative religion, dated but not negligible.

Langer, Susanne. *Philosophy in a New Key: A Study in the Symbolism of Reason, Rite, and Art*. Cambridge, Mass.: 1942.
Admirably succinct and penetrating interpretation of the role of the symbol in man's development. An original contribution, not less so because it brings together and refocusses other original contributions, such as Cassirer's.

Lanyon, W. E., and W. N. Tavolga. *Animal Sounds and Communication*. Publication No. 7, American Institute of Biological Sciences. Washington: 1960.
A new field partly opened by better recording devices that may cast a flickering sidelight on human language.

Lashley, K. S. *Brain Mechanisms and Intelligence: A Quantitative Study of Injuries to the Brain*. Chicago: 1929.
Impeccably 'objective' in method, but thereby necessarily limited to rat intelligence: the level where brain and mind are one. See Penfield, Wilder.

Leakey, L. S. B. *Adam's Ancestors*. Fourth, revised, edition. London: 1953.
A standard work by one of those who have unearthed man's hominid precursors or early collateral relatives in Africa.

The Origin of the Genus Homo. See Tax, Sol (editor).

Working Stone, Bone and Wood. See Singer, Charles (editor).

Lefranc, Georges. *Histoire du Travail et des Travailleurs*. Paris: 1957.

Lenneberg, E. H. *Language, Evolution, and Purposive Behavior*. See Diamond, Stanley (editor).

Leonardo da Vinci. *The Notebooks, Arranged, Rendered into English and Introduced by Edward MacCurdy*. New York: 1939.

Leroi-Gourhan, André. *Milieu et Techniques*. In series: *Evolution et Techniques*, part of collection: *Sciences d'aujourd'hui*. Paris: 1945.
Systematic comparative study of all aspects of material technology. Invaluable.

Prehistoric Man. New York: 1957.
Brief, witty, and within its largely French boundaries authoritative. Redresses the vulgar conception of the 'cave-man.' See Braidwood.

Préhistoire de l'Art Occidental. Paris: 1965.
Magnificent, alike in its thorough assemblage of data, its superb illustrations, its circumspect interpretations, and not least, its daring hypotheses. Leroi-Gourhan necessarily builds on Breuil's work, but challenges some of his explanations and offers alternative speculations.

Lethaby, W. R. *Architecture, Nature, and Magic*. New York: 1956.
First sketched out in 1891, in a book entitled *Architecture, Mysticism and Myth*, this remains a highly important pioneer study of the role myth has played in civilization.

Lévi-Strauss, Claude. *La Pensée Sauvage*. Paris: 1962.

Totemism. Boston: 1963.
Critique of totemism as a universal concept, indicating its varied and disparate manifestations. But see Radcliffe-Brown.

Structural Anthropology. New York: 1963.
Full of searching questions, if not always satisfactory answers. The equation of the method of linguistics with that of anthropology illuminates the whole problem of 'purpose' in all manifestations of life, and shows that it is not dependent upon consciousness.

Levy, Gertrude Rachel. *The Gate of Horn: A Study of the Religious Conceptions of the Stone Age and Their Influence upon European Thought*. London: 1948.
Often illuminating.

Lewis, M. M. *Language in Society: The Linguistic Revolution and Social Change*. New York: 1948.
Survey of the renascence of the spoken word through a variety of contemporary speech-machines from telephone to tape-recorder. But aware of the contrast between the closeness of mechanical speaking distance and social distance and alienation.

Linton, Ralph. *The Tree of Culture*. New York: 1955.
Comparative study: valuable.

Lips, Julius E. *Fallensysteme der Naturvölker*. Leipzig: 1927.
Demonstrates the importance of trapping and netting before effective hunting weapons were made.

Paleolithische Fallenzeichnungen und das Ethnologische Vergleichsmaterial. In Tagungsberichte der Deutschen Anthropologische Gesellschaft. Leipzig: 1928.

The Origin of Things. New York: 1947.
'Popular' in presentation, but rich in comparative data, with an exhaustive bibliography.

Loeb, Edwin M. *Die Institution des Sakralen Königtums*. In Paideuma: Mitteilungen zur Kulturkunde: December 1964.
Further evidence on importance of Sacred Kingship in widely scattered societies.

Wine, Women, and Song: Root Planting and Head-Hunting in Southeast Asia. See Diamond, Stanley (editor).

Lorenz, Konrad Z. *King Solomon's Ring: New Light on Animal Ways*. London: 1952.
Already a minor classic.

On Aggression. New York: 1966.
See especially the chapter on Habit, Ritual and Magic.

Lowie, Robert H. *Primitive Religion*. New York: 1924.
In essentials undated.

MacGowan, Kenneth, and Joseph A. Hester, Jr. *Early Man in the New World*. New York: 1962.
Convenient summary.

Magoun, H. W. *Evolutionary Concepts of Brain Function Following Darwin and Spencer.* See Tax, Sol (editor).

Malinowski, Bronislaw. *Myth in Primitive Psychology.* New York: 1926.
Differentiates between folk-tales, legends, and true myths: which latter are considered as attempts to justify a ritual or a moral rule preserved in myth as a statement of a primeval reality still in operation.

A Scientific Theory of Culture, and Other Essays. Chapel Hill: 1944.

Magic, Science, and Religion. New York: 1948.
Malinowski's handling of magic, myth, and religion corrects the rationalist bias and lack of field experience of Frazer without condescension or gratuitous disparagement.

Coral Gardens and Their Magic. Vol. I: Soil Tilling and Agricultural Rites in the Trobriand Islands. Vol. II: The Language of Magic and Gardening. London: 1935. Bloomington, Ind.: 1965.
Valuable for its many data and insights, alike on magic, on language, and on 'neolithic' practices.

Mallery, Garrick. *Sign language among North American Indians compared with that among other peoples and deaf-mutes.* In First Annual Report of the Bureau of Ethnology. Smithsonian Institution. Washington: 1881.
A notable contribution whose relevance to linguistics has been insufficiently appreciated.

Marett, R. R. *Faith, Hope, and Charity in Primitive Religion.* New York: 1932.
Discussion of primitive attitudes and beliefs with the kind of human insight and common sense that will outlast its data.

Sacraments of Simple Folk. Oxford: 1933.

Maringer, Johannes. *The Gods of Prehistoric Man.* New York: 1960.
Necessarily cautious but highly suggestive speculations on the religious implications of prehistoric art, well illustrated. See Bowra.

Marshack, Alexander. *On Early Calendars.* In Science: November 7, 1964.
Evidence of recorded lunar observations that "extend backward in an unbroken line from the Mesolithic Azilian to the Magdalenian and Aurignacian cultures." See Zelia Nuttall.

Marx, Karl. *Capital: A Critique of Political Economy.* Edited by Friedrich Engels. Revised by Ernest Untermann. Fourth edition. New York: 1906.

Marx, Karl, and Friedrich Engels. *Selected Correspondence: 1846–1885.* New York: 1942.
Many vivid flashes of social insight.

Mason, Otis T. *The Origins of Invention: A Study of Industry among Primitive Peoples.* New York: 1895.
An excellent early essay, now deservedly in M.I.T. paperback edition.

Mauss, Marcel. *The Gift: Forms and Functions of Exchange in Archaic Societies.* London: 1954.
Medicinal for those who imagine that affluence and freedom from economic anxiety are solely a gift of the machine. Not merely scholarly but, in its concluding chapters, wise.

Mayr, Ernst. *Animal Species and Evolution*. Cambridge, Mass.: 1963.
Recommended. See especially the passage on brain growth in man.

McCurdy, George C. *Human Origins*. 2 vols. New York: 1926.
Modified but not irremediably outdated by later finds and interpretations.

McLuhan, Marshall. *The Gutenberg Galaxy: The Making of Typographic Man*. Toronto: 1963.
Happily provocative, even in its most erratic and dubious flights.

Mead, Margaret. *Continuities in Cultural Evolution*. New Haven: 1964.
The chapter on the Paliau movement is worth the price of admission. Mead broaches some important problems against her rich background of anthropological observation among surviving primitives.

Mellaart, James. *Earliest Excavations in the Near East*. New York: 1965.

Mellink, Machteld J. *Anatolia: Old and New Perspectives*. In Proceedings of the American Philosophical Society: April 22, 1966.

Meyerowitz, Eva L. R. *The Divine Kingship in Ghana and Ancient Egypt*. London: 1960.
Sets forth the startling parallels between the cult of the divine king in Ghana and what we know of the same nest of ideas in ancient Egypt. The likenesses are so numerous that the possibility of an independent origin is extremely low.

Mitchell, Arthur. *The Past in the Present: What is Civilization?* Edinburgh: 1880.
An archeological milestone: upright though eroded. These first-hand observations on primitive houses and watermills in the Hebrides are still valuable.

Mountford, Charles. *Art in Australian Aboriginal Society*. See Smith, Marian W. (editor).

Movius, H. L., Jr. *The Old Stone Age*. See Shapiro, Harry L. (editor).

Müller, [Friedrich] Max. *Lectures on the Science of Thought*. 2 vols. New York: 1862–65.

Mumford, Lewis. *Technics and Civilization*. First edition. New York: 1934. Paperback: 1963.
Survey of modern technics from its European origins in the Middle Ages. For self-criticism see the preface to the paperback edition. The bibliography and the list of inventions are still full enough to be useful for those first approaching the subject.

The Condition of Man. New York: 1944.
See particularly the chapter on Capitalism, Absolutism, and Protestantism.

Art and Technics. Bampton Lectures in America. New York: 1952. Paperback: 1960.
Recommended in connection with Chapter Eleven.

The Transformations of Man. New York: 1956.
The best summation of my general outlook.

The City in History. New York: 1961.
The first eleven chapters intertwine with *The Myth of the Machine* and supplement it at many vital points. Likewise the bibliography.

Murray, Henry A. (editor). *Myth and Myth Making: A Symposium.* In Daedalus: Spring 1959.
An enquiry into the function and meaning of myth, by a group brought together by one of the foremost interpreters of personality and culture. Yet none of the interpretations suggests any awareness of the myths now dominating contemporary society.

Narr, Karl J. *Urgeschichte der Kultur.* Stuttgart: 1961.

Naumburg, Margaret. *Art as Symbolic Speech.* See Anshen, Ruth (editor).

Needham, Joseph, Wang Ling, and Derek Price. *Heavenly Clockwork: The Great Astronomical Clocks of Medieval China.* Cambridge: 1960.
A precious addition to our technical knowledge, pushing back by many centuries the origin of the components of the mechanical clock, and bringing the clepsydra and the European clock into closer relation through the intermediate machinery of the armillary. A further confirmation, were it needed, of the relation of the celestial order and royal authority.

Nef, John. *The Conquest of the Material World.* Chicago: 1964.
Useful for its summary of the place of mining and metallurgy in medieval society. Though the chapter on industrialism and modern science happily complements the present critique, Professor Nef places the rise of mechanical industry at too late a date to take in the critical steps already noted in my *Technics and Civilization* (1934).

Neumann, Erich. *The Origin and History of Consciousness.* German edition. Zurich: 1949. New York: 1954.
Far from impeccable from the standpoint of scientific method, yet this study is full of important intuitions that throw a sometimes sharp illumination into the fogbank of man's prehistoric experience and his later partial emergence. In its presentation of the realities of symbolic language and the difficulties of interpreting its ambiguous meanings, it seems to me superior not only to Fromm but to Jung.

Nougier, Louis-René. *Géographie Humaine Préhistorique.* Paris: 1959.
Excellent.

Nuttall, Zelia. *The Fundamental Principles of Old and New World Civilizations.* Cambridge, Mass.: 1901.
A Peabody Museum paper, by an archeologist once pre-eminent in her field whose analysis of the constellation of institutions that marked the emergence of civilization anticipated the richer data uncovered by Hocart, Frankfort, Loeb and others, which confirm it. Her hypothesis that a Polaris religion, in evidence in Maya culture, preceded the establishment of the solar calendar awaits re-examination by some rare scholar who combines equally her astronomical and archeological qualifications.

Oakley, Kenneth P. *Man the Tool-Maker.* First edition: 1949. Fifth edition. London: 1963.
Useful, despite the fact that the title begs the question.

The Earliest Firemakers. In Antiquity: June 1956.

Oppenheim, A. Leo. *Ancient Mesopotamia: Portrait of a Dead Civilization.* Chicago: 1964.
The work of a distinguished Assyriologist, cautious, skeptical, meticulous; rich in firsthand material.

earce, Roy Harvey. *The Savages of America: A Study of the Indian and the Idea of Civilization.* Baltimore: 1965.

ei, Mario. *The Story of Language.* Philadelphia: 1949.
Admirable general introduction.

eller, Lili E. *Language and Its Pre-stages.* In Bulletin of the Philadelphia Association for Psychoanalysis: June 1964.
Recommended.

enfield, Wilder, and T. Rasmussen. *The Cerebral Cortex of Man: A Clinical Study of Localization of Function.* New York: 1952.

enfield, Wilder, and L. Roberts. *Speech and Brain-Mechanisms.* Princeton: 1959.
Demonstrates the specifically human specialization of brain areas for speech and association, as distinct from Lashley's findings with rats, whose activities have not produced such specialization. See also Kurt Goldstein.

etrie, W. M. Flinders. *Egyptian Tales: translated from the papyri. First Series, IVth to XIIIth Dynasty.* London: 1895.

hillips, E. D. *The Greek Vision of Prehistory.* In Antiquity: September 1964.
As with Vico and Goethe, this essay shows how good guesses—doubtless colored by tradition—could be before knowledge was actively sought.

The Royal Hordes: The Nomad Peoples of the Steppes. See Piggott, Stuart (editor).
One of the best accounts available of the separation of cattle-raising and (later) horse-taming from mixed farming in the nomad cultures of Russia and Asia, adapting to their own purposes the traits derived from the farmers of the Near East.

iggott, Stuart. *Ancient Europe: From the Beginnings of Agriculture to Classical Antiquity.* Edinburgh: 1965.
Valuable.

iggott, Stuart (editor). *The Dawn of Civilization.* New York: 1961.
An encyclopedic survey by leading British scholars, with 940 illustrations. The most comprehensive sourcebook in English to date, with an exhaustive bibliography.

olanyi, Michael. *Personal Knowledge: Towards a Post-Critical Philosophy.* Chicago: 1958.
Both a justification of scientific knowledge and a cogent criticism of the limitations of its attempt at depersonalization, with its rejection of teleological evidence not in harmony with its current limited postulates. The chapter on Articulation supplements and strengthens my own interpretation of language.

ortmann, Adolf. *Animals as Social Beings.* Zurich: 1953. Translation. New York: 1961.
Fresh insights into animal behavior, including the role of ritual.

New Paths in Biology. New York: 1964.

ostan, M., and E. E. Rich (editors). *The Cambridge Economic History of Europe. Vol. II. Trade and Industry in the Middle Ages.* Cambridge: 1952.
See especially the chapters on trade by Postan and Robert Lopez.

Price, Derek J. de Solla. *Automata and the Origins of Mechanism and Mecha-nistic Philosophy*. In Technology and Culture: Winter 1964.
> Sets out to show that the mechanistic ideology preceded the automata, but his own data suggest that from the beginning they interacted.

Pritchard, James B. (editor). *Ancient Near Eastern Texts: Relating to the Old Testament*. Princeton: 1955.
> The relation to the Old Testament is broadly interpreted, and the texts are indispensable even to a layman, all the more because of their wide archeological range. Invaluable. There is a companion volume of illustrations.

Pumphrey, R. J. *The Origin of Language: An Inaugural Lecture*. Liverpool 1951.
> Not less interesting because it comes from a zoologist, one of the many biologists who now cast doubt on the relevance of tool-making to man's mental development. See Mayr, Ernst.

Radcliffe-Brown, A. R. *Structure and Function in Primitive Society: Essays and Addresses*. Glencoe, Ill.: 1952.
> The papers on Totemism, Taboo, and Religion are particularly relevant to the theme of this book.

Radin, Paul. *Primitive Religion: Its Nature and Origin*. New York: 1937.
> Recommended for both its historic and social insights, despite its psychological blind spots and its Marxian overweighting of primitive economic motivations.

Raven, Christian P. *Oögenesis: The Storage of Developmental Information*. In the Permanon Series on Pure and Applied Biology. New York: 1961.
> Important: particularly recommended to social scientists whose biological information too often tardily reflects the nineteenth-century neglect of historical processes and teleological functions.

Recinos, Adrian (translator). *Popul Vuh: The Sacred Book of the Ancient Quiche Maya*. English translation by Delia Goetz and Sylvanus G. Morley. Norman, Okla.: 1950.
> Illuminating and, even at third hand, valuable as 'literature.'

Redfield, Robert. *The Primitive World and Its Transformations*. Ithaca: 1953.
> Important evaluation of the essential role of morals in 'folk societies.'

How Human Society Operates. See Shapiro, Harry L. (editor).

Thinker and Intellectual in Primitive Society. See Diamond, Stanley (editor).

Révész, G. *The Origins and Prehistory of Language*. New York: 1956.
> The best discussion to date of this long-neglected subject: by a one-time professor of psychology at the University of Amsterdam. Révész's review of previous theory is admirable in its exhaustiveness, but his circular definition of man, as a creature who possesses phonetic language, causes him to dismiss peremptorily hypotheses that seek to explain the transition—or the leap—from animal to human forms.

Rivers, W. H. R. *Medicine, Magic and Religion*. New York: 1924.

Robertson, H. M. *Aspects of the Rise of Economic Individualism: A Criticism of Max Weber and His School*. Cambridge: 1935.

Robinson, J. T. *The Australopithecines and Their Bearing on the Origin of Man and of Stone Tool-making.* See Howells, William (editor).

Roe, Z., and G. G. Simpson (editors). *Behavior and Evolution.* New Haven: 1958.

Róheim, Géza. *Animism, Magic, and the Divine.* New York: 1930.

    The Riddle of the Sphinx, or Human Origins. London: 1934.
First-hand interpretation of ceremony, myth, and song among the Australian aborigines; but vitiated by its acceptance of Freud's uncritical mythology of early man, in which an imaginary event is called upon to sustain a fantastic hypothesis.

    The Eternal Ones of the Dream: A Psychoanalytic Interpretation of Australian Myth and Ritual. New York: 1945.
Based on first-hand observation; with fewer objectionable traits than the earlier books.

Sapir, Edward. *Language: An Introduction to the Study of Speech.* New York: 1921.

    Language. In *Encyclopedia of the Social Sciences.* Vol. 9. New York: 1931.
Both book and essay are still noteworthy.

Sauer, Carl O. *Agricultural Origins and Dispersals.* New York: 1952.
An outstanding summation of the available evidence on human development in relation to plant and animal species, with extraordinary resourcefulness in marshalling the contributions of paleontology, geography, climatology, and anthropology. Not less original because of Sauer's acknowledged debt to Hahn and Oakes Ames.

    Land and Life. A Selection from the Writings of Carl Ortwin Sauer, edited, with an introduction by John Leighly. Berkeley: 1963.
A notable series of papers, many of them permanent landmarks in the interpretation of man's relation with the environment, by a mind never entrapped by its own specialized knowledge, or too respectful of the authority of those who have been so caught. See especially *Environment and Culture During the Last Deglaciation,* and *Seashore—Primitive House of Man.*

    Cultural Factors in Plant Domestication in the New World. In Euphytica. Wageningen: November 1965.
Rounds out, in terms of the New World, the picture presented in *Agricultural Origins.*

Sayce, R. U. *Primitive Arts and Crafts: An Introduction to the Study of Material Culture.* Cambridge: 1933.
One of the best short introductions to date; though O. E. Mason's early work should not be altogether neglected. Sayce's consciousness of the impossibility of dealing with material culture without reference to non-material attributes, as indicated in his introductory chapter, might have made the great Singer *History of Technology* more valuable (which see). See also Leroi-Gourhan and Forde.

Schaller, George. *Year of the Gorilla.* Chicago: 1964.

Schmidt, W. *The Origin and Growth of Religion: Facts and Theories.* London: 1931.
An exhaustive survey—tendentious but judicious—by a Catholic anthropologist best known for his elaboration of Andrew Lang's insight that belief in high Gods existed in extremely primitive peoples, and was later overlaid by polytheism and animism.

Sears, Paul. *Changing Man's Habitat: Physical and Biological Phenomena.* In Yearbook of Anthropology: 1955.
Sear's application of moral principles to ecology gives him much the same role as Redfield's in anthropology.

Semenov, S. A. *Prehistoric Technology: An Experimental Study of the Tool and Artefacts from Traces of Manufacture and Wear.* Moscow: 1957 London: 1964.
An attempt by microscopic study to throw light on the function of early artifacts whose shapes give little indication of purpose.

Service, Elman E. *The Hunters.* In Foundations of Modern Anthropology Series. Englewood Cliffs, N. J.: 1966.

Shapiro, Harry L. *Human Beginnings.* See below.

Shapiro, Harry L. (editor). *Man, Culture, and Society.* New York: 1956.
Excellent survey of human origins.

Sherrington, Charles. *Man on His Nature.* New York: 1941.
An important summary: in many respects still unrivalled.

Simpson, George Gaylord. *The Meaning of Evolution: A Study of the History of Life and Its Significance for Man.* New Haven: 1949.
Summary of present data on the external pressures of adaptation and survival as the explanatory source of self-directive and teleological phenomena.

Singer, Charles. *A Short History of Science to the Nineteenth Century.* Oxford 1941.
Useful short summary.

Singer, Charles, E. J. Holmyard, and A. R. Hall (editors). *A History of Technology. Vol. I: From Early Times to the Fall of Ancient Empires.* London 1954.
Individual papers by Oakley, Childe, Forde, *et al.,* are excellent; but in certain departments there are serious omissions.

Vol. II: The Mediterranean Civilizations and the Middle Ages. c. 700 B.C. to c. A.D. 1500. London: 1958.
As with much technical history, the material presented is spotty, due to the unavailability of data. The article on machines by Bertrand Gille, with evidence from medieval monasteries, is particularly good, and relevant to the present book. (In Vol. II Trevor I. Williams joins the editors.)
Since this is the only complete history of technology in English, it is an indispensable work. But like most encyclopedic compilations, it is uneven and, also like other encyclopedias, must be taken circumspectly even in matters like dates and attributions.

Smith, Cyril Stanley. *Materials and the Development of Science.* In Science May 14, 1965.

Smith, Marian W. (editor). *The Artist in Tribal Society.* New York: 1961.

Smith, W. Robertson. *Lectures on the Religion of the Semites.* First edition 1889. Revised edition: 1907.
An outstanding classic that, like Fustel de Coulanges' *The Ancient City,* can never be entirely superseded. Particularly good on the nature and modes of sacrifice, though like all other efforts, it leaves the meaning of the impulse itself unexplained.

Sollas, W. J. *Ancient Hunters: and Their Modern Representatives.* London: 1911. Third revised edition: 1924.
> Sollas, despite his sound work, has fallen into undue disrepute because he ventured to suggest that recently existing tribes were directly connected with their paleolithic analogues. Though the odds are heavily against this, the continuity of cultural traits may be genuine; and the readings backward from the present, though risky, are the only clues we have, apart from faint historic records, to the contents of primitive life.

Sombart, Werner. *Der Moderne Kapitalismus.* 6 vols. First edition: 1902. Muenchen: 1928.
> Massive scholarship and often original interpretation of both the technical and social foundations of modern capitalism. But with a tendency to regard latter-day capitalism as a new phenomenon. Weak on those psychological implications and theological perspectives that seem important in assaying the non-rational factors that underlie economic rationalization.

Sommerfelt, Alf. *The Origin of Language: Theories and Hypotheses.* In Journal of World History: April 1954.
> An inadequate survey that shows a persistent blockage in constructing viable hypotheses. The lack of any reference to Cassirer, Tylor, or Langer is inexplicable.

Sorre, Max. *Les Fondements de la Géographie Humaine. Tome II. Les Fondements Techniques.* Paris: 1950.
> The second volume in two books covers the entire range of human activity. So far the best survey in any language in terms of place relationships.

Spengler, Oswald. *Man and Technics: A Contribution to a Philosophy of Life.* New York: 1932.
> A chamber of anthropological horrors: Spengler at his Faustian worst. Yet his *Decline of the West* was one of the earliest general interpretations of history that did justice to technics.

Spuhler, J. N. (editor). *The Evolution of Man's Capacity for Culture.* Detroit: 1959.
> Six pithy essays: partly speculative but sure-footed. Perhaps the best short summation available.

Stalin, Joseph. *Marxism and Linguistics.* New York: 1951.
> Irresistible for those who value unconscious humor.

Sudhoff, Karl. *Essays in the History of Medicine.* New York: 1926.

Taton, René (editor). *Histoire Générale des Sciences. La Science Antique et Mediévale.* Paris: 1957.

Tax, Sol (editor). *Evolution After Darwin.* 3 vols. Chicago: 1960.
> An important symposium on the occasion of the Darwin Centennial Celebration at the University of Chicago. See especially *Vol. II, The Evolution of Man,* with notable essays by various authorities, some cited in this bibliography.

Taylor, Alfred. *Mind as the Basic Potential.* In Main Currents in Modern Thought: March 1958.
> Recommended.

Theal, G. M. *History of South Africa from 1795 to 1872.* London: 1873.
> An early report on the Xosa fantasy. Edward Roux, in *Time Longer Than Rope,* gives variants.

Thomas, Elizabeth Marshall. *The Harmless People.* New York: 1959.
  Intimate picture of a small Bushman group in the Kalahari Desert by a keen observer
  whose study has rightfully won the admiration of anthropologists. One of the best
  accounts of truly primitive life available.

Thomas, William I. *Primitive Behavior: An Introduction to the Social Sciences.*
  New York: 1937.
  Useful comparative study, in a fashion that has been too completely discarded by those
  who think there is only one way of eating an orange.

Thomas, William L., Jr. (editor). *Man's Role in Changing the Face of the
  Earth: An International Symposium Under the Co-chairmanship of Carl
  O. Sauer, Marston Bates, and Lewis Mumford.* Chicago: 1956.
  A mine of important material and fresh ecological interpretations.

Thompson, Homer. *Classical Lands.* In Proceedings of the American Philosophical Society. Philadelphia: April 22, 1966.

Thompson, J. Eric S. *The Rise and Fall of Maya Civilization.* Norman, Okla.:
  1954.

Thorndike, Lynn. *History of Magic and Experimental Science During the First
  Thirteen Centuries of Our Era.* 2 vols. New York: 1923.

  Science and Thought in the Fifteenth Century. New York: 1929.
  Outstanding contributions.

Tinbergen, N. *Social Behaviour in Animals: With Special Reference to Vertebrates.* London: 1953.

Tout, Thomas Frederick. *The Collected Papers . . . With a Memoir and Bibliography. Vol. III.* Manchester: 1934.
  See chapters on Civil Service and Administration.

Toynbee, Arnold Joseph. *A Study of History.* 10 vols. London: 1934–1954.
  Magnificent despite its inevitable omissions and solecisms, and its lack of an adequate
  sociological structure. Toynbee was one of the first to point out that an 'advanced'
  technics may be a source of social retardation: a relevant fact that my interpretation
  of the megamachine substantiates.

Turney-High, Harry Holbert. *Primitive War: Its Practice and Concepts.* Colum-
  bia, S. C.: 1949.
  An able comparative survey.

Tylor, Edward B. *Primitive Culture: Researches into the Development of Mythology, Philosophy, Religion, Language, Art, and Custom.* 2 vols. First
  edition: 1865. Fourth edition. Revised. London: 1903.

  *Researches into the Early History of Mankind and the Development of Civili-
  zation.* First edition. London: 1865. Chicago: 1964. This text, with Chapters 8 and 10 omitted, edited by Paul Bohannan.
  Classic. The chapters on Gesture Language, Picture Language, and Esthetic Expression
  still valuable, and if Tylor had applied these insights into symbolism to verbal language,
  instead of taking the false trails into imitative sounds and primal roots, he would
  have anticipated Cassirer.

Usher, Abbott Payson. *A History of Mechanical Inventions.* First edition. New York: 1929. Revised edition: 1954.
An excellent scholarly work, thorough and well-illustrated: indeed a pioneer study in English.

Varagnac, André. *De la Préhistoire au Monde Moderne. Essai d'une anthropo-dynamique.* Paris: 1954.
Brief summary of both anthropological and technical development.

Civilisation Traditionelle et les Genres de Vie. Paris: 1948.
Survey of the persistent 'archaic culture,' dating back at least to neolithic times, which underlay the higher cultures till the present century. Recommended.

Varagnac, André (editor). *L'Homme Avant l'Ecriture.* Paris: 1959.
A collective effort at synthesis, but with the stress too emphatically on artifacts to even hint at the whole picture.

Veblen, Thorstein. *The Instinct of Workmanship.* New York: 1914.
Considering its date, a remarkably acute analysis of technological origins. Unfortunately, Veblen's dislike for the predatory economy which he correctly associates with divine kingship makes him impute to this institution a failure in technological adroitness and efficiency which the evidence contradicts.

The Theory of the Leisure Class. New York: 1899.
Original but somewhat perverse.

Vico, Giambattista. *The New Science.* Translated from third edition (1744) by Thomas Goddard Berin and Max Harold Fisch. Ithaca: 1948.
For all its overbold guesses and wild probings, on the insufficient factual basis available in the eighteenth century, this is a great seminal work. Many of its intuitions are now richly substantiated.

Vinci, Leonardo da. See Leonardo.

Waddington, C. H. *The Ethical Animal.* London: 1960.
Associates human development with the transmission of knowledge and a willingness to accept authority as a source of knowledge. The definition of ethics is too restricted, but the supporting argument opens up fresh channels for thought. Recommended.

The Nature of Life. New York: 1962.
An able attempt to transcend the self-imposed limits of the anti-teleologic theory derived from physics, upon which prevalent biological doctrine is based.

Wales, H. G. Quaritch. *The Mountain of God: A Study in Early Religion and Kingship.* London: 1953.
Carries the Mesopotamian symbolism and practice into Far Eastern cultures.

Wallace, Alfred Russel. *The Malay Archipelago: The Land of the Orang-Utan and the Bird of Paradise; a Narrative of Travel with Studies of Man and Nature.* New York: 1869.
Still valuable for its data on flora, fauna, and primitive human customs, not least for its interpretations of the way the population remained limited despite an abundant food supply.

Wallon, Henri. *Histoire de l'esclavage dans l'antiquité.* Paris: 1879.

Walter, W. Grey. *The Living Brain*. London: 1953.
Report on recent experiments with mechanical models simulating the autonomous activities of the brain. See Herrick, Goldstein, Lashley, and Penfield.

Washburn, Sherwood L. *Speculations on the Inter-relation of the History of Tools and Biological Evolution*. In Human Biology. Vol. 31: 1959.
A judicious re-statement.

Washburn, Sherwood L. (editor). *Social Life of Early Man*. Chicago: 1961.
Useful conspectus. See Kroeber and Shapiro.

Washburn, Sherwood L., and F. Clark Howell. *Human Evolution and Culture*. See Tax, Sol (editor).

Welby, V. (Lady Victoria). *What Is Meaning? Studies in the Development of Significance*. London: 1903.
Pioneer study still important for those seriously concerned with language and semantics. See also Peirce's correspondence with her in Peirce, Charles. *Values in a Universe of Chance*. Garden City: 1958.

Westermann, W. L. *The Slave Systems of Greek and Roman Antiquity*. Philadelphia: 1955.

White, Leslie A. *The Evolution of Culture: The Development of Civilization to the Fall of Rome*. New York: 1959.
The subtitle is misleading, for the emphasis is on the origins and development of human culture, with the author's special stress on symbolization and technics. A swing back to the evolutionary tradition of the nineteenth century, correcting the isolationism and cultural relativism of the intervening cultural school, but emphasizing, as with Kroeber and Lévi-Strauss, the integrity and persistence of culture itself.

Four Stages in the Evolution of Minding. See Tax, Sol (editor).

White, Lynn, Jr. *Medieval Technology and Social Change*. Oxford: 1962.
Acute analysis of medieval inventions in commanding power and improving agriculture as related to their social accompaniments and results.

Whitehead, Alfred North. *Symbolism: Its Meaning and Effect*. New York: 1927.

Whorf, Benjamin Lee. *Language, Thought, and Reality: Selected Writings*. Cambridge, Mass.: 1956.
Fundamental discussions by a brilliant linguist whose too early death, like that of Ventris, robbed the world of a fuller development of his essential ideas. Whorf demonstrated that each language presents, by its very structure, a metaphysical world view. Niels Bohr, from quite different premises, reached similar conclusions about the mediating functions of language in scientific observations.

Willey, Gordon R. *New World Archaeology in 1965*. In Proceedings of the American Philosophical Society: April 22, 1966.

Wilson, John A. *The Burden of Egypt: An Interpretation of Ancient Egyptian Culture*. Chicago: 1951.
Note especially chapter on The King and God.

Wilson, Richard Albert. *The Miraculous Birth of Language*. New York: 1948.
One of the few modern works that dares to speculate on the origins of language and re-examine its purposes as well as its functions. But see Whorf, Sapir, Révész, and Jespersen.

Windels, Fernand. *The Lascaux Cave Paintings*. New York: 1950.
An account, authenticated by various authorities, of the contents of these great caves, handsomely illustrated, by the photographer who in 1940 helped the Abbé Breuil to create this record.

Wolf, Eric R. *Peasants*. Englewood, N. J.: 1966.

Woolley, [Charles] Leonard. *Abraham: Recent Discoveries and Hebrew Origins*. New York: 1936.

Excavations at Ur: A Record of Twelve Years' Work. New York: 1954.
See also Hawkes, Jacquetta.

Wymer, Norman. *English Town Crafts: A Survey of Their Development from Early Times to the Present Day*. London: 1949.
Though confined to England, it presents a sympathetic picture of the wealth of craftsmanship that was still in existence all over the Old World till the twentieth century.

English Country Crafts. London: 1946.

Wynne-Edwards, V. C. *Animal Dispersion in Relation to Social Behaviour*. Edinburgh: 1962.
Exhaustive.

Young, J. Z. *Doubt and Certainty in Science: A Biologist's Reflections on the Brain*. New York: 1960.
Up-to-date scientific data embarrassed by questionable Galilean 'objectivity' and equally inadequate current communications theory, which ignores the expressive functions.

A Model of the Brain: Mechanisms of Learning and Form Discrimination. Oxford: 1964.
Though Young properly uses a simple biological model, the nervous system of an octopus, he confuses his problem and his findings by superimposing a Cartesian mechanical model, which even eliminates the organism by reducing it conceptually to a "homeostat."

INDEX

Brackets around numerals denote plate numbers in the graphic section.

Abacete, cave at, 110
Abel, 150, 238
Aborigines, Australian, 53
Abraham, 221, 223
'Absolute power,' human limitations on, 233
Absolute powers, vulnerability of, 226
'Absolute sovereignty,' 198
'Absolute weapons,' 198
Abstraction, 'disease of,' 93; primitive's powers of, 105
Abundance, economy of, 219; linguistic, 94; promise of, 283
Abydos, 200, 209
Accountability, 192
Accountancy, capitalist, 280; universal, 278
Accumulation, 141; theological and capitalist, 275
Acheulian culture, 113
Acheulian tools, improvement of, 105
Acid-alkaline balance, 34
'Acquired characters,' transmission of, 38
Acquisitiveness, 274
Acropolis, 246
Acts, ritualized, 73
Adams, Henry, 292
Adaptability, human, 107
Adaptive mechanism, brain as, 39
Advance, technological medieval, 271
Advances, Aurignacian and Magdalenian, 33
Affluence, as anxiety breeder, 178; early dreams of, 241
'Affluent society,' 42, 96
African Bushmen, 102
African Pygmies, 118
Age of Poetry, 93
Aggression, capitalist, 279; 'civilized,' 257; early limitations on, 217
Agricultural advance, gradualness of, 129
'Agricultural revolution,' 15, 130, 163

Agriculture, neolithic practice of, 155
Air-conditioning, [30–31]
Akkad, 175
Akkadian poem, 66, 152
Albertus Magnus, 282
Albright, W. F., 200
Alcheringa, 53
Alexander the Great, 42, 187, 210
Alienation, 42
Allier, R., 95
Altamira caves, 8, 119, 128
Altar, sacrificial, 150
Alteration, sexual, 110
al' Ubaid, 170
Amana, Iowa, 161
Amaranth, 130
Amazonite beads, 165
American home, mechanization of, 256
Ames, Adelbert, 44
Ames, Oakes, 102, 106, 129, 132, 144
Ammann, Jost, [22–23]
Amok, 55
Amos, 258
Anabaptists, 276
Anabolism, woman's, 140
Ancestor, totem, 69
Ancestor-images, archetypal, 50
Ancestors, man's early, [7]
Ancestors, the, 50, 98
Ancestral past, inviolability of, 98
'Ancient Hunters,' 75
Andersen, Hans Christian, 226
Anderson, Edgar, 144
Animal, man as dreaming, 49
Animal aggression, 216
Animal domestication, 151
Animalhood, man's emergence from, 26
Animal-hunt, 218
Animal signals, 86
Animals, backboned, 32; cultural influence of, [2–3]; domestication of, 130, 152; draught, 23; magic capturing of, 119; veneration of, 151

Animus, woman's masculine, 148
Anna Karenina, 160
Annuals, cultivated, 129
Ant-hill, lesson of, 266
Anticipation, imaginative, 44; mechanical, 283
Antipater of Thessalonica, 247
Anu, 179
Anxiety, 44, 177; as source of war, 220
Apprehensiveness, prophetic, 44
'Arabian Nights,' 204, 282
'Archaeology and Society,' 217
Archaic man, 68
Archeologists, [12]; nineteenth-century, 8
Archimedes, 227, 245
Archytas, 245
Aristophanes, 148
Aristotle, 39, 271
Army, as archetypal machine, 192; organization of, 218; wrecking functions of, 223
Arrow-straightener, 24
'Art and Technics,' 285
Art, as cult-secret, 120; democratization of, 273; paleolithic, [2–3], 118; primacy of, 252; ritual practice of, 119; sacrifices for, 120
Artifacts, cave, 112; material, deceptiveness of, 23; paleolithic, 20
Artist, the paleolithic, 127
Arts, folk, 253; subjective, 254
Arunta, 79
Assyrians, [19]
Astrologers, 283
Astronauts, 34
Astronomy, Babylonian, 166
Atum, 167, 171; true name of, 94
Audubon, Egyptian, [14]
Audubon, John James, 147
Aurignacian man, 113
Ausonius of Bordeaux, 248

325